Calendars in the Making

Time, Astronomy, and Calendars

TEXTS AND STUDIES

VOLUME 10

The titles published in this series are listed at *brill.com/tac*

Calendars in the Making

The Origins of Calendars from the Roman Empire to the Later Middle Ages

Edited by

Sacha Stern

BRILL

LEIDEN | BOSTON

Cover illustration: Detail from Berlin Staatsbibliothek, or. quart. 211, fol. 1r. Arabic Ephemeris for the year 1804 CE with dates in the Arabic, Coptic, Syriac, Julian, Persian (Yazdgird), Hebrew, and Jalālī calendars (Public Domain Mark 1.0; image: Staatsbibliothek zu Berlin – PK).

Library of Congress Cataloging-in-Publication Data

Names: Stern, Sacha, editor.
Title: Calendars in the making : the origins of calendars from the Roman Empire to the later Middle Ages / edited by Sacha Stern.
Other titles: Origins of calendars from the Roman Empire to the later Middle Ages
Description: Leiden ; Boston : Brill, [2021] | Series: Time, astronomy, and calendars, 2211-632X ; volume 10 | Includes bibliographical references and index.
Identifiers: LCCN 2021010841 (print) | LCCN 2021010842 (ebook) | ISBN 9789004459632 (hardback) | ISBN 9789004459694 (e-book)
Subjects: LCSH: Calendar – Europe – History – To 1500. | Calendar – Rome – History.
Classification: LCC CE57 .C35 2021 (print) | LCC CE57 (ebook) | DDC 529/.3094 – dc23
LC record available at https://lccn.loc.gov/2021010841
LC ebook record available at https://lccn.loc.gov/2021010842

Typeface for the Latin, Greek, and Cyrillic scripts: "Brill". See and download: brill.com/brill-typeface.

ISSN 2211-632X
ISBN 978-90-04-45963-2 (hardback)
ISBN 978-90-04-45969-4 (e-book)

This book is printed on acid-free paper and produced in a sustainable manner.

Contents

Figures

Notes on Contributors

Ilaria Bultrighini

is Honorary Research Fellow at UCL. She is an ancient historian specializing in epigraphy and the social, religious, and cultural history of Greek and Roman Antiquity. After completing her studies in Italy, she held fellowships at the Harvard University Center for Hellenic Studies, the Deutsches Archäologisches Institut, and the Einstein Center Chronoi at Freie Universität in Berlin. She also worked as Research Fellow on major research projects at UCL and the Institute of Classical Studies (London). Her research interests include the history of non-urban areas of the Athenian polis, time reckoning, astrology, and the Greek East under Roman rule.

François de Blois

is a scholar of Semitic and Iranian languages and literatures, and of the history and religious history of the Near East and Central Asia in pre-modern times. He studied at Tübingen University, and has taught and researched in London (mainly SOAS and UCL), Cambridge, Hamburg, and Princeton. He has worked on the history of calendars and eras of Iran (ancient Iranian, Zoroastrian, Bactrian) and the Arabian peninsula (ancient South Arabian and Islamic). His most recent book is *Studies in the Chronology of the Bactrian documents from Northern Afghanistan*, co-authored by Nicholas Sims-Williams (Vienna 2018). He is currently completing a new edition and translation of al-Biruni's *Chronology of Ancient Nations*.

Sacha Stern

DPhil. Oxon. 1992, is Professor of Jewish Studies at University College London. An ancient historian by training, he has published several books on calendars and time reckoning in Antiquity and the Middle Ages, including *Calendar and Community: A History of the Jewish Calendar* (2001), *Time and Process in Ancient Judaism* (2003), *Calendars in Antiquity* (Oxford 2012), and *The Jewish Calendar Controversy of 921/2* (Leiden 2019). Since 2008 he has been Principal Investigator of five research projects at UCL on calendars in late Antiquity and the Middle Ages.

Johannes Thomann

Ph.D. (1992), researches at the Institute of Asian and Oriental Studies, University of Zürich. He specializes in the history of sciences in the premodern Islamic world, particularly the history of astronomy. In recent years, his main occupation has been the edition of Arabic astronomical documents, but he has also published on Arabic and Turkish folk literature. Recent publications on chronology include 'Tools of Time' in A. Pelliteri et al. (ed.), *Re-defining a space of encounter* (Leuven 2019), and 'Ritual Time, Civil Time, and Cosmic Time', in *KronoScope* 20 (2020).

Nadia Vidro

is Senior Research Fellow at UCL. Her main research interests include the intellectual history of the Jews in the medieval Near East, and Jewish manuscripts. Her PhD (University of Cambridge, 2010) was on the history of medieval grammatical traditions. In her current research she focuses on the history of Jewish calendars and the socio-historical implications of calendar diversity. Her publications on Jewish calendars include 'The origins of the 247-year calendar cycle', *Aleph: Historical Studies in Science and Judaism*, 17/1 (2017), and 'R. Naḥshon Gaon – calendar scholar or pseudo-author?', *Jewish Studies Quarterly*, 26/1 (2019).

Immo Warntjes

is Ussher Assistant Professor in Early Medieval Irish History at Trinity College Dublin. His research interests include the Easter controversy of the first eight centuries of Christianity, and the transmission of ideas and scientific thought in the early medieval Latin West. He is the author of *The Munich Computus: text & translations. Irish computistics between Isidore of Seville and the Venerable Bede and its reception in Carolingian times* (Stuttgart, 2010), and co-editor of a series of conference proceedings on the science of computus in Late Antiquity and the early Middle Ages.

Abbreviations

ACO	E. Schwartz, J. Straub, and R. Schieffer (eds), *Acta conciliorum oecumenicorum* (Berlin: Walter de Gruyter, 1914–1984).
AE	*L'Année Épigraphique* (Paris: Presses Universitaires de France, 1888–).
BGU	*Aegyptische Urkunden aus den Königlichen (later Staatlichen) Museen zu Berlin, Griechische Urkunden* [illegible] 1895–2014).
CAG	*Carte Archéologique de la Gaule* (Paris: Académie des inscriptions et belles lettres, 1988–).
CCAG	*Catalogus Codicum Astrologorum Graecorum* (Brussels: Lamertin, 12 vols, 1899–1953).
CEL	P. Cugusi, *Corpus epistularum Latinarum papyris tabulis ostracis servatarum*, 3 vols (Florence: Gonnelli, 1992–2002).
C. Gloss. Biling.	J. Kramer (ed.), *Glossaria bilinguia in papyris et membranis reperta* (Bonn: Habelt, 1983).
ChLA	*Chartae Latinae Antiquiores*, 49 vols (Basel: Dietikon, 1954–1998).
CIGP	P. Kovács, *Corpus inscriptionum Graecarum Pannonicarum* (Hungarian Polis Studies 8; Debrecen: University of Debrecen, 2001).
CIL	*Corpus Inscriptionum Latinarum* (Berlin: Walter de Gruyter, 1862–).
CLA	E.A. Lowe et al., *Codices Latini Antiquiores: a Palaeographical Guide to Latin Manuscripts Prior to the Ninth Century* (Oxford: Clarendon Press, 1934).
CPJ	V.A. Tcherikover (ed.) *Corpus Papyrorum Judaicarum* (*Cambridge*, Mass.: Harvard University Press: 1957–1964).
CPL	R. Cavenaile (ed.), *Corpus Papyrorum Latinarum* (Wiesbaden: O. Harrassowitz, 1958).
DNP	*Der Neue Pauly: Enzyklopaedie der Antike* (Stuttgart: J.B. Metzler, 1996–).
FIM	W. Boppert, *Die frühchristlichen Inschriften des Mittelrheingebietes* (Mainz am Rhein: P. v. Zabern, 1971).
FITrier	H. Merten, *Katalog der frühchristlichen Inschriften des bischöflichen Dom- und Diözesanmuseums Trier* (Trier: Bischöfliches Dom- und Diözesanmuseum Trier, 1990).

ICatania	K. Korhonen, *Le iscrizioni del Museo Civico di Catania: storia delle collezioni – cultura epigrafica – edizione* (Helsinki: Societas Scientiarum Fennica, 2004).
ICG	C. Breytenbach, K. Hallof, et al., *Inscriptiones Christianae Graecae (ICG): A Digital Collection of Greek Early Christian Inscriptions from Asia Minor and Greece* (Berlin: Edition Topoi, 2016).
ICI	*Inscriptiones Christianae Italiae septimo saeculo antiquiores* (*nova series*) (Bari: Edipuglia, 1985–<2016>).
ICKarth	L. Ennabli, *Les inscriptions funéraires chrétiennes de Chartage. 3 – Carthage intra et extra muros* (Rome: École Française de Rome, 1991).
ICUR	*Inscriptiones Christianae Urbis Romae Septimo Saeculo Antiquiores* (Rome: Ex Officina Libraria Pontificia, 1857–).
IG	*Inscriptiones Graecae* (Berlin: Walter de Gruyter, 1873–).
IGerasa	C.B. Welles, in: C.H. Kraeling, *Gerasa: City of the Decapolis. An Account Embodying the Record of a Joint Excavation Conducted by Yale University and the British School of Archaeology in Jerusalem (1928–1930), and Yale University and the American Schools of Oriental Research (1930–1931, 1933–1934)* (New Haven: American Schools of Oriental Research, 1938).
IGLS III.1	L. Jalabert and R. Mouterde, *Inscriptions grecques et latines de la Syrie*, vol. III.1: *Région de l'Amanus. Antioche* (Paris: Geuthner, 1950).
IGLS VI	J.-P. Rey-Coquais, *Inscriptions grecques et latines de la Syrie, VI. Baalbek et Beqaʿ* (Paris: P. Geuthner, 1967).
IGR	E. Leroux (ed.), *Inscriptiones graecae ad res romanas pertinentes* (Paris: E. Leroux, 1906–1927).
IGUR	L. Moretti, *Inscriptiones graecae urbis Romae*, 4 vols (Rome: Istituto italiano per la Storia antica, 1968–1990).
IIt	*Inscriptiones Italiae* (Rome: Liberia dello stato, 1931–).
IKaunos	C. Marek, *Die Inschriften von Kaunos* (Vestigia. Beiträge zur Alten Geschichte 55; Munich: C.H. Beck, 2006).
IK *Estremo oriente*	F. Canali De Rossi, *Iscrizioni dello estremo oriente greco. Un repertorio* (Inschriften griechischer Städte aus Kleinasien 65; Bonn: Habelt, 2004).
ILAlg 1	S. Gsell, *Inscriptions latines de l'Algerie, Tome 1* (Paris: E. Champion, 1922).

ILaodikeia	T. Corsten, *Die Inschriften von Laodikeia am Lykos* (Inschriften griechischer Städte aus Kleinasien 49; Bonn: Habelt, 1997).
ILCV	E. Diehl, *Inscriptiones Latinae Christianae veteres* (Berlin: Apud Weidmannos, 1925–1931).
ILS	H. Dessau, *Inscriptiones Latinae Selectae*, 3 vols (Berlin: Berolini Apud Weidmannos, 1892–1916).
ILTun	A. Merlin, *Inscriptions Latines de la Tunisie* (Paris: P.U.F., 1944).
INikaia	*Katalog der antiken Inschriften des Museums von Iznik (Nikaia) 1: Stadtgebiet und die nächste Umgebung der* [illegible] 1979).
IOSPE I²	V. Latyshev, *Inscriptiones antiquae orae septentrionalis Ponti Euxini graecae et latinae*, 3 vols (St. Petersburg, 1885–1901).
IPalaestina Tertia Ia	Y.E. Meimaris and K.I. Kritikakou-Nikolaropoulou, *Inscriptions from Palaestina Tertia, Ia: The Greek Inscriptions from Ghor Es-Safi (Byzantine Zoora)* (Meletemata 41; Athens: National Hellenic Research Foundation, 2005).
IPalaestina Tertia Ib	Y.E. Meimaris and K.I. Kritikakou-Nikolaropoulou, *Inscriptions from Palaestina Tertia. Vol. 1b, The Greek inscriptions from Ghor Es-Safi (Byzantine Zoora) (Supplement), Khirbet Qazone and Feinan* (Meletemata 57; Athens: National Hellenic Research Foundation, 2008).
IPergamon 2	M. Fränkel, *Die Inschriften von Pergamon* (Altertümer von Pergamon 8,1–2; Berlin: W. Speman, 1890–1895), vol. 2, nos. 251–1334, 'Römische Zeit'.
IPriene (2014)	W. Blümel and R. Merkelbach, *Die Inschriften von Priene* (Inschriften griechischer Städte aus Kleinasien 69; Bonn: Habelt, 2014).
IRT	J.M. Reynolds and J.B. Ward-Perkins, *Inscriptions of Roman Tripolitania* (Rome and London: British School at Rome, 1952), enhanced electronic reissue by G. Bodard and C. Roueché (London: King's College, 2009), available: http://inslib.kcl.ac.uk/irt2009/ (accessed 22 September 2020).
IStratonikeia	M. Sahin, *Die Inschriften von Stratonikeia, 2.1* (Bonn: Habelt, 1982).
ITrebula	H. Solin, *Le iscrizioni antiche di Trebula, Caiatia e Cubulteria* (Caserta: Associazione Storica del Caiatino, 1993).
MAMA VI *List*	W.H. Buckler and W.M. Calder, *Monuments and Documents from Phrygia and Caria* (Monumenta Asiae Minoris Antiqua 6; Manchester: Manchester University Press, 1939).

MGH	G.H. Pertz *et al.*, *Monumenta Germaniae Historica Inde Ab Anno Christi Quingentesimo Usque Ad Annum Millesimum Et Quingentesimum* (Auspiciis Societatis Aperiendis Fontibus Rerum Germanicarum Medii Aevi; Hannover: Hahn, 1826–).
Auct. Ant.	*Auctores antiquissimi*
Conc.	*Concilia*
SS rer. Merov.	*Scriptores rerum Merovingicarum*
P.Abinn.	H.I. Bell, V. Martin, E.G. Turner, and D. van Berchem (eds), *The Abinnaeus Archive: Papers of a Roman Officer in the Reign of Constantius II* (Oxford: Clarendon Press, 1962).
P.Acad.	J.-L. Fournet, J. Gascou, *et al.* (in preparation).
P.Bagnall	R. Ast, H. Cuvigny, T.M. Hickey, and J. Lougovaya, *Papyrological Texts in Honor of Roger S. Bagnall* (American Studies in Papyrology 53; Durham, NC: The American Society of Papyrologists, 2013).
P.Cair.Isid.	A.E.R. Boak and H.C. Youtie (eds), *The Archive of Aurelius Isidorus in the Egyptian Museum, Cairo, and the University of Michigan* (Ann Arbor: University of Michigan Press, 1960).
P.Cair.Zen.	C.C. Edgar (ed.), *Zenon Papyri, Catalogue général des antiquités égyptiennes du Musée du Caire* (Cairo: impr. de l'Institut Français d'Archéologie Orientale, 1925–40).
P.Dura	C.B. Welles, R.O. Fink, and J.F. Gilliam (eds), *The Excavations at Dura-Europos conducted by Yale University and the French Academy of Inscriptions and Letters, Final Report V, Part I, The Parchments and Papyri* (New Haven: Yale University Press, 1959).
P.Euphrates	D. Feissel, J. Gascou, and J. Teixidor, 'Documents d'archives romains inédits du Moyen Euphrate (IIIe s. après J.-C.)', *Journal des savants* (1997), pp. 3–57.
P.Fay.	B.P. Grenfell, A.S. Hunt and D.G. Hogarth (eds), *Fayum Towns and their Papyri* (Egypt Exploration Society, Graeco-Roman Memoirs 3; London: Offices of the Egypt exploration fund, 1900).
P.Flor.	*Papiri greco-egizii, Papiri Fiorentini* (Supplementi Filologico-Storici ai Monumenti Antichi; Milan: Hoepli, 1906–1915).
P.Genova	M. Amelotti and L. Zingale Migliardi (eds), *Papiri dell'Università di Genova* (Milano: A. Giuffrè, 1974–1980).
P.Giss.	O. Eger, E. Kornemann, and P.M. Meyer (eds), *Griechische Papyri im Museum des oberhessischen Geschichtsvereins zu Giessen* (Leipzig: Teubner, 1910–1912).

PGM	K. Preisendanz and A. Henrichs, *Papyri Graecae magicae. Die griechischen Zauberpapyri*, 2 vols (Stuttgart: B.G. Teubner, 1973–74).
P.Herm.	B.R. Rees (ed.), *Papyri from Hermopolis and Other Documents of the Byzantine Period* (Graeco-Roman Memoirs 42; London: Egypt Exploration Society, 1964).
P.Iand.	C. Kalbfleisch et al. (ed.), *Papyri Iandanae* (Leipzig: Teubner, 1912–1938).
*PIR*2	*Prosopographia Imperii Romani* (Berlin: Walter de Gruyter, 2nd ed., 1933–).
P.Kellis	*Papyri from Kellis* (Oxford: Oxbow Books, 1995–2014).
PL	J.-P. Migne, *Patrologia Latina*.
P.Lond.	*Greek Papyri in the British Museum* (London: British Museum, 1893–1974).
PLS	J.-P. Migne, *Patrologia, series latina. Supplementum* (ed. A. Hamman; Paris: Garnier Fratres, 1958–1974).
P.Merton	*A Descriptive Catalogue of the Greek Papyri in the Collection of Wilfred Merton* (London: Institute of Classical Studies, 1948–1967).
P.Mur.	P. Benoit, J.T. Milik and R. de Vaux (eds), *Les grottes de Murabba'ât* (Discoveries in the Judaean Desert of Jordan 2; Oxford: Clarendon Press, 1961).
P.Ness.	*Excavations at Nessana* (London: British School of Archaeology in Jerusalem, 1950–1962).
P.Oxy.	*The Oxyrhynchus Papyri* (Graeco-Roman Memoirs; London: Egypt Exploration Society, 1898–).
P.Petra	*The Petra Papyri* (Amman: ACOR, 2002–2013).
P.Rain.Cent.	*Festschrift zum 100-jährigen Bestehen der Papyrussammlung der Österreichischen Nationalbibliothek, Papyrus Erzherzog Rainer* (Vienna: Hollinek, 1983).
P.Ross.Georg.	*Papyri russischer und georgischer Sammlungen* (Tiflis: Universitätslithographie, 1925–1935).
P.Ryl.	*Catalogue of the Greek and Latin Papyri in the John Rylands Library, Manchester* (Manchester: Manchester University Press, 1911–1952).
P.Sakaon	G.M. Parássoglou (ed.), *The Archive of Aurelius Sakaon: Papers of an Egyptian Farmer in the last Century of Theadelphia* (Bonn: Habelt, 1978).

PSI	*Papiri greci e latini* (Pubblicazioni della Società Italiana per la ricerca dei papiri greci e latini in Egitto; Florence: E. Ariani, 1912–).
PSI.Congr.XVII	*Trenta testi greci da papiri letterari e documentari editi in occasione del XVII Congresso Internazionale di Papirologia* (Florence: Istituto papirologico G. Vitelli, 1983).
P.Thead.	*P. Jouguet (ed.), Papyrus de Théadelphie (Paris: Fontemoing, 1911).*
PXHev/Se	H.M. Cotton and A. Yardeni (eds), *Aramaic, Hebrew and Greek Documentary Texts from Nahal Hever and Other Sites, with an Appendix containing Alleged Qumran Texts (The Seiyâl Collection II)* (Discoveries in the Judaean Desert 27; Oxford: Clarendon Press, 1997).
P.Yadin	*The Documents from the Bar Kochba Period in the Cave of Letters* (Jerusalem: Israel Exploration Society, 1989–2002).
RICG	N. Gauthier, *Recueil des inscriptions chrétiennes de la Gaule I. Première Belgique* (Paris: Centre national de la recherche scientifique, 1975).
RICM	D. Feissel, 'Recueil des inscriptions chrétiennes de Macédoine, du IIIe au VIe siècle', *Bulletin de correspondance hellénique. Supplément 8* (Athens: Ecole française d'Athènes, 1983).
SB	*Sammelbuch griechischer Urkunden aus Ägypten* (Strassburg: K.J. Trübner, 1915–).
SEG	*Supplementum Epigraphicum Graecum* (Leiden: Brill, 1923–).
TAM V.3	G. Petzl, *Tituli Asiae Minoris. Vol. 5, Tituli Lydiae linguis Graeca et Latina conscripti. Fasc. 3, Philadelpheia et Ager Philadelphenus* (Vienna: Verlag der Österreichischen Akademie der Wissenschaften, 2007).
Terrien-2007	M.-P. Terrien, *La christianisation de la région rhénane du IVe au milieu du VIIIe siècle. Corpus et synthèse* (Besançon: Presses universitaires de Franche-Comté, 2007).
Thylander	H. Thylander, *Inscriptions du Port d'Ostie* (Lund: C.W.K. Gleerup, 1952).

Introduction

Sacha Stern

Sed ut hos annos omittam caligine iam profundae vetustatis obductos, in his quoque, qui sunt recentioris memoriae et ad cursum lunae vel solis instituti, quanta sit varietas, facile est cognoscere.

But let me leave aside those years that are concealed in the darkness of deep antiquity. Even among those of more recent memory, which are regulated by the course of the moon or sun, it is easy to recognize how much variety there is.

CENSORINUS, *De Die Natali*, 20:1

In his short calendar treatise of 238 CE, in a passage discussing the variety of calendars, Censorinus decides to ignore ancient methods of reckoning years because, he says, they are 'concealed in the darkness of deep antiquity'; instead, he will focus on those 'of more recent memory'. His avoidance of the ancient 'years' – which we would probably call 'calendars' – is partly a profession of ignorance, but also reflects the difficulty Censorinus might have had explaining the little he knew about their strange features. Following an established tradition attested also elsewhere, Censorinus writes in the preceding passage (ibid. 19:4–7) that the ancient civil years were originally very short, and gradually evolved through increasingly longer periods, eventually to reach the length of the natural solar year.[1] The Egyptian year originally consisted of only one month; it later grew to four months; and only later, to a year of twelve months and five days. In Achaea, the Arcadians originally reckoned a year of three months. The Carian and Acarnanian years counted six months, a pair of which made up a 'great year'. For Censorinus, these somewhat primitive, aberrant calendars, not 'regulated by the course of the moon or sun' (at least, not by the sun), were bizarre and unknown, and best left to the darkness of their deep antiquity.

1 'Civil years' (*anni civiles*) means the calendar years that were reckoned by cities, in contrast to the 'natural', solar year (*annus naturalis*: ibid. 19:1–3).

 | DOI:10.1163/9789004459694_002

The calendars and other time reckoning methods that are studied in this present volume were all of recent origin in Censorinus' day, or from a later period; he would certainly not have regarded them as 'ancient'. The Julian calendar, whose model was being adopted and adapted, in Censorinus' day, in the eastern provinces of the Roman Empire, had only been instituted in 46 BCE, replacing the structurally quite different Roman Republican calendar. The seven-day week, although of ancient Biblical origin, only came to be used as a time reckoning device, by Jews and increasingly by pagans, in the first century CE – surprisingly late, as will be shown later in this volume; Dio Cassius, Censorinus' contemporary, readily acknowledged that the week was a rather recent institution (Dio 37:18–19). The computation of the date of Easter, leading to a distinctly Christian calendar, was being invented exactly in Censorinus' time; the first Easter cycle, of 112 years, was designed by Christians in Rome in 222 CE. Islamic and Jewish (rabbinic) calendars were to follow some time later, and the Zoroastrian calendar was also about to undergo several transformations in late Antiquity and the early Middle Ages. All these calendars, which have come to dominate the modern world today – in particular the Julian and Gregorian Christian calendars, and the Islamic calendars – originated in the Roman imperial and the early medieval periods.

Yet although Censorinus would have considered these calendars as 'of more recent memory', and all broadly regulated by the moon and/or the sun, it is remarkable how little we know about their origins or how, exactly, they came into existence. The Islamic calendar, the Latin Christian calendars, and the fixed rabbinic calendar, are all very well known, but the processes that brought about their institution and gave them their distinctive structures – even the dates when they were instituted – remain obscure and concealed in darkness, open to much speculation and debate in modern scholarship.[2]

The purpose of this present volume is to address the question of the origins of calendars that we are most familiar with, but whose origins are still very poorly known or understood. Chapter 1 describes the sudden rise and diffusion of the seven-day week as a method of time reckoning among Jews and others in the Roman Empire, starting from the first century CE – a surprisingly obscure phenomenon, too easily taken for granted, that has not been systematically investigated since Colson (in 1926).[3] This chapter attempts to unpack

---

2 For example, the date, author(s), and circumstances of the institution of the Alexandrian 19-year cycle, which was later to become dominant in the computation of the date of Easter (see Chapter 3), are completely unknown; in spite of many scholarly theories, any date in the late third century or first half of the fourth remains possible (Stern, *Calendars in Antiquity*, pp. 402–4).

3 Colson, *The Week*.

and explain the relationship between the Jewish week, its Christian adaption, and its Roman, planetary interpretation, which went on co-existing in complex ways through late Roman and medieval society. Chapter 2 considers the spread of the Julian calendar, in various adaptations, in the Roman provinces of the East, mainly as evidenced in the late antique multi-ledger calendars called *Hemerologia* which tabulate and compare a wide range of provincial calendars. Since Kubitschek (in 1915), very little research has been carried out on the *Hemerologia* and on epigraphic and other evidence of provincial calendars, in spite of the fact that these calendars must have constituted a fundamental structure of social organization in the provinces of the Roman Empire.[4] Chapter 3 challenges the modern scholarly belief in a centralized origin and a linear development of the medieval Christian calendar in the Latin West, stressing instead the significance of independent initiative, local variety, and complex lines of cultural exchange. Chapter 4 proposes a new explanation of the origins of the Islamic calendar, which in spite of its historical importance, has never been satisfactorily accounted for or resolved. Chapter 5 documents the eleventh-century institution of the Jalālī calendar, a little known yet major development in the history of the Persian Zoroastrian calendar, and reflects on the implications of the resulting simultaneous use, in Persian (or any other) society, of different time reckoning systems. Chapter 6 presents, for the first time, a study of the origins, diffusion, diversity, and social significance of a cyclical calendar that pervades later medieval Jewish manuscripts in spite of deviating from the normative rabbinic calendar.

Our research is innovative; our approach, through the book, is largely empiricist. The majority of the chapters are based on original compilation and analysis of large corpora of primary sources, mainly manuscripts and inscriptions, that supply evidence of dating practices in the provinces of the Roman Empire and of the large-scale medieval production of Jewish, Christian, and other calendar texts and calendars (especially Chapters 1, 2, 3, 6; some of the corpora are listed in appendices to individual chapters).

The background to this book is an extensive research project funded by an Advanced Grant of the European Research Council that ran at UCL from 2013 to 2018 with a team of six researchers. The project was entitled 'Calendars in Late Antiquity and the Middle Ages: Standardization and Fixation', and was concerned, as the title indicates, with the processes through which late antique and medieval calendars became increasingly standard and universal,

4 Kubitschek, *Kalenderbücher*. See more recently Samuel, *Greek and Roman Chronology*; Meimaris, *Chronological Systems*; and for more specific studies, references below in Chapter 2.

determined and fixed.[5] The majority of contributors to this volume were part of this team, and their contribution in this volume derives directly from their ERC project research (the only notable exception is François de Blois, whose paper was an unexpected side-line of his substantial work, in the ERC project, on al-Biruni's *Chronology of the Ancient Nations*). Immo Warntjes and Johannes Thomann, from outside the team, participated in the project's workshops and in the final conference in July 2017, when most of the present papers were presented; I am grateful for their distinguished contribution to this volume. Two of the ERC team, Israel Sandman and Ilana Wartenberg, worked mainly on Isaac Israeli's *Yesod Olam* (Toledo, 1310) and will be publishing the results of their research separately, with relevant contributions from other members of the team. The theme of this present volume, the origins of calendars, is slightly different from that of the ERC project as a whole, but it reverts to the more general question of standardization and fixation, as these processes were closely involved in the origins, formation, making, and diffusion of new calendars and reckoning methods in the Roman period and the early Middle Ages.

As an edited book rather than a multi-authored monograph, it makes no claim of being comprehensive. There is no chapter, for example, on the institution of the Julian calendar in 46 BCE, although much attention is given to the subsequent diffusion of the Julian calendar in the eastern part of the Roman Empire (Chapter 2) and its transmission in the medieval Latin West (Chapter 3). There are no chapters on the origins of the Christian Easter computation (though early medieval Latin Computus is discussed in Chapter 3), or on the late antique and early medieval Jewish calendars (early rabbinic, fixed rabbinic, and Qaraite – but the late medieval Jewish calendar is treated in Chapter 6). The origins of the Julian calendar and the Christian Easter calendars have been dealt with at length elsewhere,[6] but those of the early medieval Jewish calendars remain, in spite of much recent research, largely unknown and in need of further investigation.[7] This volume does not deal with short time (like sub-divisions of the day), nor with long time (like eras).[8] The term

5 https://www.ucl.ac.uk/hebrew-jewish/research/research-projects/calendars-late-antiquity-and-middle-ages-standardization-and-fixation-0, accessed 17 May 2020.

6 Julian calendar: for example Rüpke, *Kalender und Öffentlichkeit*; Feeney, *Caesar's Calendar*; Stern, *Calendars In Antiquity*, pp. 211–27. Easter cycles and computation: for example, Jones, *Bedae Opera de Temporibus*; Lejbowicz, 'Des tables pascales'; Holford-Strevens, 'Paschal lunar calendars'; Mosshammer, *Easter Computus*; Stern, *Calendars in Antiquity*, pp. 326–30 and 388–411; Mosshammer, *Prologues on Easter*, pp. 1–29.

7 As emerges from Stern, *Calendar and Community*, chs. 4–5.

8 Short time in Antiquity: see now Miller and Symons (eds), *Down to the Hour*. Ancient eras: Samuel, *Greek and Roman Chronology*; Meimaris, *Chronological Systems*; Mosshammer, *Easter Computus*.

'calendar' is used, in this volume, in a variety of senses. The seven-day week, to begin with, is not properly speaking a 'calendar', although it is a fundamental time-reckoning method that became integrated in broader calendar frameworks. For most contributors in this volume, 'calendar' is primarily used in the sense of the formal structure of time reckoning, within a medium time-scale of typically one year; but some contributors use 'calendar' more in the sense of the design and way through which time-reckoning is represented (e.g. the *Hemerologia*, Chapter 2, and the 'calendars' of the medieval Latin West, Chapter 3).

Although not fully comprehensive or consistent in its conception of what classifies as 'calendars', this volume is not a random collection of articles or conference proceedings. It attempts, above all, to make a statement about the 'making' and 'origins' of calendars. In spite of their claim to accuracy and exactitude, it is evident that calendars have always been man-made and the outcome of a process of making. The creativity of scribes and scholars in fashioning calendars, and thereby determining how time should be reckoned, comes out very clearly in Chapters 3 (Christian calendars in the West) and 6 (medieval Jewish calendar cycles). Creativity is also evident in the production of tables that bring together a selection of calendars and compare them side by side, as in the late Roman *Hemerologia* (Chapter 2) and the yearbooks of the Islamic world (Chapter 5).

'Making' is inherently related to the question of 'origins'; and it is appropriate, at this point, to reflect on the meaning of this term. 'Origins' are often associated, as in Censorinus, with 'deep antiquity', a vague historical notion which can be used to mask one's ignorance, but also to conjure up romanticized views about the birth of human civilization. For example, the calendar of Coligny, a Gallic five-year lunar calendar inscribed on a monumental copper plate of the second century CE, has been described by some scholars as going back to the 'dawn of time',[9] implying a romantic and possibly nationalistic claim that long before the arrival of the Romans, the ancient Celts were versed in astronomy and capable of constructing highly sophisticated calendars.[10] In actual fact, the solar dates in this calendar – its most sophisticated element – were most likely derived from the Julian calendar; the Coligny calendar was far more plausibly a Gallo-Roman product of the Roman imperial period, a

9 '*La nuit des temps*': Duval and Pinault, *Recueil des inscriptions gauloises, iii: Les Calendriers*, pp. 19–21 and 399–400. The second-century CE dating of the inscription is palaeographical.

10 A claim developed at length by Olmsted, *The Gaulish Calendar*.

subversive, lunar Gallification of the Roman monumental *fasti*.[11] Similarly, claims of antiquity and myths of origins have long been used for purposes of religious legitimization: thus, for example, the later medieval attribution of the institution of the fixed rabbinic calendar to a fourth-century sage named Hillel – still assumed today in many general, popular works – was intended to support the halakhic (legal) legitimacy of the post-Talmudic Rabbanite calendar; historians now know that it was not finalized and adopted before the ninth century.[12] It goes without saying that the search of 'origins', in this present volume, needs to be de-mystified and stripped of ideological agendas. Even if the institution of the Islamic calendar turns out, as proposed by de Blois (Chapter 4), to go back to the time of the *hijra*, this possible vindication of Islamic sacred chronology only serves for us the purpose, in de Blois' words, of 'bringing the early history of Islam just a bit closer to documented world history'.

The concept of origins, however, is complex and needs to be problematized. In contrast perhaps to the Islamic calendar, the origins of many calendars were not punctual or the result of a single event or act of invention and institution. This emerges very clearly, for example, with regard to the seven-day week (Chapter 1) and the early medieval Latin Christian calendars (Chapter 3), whose making, adoption, and diffusion were non-linear, complex processes, driven by agents and motives that are often unclear, and through a time-line, dates, and stages that are equally unclear. The Jewish, fixed rabbinic calendar, distinctive in the complexity of its calculations and rules, was also the outcome of a slow, cumulative process, which can be partially reconstructed as stretching from the third to ninth centuries CE.[13]

Many calendars, however, are likely to have originated from a single, deliberate act of institution and making, even if their subsequent diffusion and adoption could have been drawn out and somewhat more complex. The institution of the Julian calendar in 46 BCE is a well-known and well documented example; to this category can also be added the institution of the Jalālī calendar (Chapter 5). Several other calendars can be assumed to belong to this category, even if the historical details of their institution remain unclear: the Islamic calendar, and much earlier in history, the Egyptian, Zoroastrian, and Roman Republican calendars, are all likely to have been formally instituted at a specific point in time. The Egyptian calendar, to begin with, is highly abstract,

---

11 McCluskey, *Astronomies and Cultures in Early Medieval Europe*, pp. 54–66; Stern, *Calendars in Antiquity*, pp. 303–13.

12 Stern, *Calendar and Community*, pp. 175–81.

13 Ibid. ch. 4.

simple, and schematic, and unlikely to have been created through a lengthy, cumulative historical process: whenever it was conceived, sometime in the third millennium BCE, it must have been instituted and designed as a whole piece. The Zoroastrian calendar, a near copy of the Egyptian one, was most likely adopted and instituted as a single political act by one of the Achaemenid kings, not too long after the Persian conquest of Egypt in 525 BCE.[14] The Roman Republican calendar was almost certainly instituted by senatorial decision, even if the date and circumstances of this remain contentious among modern historians.[15]

The institution of calendars is almost always a political act, often attributed to a political ruler (the known cases include Julius Caesar for the Julian calendar, Malikshāh for the Jalālī calendar), to which is often added a religious dimension, especially if instigated by a priest or priesthood (e.g. in Rome the pontifical college and the pontifex maximus, which Julius Caesar was, or the Prophet in early Islam). The social importance of calendars is such that any calendar institution is bound to constitute an exercise of power, authority, and social control. It is political authority, moreover, that underpinned the subsequent stability of the calendar, its diffusion, and its standardization.[16] Even calendars that were not formally instituted, and that resulted from more complex historical processes, must often be associated with an element of authority. The Christian scholars involved in the production of liturgical calendars in Ireland and other parts of the early medieval Latin West were undoubtedly exercising their social, religious authority over the development of liturgical practices in local Christian communities.

Most of the calendars under discussion in this book were designed as fixed schemes with little room for flexibility or deviation. The Islamic month, depending for most Muslims on the appearance of the new moon crescent, could be subject to a small margin of unpredictable variation; and the 247-year Jewish calendar cycle, not being truly cyclical, could be designed unpredictably with different data and in different ways (Chapter 6). But the seven-day week, the Julian and Roman provincial calendars, the normative Christian Easter computations, the Persian Zoroastrian calendar, the fixed rabbinic calendar, as well as the annual sequence of Islamic months, were all based on schemes, structures or calculations that were predictable and fixed, and that ran unchanged

14 De Blois, 'Persian Calendar', pp. 39–54; Stern, *Calendars in Antiquity*, pp. 174–91.

15 The literature is vast. See for example Rüpke, *Kalender und Öffentlichkeit*, pp. 319–30, and id., *Zeit und Fest*, pp. 26–27. For a summary of the scholarship, see Stern, *Calendars in Antiquity*, pp. 207–11.

16 As I have generally argued in Stern, *Calendars in Antiquity*.

through long periods of history. Their historical stability and longevity are what make the question of their origins important.

Equally important, at the end of these long periods of changelessness, are processes of termination, transformation, or reform, which are themselves the origins of something new: so, for example, the Gregorian reform, and modern proposals of seven-day week reform. This draws out the point that origins are not necessarily primordial or ancient. Origins are only moments of significant historical change; what gives them the quality of 'origins' is only our retrospective, often distant, gaze upon them, and the explanatory function that origins are deemed to fulfil with regard to what came next. It is important to recognize, in fact, that our interest in origins and our interpretation of them are unavoidably conditioned by our retrospective gaze, our knowledge (or beliefs) of what came next, and our expectations with regard to what these origins should be coming to explain.

An open-minded perspective on the origins of ancient and medieval calendars, and on the processes of their making, is the challenge that lies ahead of us in this volume.

References

F. de Blois, 'The Persian Calendar', *Iran* 34 (1996), pp. 39–54.

F.H. Colson, *The Week: An Essay on the Origin and Development of the Seven-day Cycle* (Cambridge: The University Press, 1926).

P.-M. Duval and G. Pinault, *Recueil des inscriptions gauloises, iii: Les Calendriers* (Paris: Éditions du CNRS, 1986).

D. Feeney, *Caesar's Calendar: Ancient Time and the Beginnings of History* (Berkeley and Los Angeles: University of California Press, 2007).

L. Holford-Strevens, 'Paschal lunar calendars up to Bede', *Peritia* 20 (2008), pp. 165–208.

C.W. Jones, *Bedae Opera de Temporibus* (Cambridge, Mass.: Medieval Academy of America, 1943).

W. Kubitschek, *Die Kalenderbücher von Florenz, Rom und Leyden* (Vienna: Denkschriften der Kaiserlichen Akademie der Wissenschaften in Wien, Philosophisch-historische Klasse 57, Abhandlung 3, 1915).

M. Lejbowicz, 'Des tables pascales aux tables astronomiques et retour: Formation et réception du comput patristique', *Methodos*, 6, http://journals.openedition.org/methodos/538, DOI 10.4000/methodos.538, 2006.

S. McCluskey, *Astronomies and Cultures in Early Medieval Europe* (Cambridge: Cambridge University Press, 1998).

Y.E. Meimaris, *Chronological Systems in Roman-Byzantine Palestine and Arabia: The Evidence of Dated Greek Inscriptions* (with K. Kritikakou and P. Bougia; Meletemata, 17; Athens: Diffusion de Boccard, 1992).

K.J. Miller and S. Symons (eds), *Down to the Hour: Short Time in the Ancient Mediterranean and Near East* (Time, Astronomy, and Calendars 8; Leiden: Brill, 2020).

A.A. Mosshammer, *The Easter Computus and the Origins of the Christian Era* (Oxford: Oxford University Press, 2008).

A.A. Mosshammer, *The Prologues on Easter of Theophilus of Alexandria and [Cyril]* (Oxford: Oxford University Press, 2017).

G. Olmsted, *The Gaulish Calendar* (Bonn: Rudolf Habelt, 1992).

J. Rüpke, *Kalender und Öffentlichkeit: Die Geschichte der Repräsentation und religiösen Qualifikation von Zeit in Rom* (Berlin and New York: Walter de Gruyter, 1995).

J. Rüpke, *Zeit und Fest: Eine Kulturgeschichte des Kalenders* (Munich: Beck, 2006).

A.E. Samuel, *Greek and Roman Chronology: Calendars and Years in Classical Antiquity* (Munich: C.H. Beck, 1972).

S. Stern, *Calendar and Community: A History of the Jewish Calendar 2nd century BCE–10th century CE* (Oxford: Oxford University Press, 2001).

S. Stern, *Calendars in Antiquity: Empires, States, and Societies* (Oxford: Oxford University Press, 2012).

CHAPTER 1

The Seven-Day Week in the Roman Empire

Origins, Standardization, and Diffusion

Ilaria Bultrighini and Sacha Stern

The seven-day week is a most familiar way of reckoning days, as it is used nowadays in all parts of the globe, and binds humanity in an almost complete consensus.[1] Beyond its significance in the world religions of Christianity, Islam, and Judaism, the seven-day week functions in most modern societies as a fundamental temporal structure that governs all work and human activity. It is arguably more significant, as a social structure, than the counting of the days of the months, the months, and the years.

But for all its familiarity, the origins of the seven-day week are poorly understood and known. The seven-day week is not derived from any natural phenomenon (e.g. astronomical), nor is it the divider of any natural period such as the lunar months (29 or 30 days) and solar year (365 or 366 days – none of which are multiples of seven). The week is a purely abstract scheme. This entails that its origins must be conventional and cultural. As we shall see, the week as presently known was in fact the result of the convergence of several independent cultural traditions.

In spite of its much earlier origins, the seven-day week as a method of time reckoning was, as we shall see, the invention of the early Roman imperial period. It was standardized in the course of the first or second centuries CE, with the merger of the traditions of the Jewish, Biblical week and the planetary week. The seven-day week was massively diffused in the Roman Empire as a result of Christianization in the fourth century, but its mixed cultural

1 'Almost', because of the continually unresolved debates about the location of the international dateline, which defines, inter alia, the global reckoning of the seven-day week: see Bartky, *One Time Fits All*; Barrows, *Cosmic Time of Empire*; S. Stern, 'Problem of the date line'; and Birth, 'Standards in the shadows'. Research for this article was completed within the ERC Advanced Grant project 'Calendars in Late Antiquity and the Middle Ages: Standardization and Fixation' (2013–18), on which see https://www.ucl.ac.uk/hebrew-jewish/research/research-projects/calendars-late-antiquity-and-middle-ages-standardization-and-fixation-0, accessed 17 May 2020. The article was partly written up when Ilaria Bultrighini was a Research Fellow of the Einstein Center Chronoi of the Freie Universität Berlin, funded by the Einstein Foundation Berlin (September–December 2019).

 | DOI:10.1163/9789004459694_003

origins – Judeo-Christian and astrological – persisted as a challenge to pagans and Christians of Late Antiquity and the early Middle Ages.[2]

1 The Jewish Sabbath Week

1.1 *Early Biblical Origins*

Various claims, or rather assumptions, have been casually made about the ancient Egyptian or ancient Mesopotamian origins of the seven-day week.[3] These assumptions are unfounded: there is actually no evidence of any seven-day cycle in all the rich corpora of literary and astronomical sources of ancient Egypt and Mesopotamia. The only possible exception is the term *hamuštum* in early-second-millennium Assyrian documents, which some have identified as a seven-day week; this interpretation, however, is problematic, as the term more likely means a quarter of the lunar month (thus variably seven or eight days), or even another period such as six or ten days.[4]

The only early source of the seven-day week, which most people know, is the Hebrew Bible. The book of Genesis (1–2:3) builds it into the narrative of the Creation, thus defying in a certain way its abstract identity, and turning it into a fundamental structure of created reality. The seven-day week, conceived of as a cycle of six days of work and one day of Sabbath or rest, then makes many appearances in the Hebrew Bible, not only in the Torah where the law of

2 The only comprehensive study of the seven-day week in Antiquity is Colson, *The Week*. Zerubavel, *The Seven day Circle*, has a detailed chapter on the week in Antiquity, but like much of what has been written on the subject, it is largely derived from Colson. Colson's work has become outdated because of the wealth of epigraphic and documentary evidence that has been subsequently discovered: on the Hebrew and Aramaic side, the Dead Sea Scrolls and Judaean ostraca and documents; on the Latin side, a number of Fasti, parapegmata, and inscriptions; on the Greek side, *P.Harris* I.60, the Tremithus inscription, etc. Colson's analysis of the evidence in his first chapters is cautious and well balanced, but in later chapters he drifts into speculative theories which have now lost their relevance in the light of the newly discovered materials. He should be acknowledged for his general insight, with which we concur in this article, that the Roman Empire was critical to the diffusion of the seven-day week (Colson, *The Week*, pp. 6–9).

3 Rüpke, *Kalender und Öffentlichkeit*, pp. 456–57; id., *Roman Calendar*, pp. 162, 176–77; also Colson, *The Week*, pp. 55–59. The claim of Egyptian origins was first made, in the third century CE, by Cassius Dio (37:18:1), although he concedes in the same passage that its institution was only recent (see further below).

4 Six days would be based on the meaning of *hamuštum*, 'one fifth' (of a standard, 30-day month); other interpretations are based on the documentary evidence, which sometimes also attests the usage of the term for a whole month. See in detail Dercksen, 'Weeks', against the seven-day period interpretation of Veenhof, 'The Old Assyrian Hamuštum period'.

Sabbath is repeated in narratives (Ex. 16:22–30) and in legal passages (e.g. the Ten Commandments: Ex. 20:8–11, also Lev. 23:1–3, etc.), but also in later Biblical books (Ezek. 46:1).[5]

Yet in spite of its importance in Mosaic law, the seven-day week is never used in the Hebrew Bible as a method of time reckoning. Events in the Bible are dated by the day of the month, the number or name of the month, and/or the number of the year, but never by the day of the week.[6] No event in the Bible is said, for example, to have occurred on the Sabbath. The same applies to the Persian period, Judaean documents from Elephantine, many of which are precisely dated according to the Babylonian or Egyptian calendars, but never with a mention of the day of the week.[7]

The absence of the seven-day week as a chronological or dating device raises questions about the consistency of its reckoning in the Hebrew Biblical period. Even if the Sabbath was, to a certain degree, observed, and observance of the Sabbath was presumably synchronized within single cities and communities, it may be asked whether the seven-day week was in phase – i.e. beginning on the same day – across all Sabbath observant communities before the Roman period. We shall return to this important question further below.

1.2 *Sabbath as a Date: Maccabees and Ptolemaic Papyri*

The first known events to be dated by a day of the week appear very late, in the first book of Maccabees (the late second century BCE). At this stage, however, the days of the week are still not used as part of standard dating formulae: thus in 1Maccabees, Seleucid calendar datings do not include the day of the week (e.g. 1Macc. 4:52). Moreover, the use of the week for dating events is limited: 1Maccabees does not make use of the full range of the seven-day week, but only – and significantly – of the Sabbath. The references are to battles that took place on the Sabbath, when the Seleucids sought to take advantage of Jewish reluctance to fight (1Macc. 2:32–8, 9:34,43; also 2Macc. 5:25, 8:26–8, 12:38, 15:1–4). The identification of these days as Sabbath does not serve in these passages a chronological or time-reckoning purpose – indeed, no other information is provided about the dates of the events – but rather to make a point about Jewish religious practices.

5 Many other passages refer to Sabbath, but the meaning of this term – whether the seventh day, or another holy day – is not always clarified (see Doering, *Schabbat*, pp. 18–22). Only Ezek. 46:1 contrasts explicitly the Sabbath day to the 'six days of work'.

6 The absence of days of the week in Biblical dating was already noted in early rabbinic literature, in the Palestinian and Babylonian Talmuds: *yRosh Ha-Shanah* 1:1, *bRosh Ha-Shanah* 3a.

7 See Porten, *Archives from Elephantine*, pp. 126–27; Doering, *Schabbat*, pp. 23–42; id., *Ancient Jewish letters*, pp. 36–37.

The same applies to a small number of papyri from Ptolemaic Egypt which mention the Sabbath. Thus *P.Cair.Zen.* IV 59762 (= *CPJ* 10), a third-century BCE document from Philadelphia (in the Arsinoite nome or modern Fayūm), presents a daily account of brick deliveries in a month of Epeiph (the year is not given). The entry for day 7 (of the month, on col. 1, line 6) is not a number of bricks but instead the word Σάββατα, which seems to mean that no bricks were delivered as it was the Sabbath. The term *Sabbata* should not be construed as a part of the date (together with the digit ζ, for day 7: 'Sabbath the 7th of Epeiph'), but rather as the text of the entry. It is not a time marker, but only the indication of a certain religious observance on that day. Likewise, *PSI.Congr. XVII* 22, another account from the same region but dating to the late second or early first centuries BCE, enters for day 1 of the month of Hathyr the same term Σάββατα, in contrast to the next entry on day 2 which refers to a delivery of beer (recto col. 3, ll. 2–3). A rather different document is *BGU* 20 2847 (first published as *BGU* 8 1763), from the Herakleopolite nome, a letter about a military incident that occurred in year 3, 27 Epeiph (27 July 49 BCE), 'which was ̣προσάμβατον', the day before Sabbath (this date was indeed a Friday). Again, the purpose of mentioning Friday is not to provide a full date, but apparently to make a point about the availability of Jewish fighters before the oncoming Sabbath.[8] All this suggests that even in the Hellenistic period, the Sabbath – let alone the other days of the seven-day week – was not yet conceived as a structure of time reckoning or as a method of dating events.

1.3 *The Seven-Day Week as a Temporal Structure: Dead Sea Scrolls*

The earliest evidence of the week as a temporal structure appears in the same period, and in full force, in the Dead Sea Scrolls. The calendar texts from Qumran Cave 4, which date from the second-first centuries BCE, assume as the basis of their annual calendar a year of 364 days, which is divisible exactly into 52 weeks. This calendar appears earlier on in the Astronomical Book of Enoch, which is thought to date from the third century BCE; but the book of Enoch, in its Ethiopic and other versions, does not refer to weeks or ever remark that its year of 364 days is divisible into a whole number of weeks.[9] The book of Jubilees, usually dated to the second century BCE, does notice in a very short aside that each quarter of its 364-day year consists of 13 weeks (Jub. 6:29–30).

---

8 Maresch in Reiter (ed.), *Dokumentarische Texte*, pp. 67–76, with reference to the related text, *BGU* 20 2846. We are grateful to Tal Ilan for this reference; these newly discovered papyri will appear in the new *CPJ* volume edited by Tal Ilan as nos. 581–82.

9 There is only one fleeting reference to weeks in Ethiopic Enoch 79:4, but not in relation to the 364-day year.

But still, the week plays a very minor part in this work. Jubilees does not correlate any of its dates with specific days of the week.

In contrast, Qumran calendar texts make extensive use of the week and days of the week, most visibly by allocating every week of the year in rotation to one of the 24 priestly divisions, also known as 'watches' and 'courses', that are listed in 1Chron. 24:1–19 (the book of Chronicles does not relate these divisions to weeks, but the allocation of the priestly courses to weeks is well established much later in rabbinic literature). Since the 52 weeks of the year are not a multiple of the 24 courses, a six-year cycle needs to be constructed in Qumran sources, at the end of which whole number of priestly courses is completed. This six-year cycle of weekly, priestly courses constitutes the fundamental structure of most of the Qumran calendars.

Days of the week are also prominently used, in all the Qumran calendar texts, to identify specific dates in the calendar: thus, the New Year is said always to fall on the fourth day of the week, i.e. a Wednesday (4Q319 and 4Q320), and likewise other significant events of the year, liturgical as well as astronomical (lunar).[10] Weeks are also mentioned in various ritual texts, such as 4Q284 frg. 1 and 4Q512 frg. 33, and the liturgy of 4Q504 ('Words of the Heavenly Lights') appears to be allocated to specific days of the week.[11] The Genesis Commentary 4Q252 – in contrast to Jubilees – dates events such as the Flood according to day of the month as well as day of the week; and so does 4Q317, an astronomical, lunisolar text.

The Hebrew terms for 'week' in the Dead Sea Scrolls are, as in later Hebrew, either שבוע (*shavua*, lit. 'period of seven': a term attested already in Deut. 16:9, though not in the sense of a perpetually recurring week; also in Jeremiah 5:24) or שבת ('Sabbath'; perhaps in this sense already in Lev. 23:16–17). The latter is perhaps a synecdoche, but it also expresses the idea that the Sabbath is the focus and, literally, the *telos* or ultimate purpose of the Jewish week. As in later Hebrew, the days of the week are numbered, except for the seventh day which is just called 'Sabbath'; in the Dead Sea Scrolls, numbers are given as numerals (4Q320), as cardinal numbers (4Q321), or as ordinal numbers (4Q326).

---

10 The main Qumran calendar texts are 4Q319–30, 4Q337, and 4Q394 1–2 (= 4Q327), in Talmon, Ben-Dov, and Glessmer, *Qumran Cave 4 (XVI)*. Priestly courses are used in nearly all these texts (in 4Q319–30). On Qumran calendars, see generally VanderKam, *Calendars in the Dead Sea scrolls* and Stern, 'Qumran calendars'. See also Ben-Dov, *Head of All Years*, esp. pp. 52–66, on the seven-day week.

11 4Q504 2:vii:4 (Sabbath) and 3:ii:5 (Wednesday): Baillet, *Qumrân grotte 4 3 (4Q482–4Q520)*, pp. 137–68, with the passages on pp. 150 and 152. Baillet dates the manuscript to the Hasmonean period and possibly 150 BCE, but it could be later in the century.

1.4 *Days of the Week in the Septuagint*

Another literary source from this period that mentions days of the week is the Greek, so-called 'Septuagint' translation of Psalms, which has been identified as Judaean and from the early Hasmonean period, or second half of the second century BCE.[12] A number of Psalms are associated, in their opening verse, with a particular day of the week: Psalm 23 with Sunday, 47 with Monday, 93 with Wednesday, 92 with Friday, and 91 with Sabbath. The association of these Psalms with these days of the week became a tradition that is preserved much later in early rabbinic sources, and suggests a liturgy structured by the seven-day week.[13]

The days of the week, in the Greek Psalms, present themselves as a close translation of the Hebrew usage in the Dead Sea Scrolls: they are numbered, together with the word 'Sabbath' in the genitive (singular or plural) which may be construed literally as 'Sabbath' (e.g. 'first day counted from the Sabbath', or 'first day of the days leading to the Sabbath'), or in a looser sense, perhaps by synecdoche, as meaning 'week'.[14] Friday is called: the 'day before the Sabbath'.[15]

1.5 *The Significance of the Second Century BCE*

All the evidence converges, therefore, to suggest that the second century BCE marked a critical phase in the origins and development of the seven-day week. Its use as a time-reckoning structure, with the numbering of its seven days, in the Dead Sea Scrolls and (in a much more limited way) in the Greek Psalms, was a complete innovation; it could almost be said that it is at Qumran that the week, as a time-reckoning structure, was invented.

The historical context and significance of this invention are unclear. Qumran calendars have a reputation, in modern scholarship, of being 'sectarian',

---

12 Schaper, 'Septuagint Psalter', pp. 174–75.

13 The tradition associating these Psalms with the same days of the week is found in the Mishnah (early third century CE), *mTamid* 7:4. This text, at the end of the tractate, is certainly a later addition to the Mishnah, but it is evidently early, and probably still from the third century. One liturgical text from Qumran, 4Q504 frg 3 (Words of the Luminaries), refers perhaps to the 'fourth day', which may mean Wednesday, and which may signify the day on which the liturgy is recited.

14 Ps. 23:1: τῆς μιᾶς σαββάτων; 47:1: δευτέρᾳ σαββάτου; 93:1: τετράδι σαββάτων. Pelletier argues that the form σάββατα was originally not a plural, but a transcription of the Aramaic singular, which was then construed in Greek as a neutral plural; Pelletier, 'Σαββατα', pp. 436–77.

15 Ps. 92:1: τὴν ἡμέραν τοῦ προσαββάτου. Also slightly later, in the book of Judith 8:6: προσαββάτων καὶ σαββάτων. The word προσάββατ(ον) is inscribed on a bronze icosahedron from Egypt, possibly dating from around 200 BCE, where it is associated with the numeral six (ϛ'), which confirms its reference to the sixth day of the Biblical week, Friday (Perdrizet, 'Le jeu alexandrin de l'icosaèdre').

inasmuch as its 364-day year (also attested in the books of Enoch and Jubilees) was radically different from the lunar calendar that is attributed in this period to the Jewish 'mainstream'.[16] Whilst it is likely that the 364-day calendar, and its sub-division into a whole number of weeks, was only promoted in limited circles in society (of which Qumran was one), whereas the Judaean province (and later ethnarchy and kingdom) as a whole used a lunar calendar that was similar or identical to the calendar of the Seleucids, the appropriateness of labels such as 'sectarian' is problematic in general theoretical terms, as well as in the particular context of Qumran and Judaean history.[17] The extent to which the use of weeks – not to speak of the 364-day calendar – would have been restricted to insular groups such as the 'Qumran sect', and not shared in fact in broader segments of Judaean society, remains a matter of speculation. It may be asked, for example, whether the system of weekly priestly courses – which the Qumran calendars exploit – was actually in use in the Jerusalem Temple in this period.

Another uncertainty is whether Qumran calendars were ever actually used in practice, even by the authors of the Qumran calendar texts. The observance of a 364-day calendar would have led to a gradual deviation from the seasons, as this year length is 1 1/4 days shorter than the solar year. As a result, Passover (for example) would have fallen eventually in the winter, which would surely have been viewed as a violation of the Mosaic injunction of observing it in the month of *aviv* or spring, and would generally have disrupted the agricultural associations of the major Biblical festivals. It has been argued, on that basis, that the Qumran calendars may have been no more than ideal, theoretical schemes, that presented an abstract view of perfected time which could not, however, be used or applied in lived reality. Although this interpretation is contentious – especially as Qumran literature knows no other calendar but the 364-day year – it does raise questions as to whether Qumran sources reflect a real-life use of the seven-day week (and maybe even attest its use in contemporary Judaean society), rather than representing a purely abstract, theoretical scheme.[18]

In addition, the Dead Sea Scrolls, like the Greek Psalms where the seven-day week is occasionally used, are no more than literary works. No documents from the second or first centuries BCE, in Judea or (as we have seen) in Ptolemaic Egypt, are dated by the seven-day week. In the absence of documentary

16 Talmon et al., *Qumran Cave 4 (XVI)*, esp. pp. 3, 6.
17 Stern, 'The "Sectarian" calendar of Qumran'.
18 Id., 'Qumran calendars'.

evidence, it cannot be proved that the seven-day week reckoning had yet become a social reality in this period.

Nevertheless, it is probably significant that the second century BCE is precisely when the seven-day week makes its first appearance in a number of literary contexts, ranging from events dated to the Sabbath in the books of Maccabees and days of the week in the Greek Psalms, to fully fledged calendar schemes in Qumran sources. The contrast between second-century BCE Jubilees and Qumran calendar texts, that subdivide the 364-day year into whole weeks, and the third-century BCE Enoch, that does not (even in the later and only versions that have been transmitted to us), is particularly telling. The second-century BCE invention of the seven-day week as a time reckoning system – even if only theoretical or literary – may well have been related to the revival and promotion of the observance of the Sabbath, which is credited to a Maccabean rebels of the 160s,[19] but was surely also shared and promoted by other Judaean groups at the time, such as the communities described in Qumran literature (and the authors of this literature),[20] and may have percolated further on to the Diaspora in Egypt (as the papyri above mentioned possibly suggest). Promotion of Sabbath observance in this period may have elicited the conceptualization of the week as a recurring sequence of seven numbered days and as a fundamental structure of time reckoning and calendars.

1.6 *The Days of the Week in Dating Formulas: First-Century CE Judaean Ostraca*

As stated, the Jewish seven-day week as a time reckoning scheme only begins to appear in the second century BCE in literary sources, and in a context that may have been largely theoretical. The first attested, documentary use of days of the week as a dating method is considerably later, and marks a further stage in the development of the seven-day week. It is in the early first century CE that one first finds, in Judaean ostraca, the day of the week as part of a date. These ostraca indicate that the day of the week had become, by then, an established element of time reckoning in daily life.

19 In addition to the sources cited above, 1Macc. 1:41–8, 2Macc. 6:6, 6:11. The reluctance of the Maccabees to fight on the Sabbath has a precedent, however, in a narrative relating to the reign of Ptolemy I Soter ('son of Lagus'), probably in the late fourth century BCE, whereby Ptolemy entered Jerusalem on the Sabbath and met no resistance; the narrative is told by Agatharchides (of Cnidus, second century BCE) as quoted by Josephus, *Contra Apionem* 1:209–11. See generally Doering, *Schabbat*, pp. 539–65.

20 On Sabbath observance in the Dead Sea Scrolls, see Doering, *Schabbat*, pp. 119–282; in Jubilees, ibid., pp. 43–118.

Only a small number of Judaean Aramaic ostraca from this period has been discovered and published, by Ada Yardeni (first in 1990, then in an augmented and revised edition in 2012). The ostraca are fully dated but their preservation is fragmentary, and moreover, the year numbers are not identified by any reign or era; Yardeni's dating to the first century CE is therefore only palaeographical.

Ostracon 1 in this series, dated palaeographically to the early first century CE, records a series of deliveries of fig cakes, on 'day of Sabbath 29 Tishri', 'day of Sabbath 13 Marheshvan', and 'day one of the week'.[21] Ostracon 2, dated less precisely by Yardeni to the 'first century', refers to '4 of the week, 24 Iyyar (?)' and '5 of the week'.[22] Ostracon 4, also 'first century', is dated 'day (?) four 20 Marheshvan year 104 (?) ... eve of Sabbath on 9 Kislev'; 'eve of Sabbath' is presumably the technical term for Friday.[23] This ostracon may be close in time to early first-century CE ostracon 1, which is similarly dated 'year 102'.

What these ostraca reveal is not only the use of the day of the week, but also its formal incorporation into the dating formula, positioned before the day of the month. The terminology, moreover, appears to be fairly consistent, with the use of either numerals or cardinal numbers (except for Friday and Saturday, both named with reference to the Sabbath) and the term 'Sabbath' for the week.

1.7 *Later Aramaic Evidence*

This Aramaic (and Hebrew) usage is likely to have maintained itself in subsequent centuries, although few documents have survived to be able to confirm this. Close in time to the first-century CE ostraca are two Judaean desert documents, the earliest of which is *P.Yadin* 3, a Nabatean document dated to the year 28 of Rab'el (II) (97/8 CE) which refers to the 'first day of the week'. The next is an Aramaic deed of sale dated 120 CE, *P.Yadin* 7, with several references to the day of the week.[24] In these documents, the week is not used as part of the

21 Yardeni, 'Twelve published and unpublished Jewish Aramaic ostraca': יום שבתא 29 לתשרי ... יום שבתא 13 למרחשון ... יום חד בשבה .... The term for 'week' is 'Sabbath', as in the Qumran calendar texts (see above, n. 10), and in this text a cardinal number is used ('day one'). Yardeni remarks that the author of this ostracon must have worked on the Sabbath, as he delivered goods on this day.

22 Ibid., pp. 218–20: 4 בשבה 24 לאיר ... 5 בשבה. Numerals are used for the day numbers, and therefore cannot be determined as cardinal or ordinal. Yardeni does not specify if she means first century BCE or CE.

23 Ibid., pp. 221–26: יום ארבעה 20 למרחשון שנת 104 ... ערובת שבתא ב 9 לכסלו. Again, a cardinal number is used. The year number could be 14, and thus a regnal year (year 104, the reading favoured by Yardeni, would imply an era which neither Yardeni nor we can identify). The same uncertainty surrounds the dating of ostracon 1, 'year 102'.

24 *P.Yadin* 3.25 (Nabatean) and *P.Yadin* 7.6–7, 12, 43, 46, 47 (Aramaic): Yadin, Greenfield, Yardeni, and Levine, *Documents*, pp. 238–9 and 80–87 respectively. Here also, a cardinal

dating formula, but only as a scheme for the allocation of irrigation rights. The use of the week as a scheme for time distribution, in these documents, possibly marks a new development in the socio-economic uses of the seven-day week. Furthermore, as Guy Stiebel has rightly pointed out, the day and the night are treated here as separate time units, so that the week is effectively divided in seven numbered days and seven numbered nights – a practice still attested, much later, in Jewish texts of the medieval period.[25]

The next extant Aramaic documents dated by the day of the week are the Jewish tombstones from Zoar (south of the Dead Sea), of the late fourth to early sixth centuries. The earliest of these is dated '1 in the Sabbath, 7 days in the month Tammuz (?) ... in the year 300 (?)' (from the destruction of the temple, i.e. 369 CE); '1 in the Sabbath', a literal translation, means first day of the week.[26] In all subsequent tombstones, the day of the week is given as a mere number, without 'in the Sabbath': for example, 'day two ... three Iyyar ... year three hundred and twenty three from the destruction of the Temple' (c. Monday 12 April 392).[27] The omission of this clause, from the end of the fourth century onwards, suggests perhaps a normalisation of the seven-day week, whereby it becomes no longer necessary to explain that these day numbers are 'in the Sabbath' or the week. The day of the week is given in the first example above as an alpha-numeral,[28] and in the second, as a cardinal number;[29] elsewhere in the

---

number is used with an Aramaic lexeme for 'Sabbath', שבה. For a general discussion, see Katzoff and Schreiber, 'Week and Sabbath', pp. 102–6.

25 Stiebel, 'Meager bread', p. 297 n. 77. Stiebel argues that the phrase לילא חמשא בשבה (*P.Yadin* 7.6–7), for example, does not mean 'at night, on Thursday' (as translated by Yadin et al., *Documents*) but 'the fifth night of the week'; similarly, יום ארבעה בשבה (*P.Yadin* 7.46) means 'the fourth day of the week' ('day' in the sense of daytime). For similar tenth-century Hebrew examples, see Stern, *Jewish Calendar Controversy*. It is also interesting to find, in the context of irrigation practices in *P.Yadin* 7, that the night is considered to follow the day (thus the fourth night of the week, *P.Yadin* 7.47, comes after the fourth day of the week in the previous line), which is not what one would normally expect in a Jewish calendar.

26 Meimaris and Kritikakou-Nikolaropoulou, in collaboration with Brock, *Inscriptions from Palaestina Tertia*, IC, no. 7: א בשבה (Sunday), but readings of the month's name and the year are uncertain; Stern, 'Jewish Aramaic tombstones', p. 165. On the era of Destruction at Zoar, see Stern, *Calendar and Community*, pp. 88–91. On the days of the week, see Meimaris et al., *Inscriptions from Palaestina Tertia*, IC, p. 34.

27 Ibid., no. 12 (= Naveh, 'Aramaic tombstones', no. 7): ביום [ת]ריה ... בתלתה יומין בירח אייר.

28 Alpha-numerals: Meimaris et al., *Inscriptions from Palaestina Tertia*, IC, no. 7, no. 16 (= Naveh, 'Seven new epitaphs', no. 20, on which see Stern, 'Jewish Aramaic tombstones', pp. 166), no. 20, no. 37 (= Stern, 'New Tombstones', no. 15), no. 40 (= Stern, 'New Tombstones', no. 16).

29 Cardinal numbers: Meimaris et al., *Inscriptions from Palaestina Tertia*, IC, no. 12, no. 39 (= Bitton, Dweck, and Fine, 'Yet another Jewish tombstone', no. 59).

Zoar tombstones, ordinal numbers can also be used.[30] Friday is now called 'day of eve' (יום ערובתה),[31] and in one instance numbered 6.[32] Saturday is 'day of Sabbath'.[33] Also from this period, but in Egypt, the date of the Jewish Aramaic marriage contract of Antinoopolis (417 CE) includes the weekday of 'four in the Sabbath' (Wednesday).[34] This inconsistent use of cardinals and ordinals shows perhaps that even in this period, the nomenclature of the seven-day week was not yet fully standardized.[35] Alternatively, this may be seen as a transitional period from cardinal numbers (which seem to dominate in the first and early second centuries) to ordinal numbers (which became standard in the Middle Ages and until today).

By this late period, however, the seven-day week and its numbered and Sabbath-related days are already well attested in Aramaic and Hebrew sources. They are also well attested in Syriac, the earliest example of which might be the Peshitta version of the New Testament, where days of the week occasionally appear: Friday is thus 'the eve', and Sunday 'one of the Sabbath'[36] (on the original Greek version, see below). In early rabbinic sources, the full range of the days of the week appears, in Hebrew, already in the Mishnah (early third century), often just as an ordinal number. The days of the week are fully embedded in Mishnaic time reckoning: for example, witnesses in court are expected to know the day of the week when an event was witnessed.[37]

---

30 Ordinal numbers: Meimaris et al., *Inscriptions from Palaestina Tertia*, Ic, no. 21 (= Naveh, 'Aramaic tombstones', no. 10), no. 25 (= Stern and Misgav, 'Four additional tombstones', no. 28), no. 28, no. 34 (= Stern, 'New Tombstones', no. 14, on which see id., 'Jewish Aramaic tombstones', p. 167), no. 29 (= Naveh, 'Seven new epitaphs', no. 22), no. 36, no. 45 (= Naveh, 'Aramaic tombstones', no. 3), no. 47, no. 61. Ordinal numbers are more numerous, but the corpus is too small for this to be considered significant. Alpha-numerals, cardinal and ordinal numbers seem to be distributed fairly evenly among all weekdays (Sunday to Thursday).

31 Meimaris et al., *Inscriptions from Palaestina Tertia*, Ic, no. 30 (= Naveh, 'Aramaic tombstones', no. 4 as corrected in id., 'More on the tombstones', p. 585 n. 9), no. 35 (= Naveh, 'Aramaic tombstones', no. 11), no. 38 (= Naveh, 'Seven new epitaphs', no. 24).

32 Meimaris et al., *Inscriptions from Palaestina Tertia*, Ic, no. 20: יום ו (day 6).

33 Ibid., no. 23 (= Misgav, 'Two Jewish tombstones', no. 30), no. 50: יום שובתה.

34 With a cardinal number: בארבעה בשו[בתה]: Sirat, Cauderlier, Dukan, and Friedman, *La Ketouba de Cologne*.

35 As we shall later see, terminological inconsistency is even more pronounced in Greek inscriptions from this period, not least from the same site of Zoar, where planetary and Christian designations of the days of the week are freely mixed.

36 Matthew 27:62 ܥܪܘܒܬܐ and 28:1 ܚܕ ܒܫܒܐ.

37 Full range of weekdays: e.g. *mTaanit* 2:7, 2:9; *mMegillah* 1:2. Witnesses: *mSanhedrin* 5:1.

1.8 *The days of the Week for Dating Purposes: Early Greek Sources*

The days of the Jewish week also appear for dating purposes, in the first century CE, in Greek sources, but the sources are mainly literary and, not surprisingly, nearly all Jewish or Jewish-related.

As in the earlier Greek sources (the books of Maccabees), Josephus' references to days of the week – mostly Sabbath, which he often calls 'the seventh' (ἑβδομάς: e.g. Jewish War 2.517) – are generally not for the purpose of dating events, but rather to highlight Sabbath observance. Thus in Antiquities 12.4, he explains that Pompey entered Jerusalem unopposed because that day was Sabbath. In Antiquities 13.250–2, he quotes a passage of Nicolaus of Damascus whereby in 130 BCE, during his campaign against the Parthians, Antiochus VII remained stationed for two days by the river Lycus at the request of his ally John Hyrcanus, because of some ancestral festival during which it was forbidden for the Jews to march out; Josephus explains that Pentecost occurred in that year after the Sabbath, and that it would have been forbidden to travel on these consecutive holy days.[38] In Antiquities 16.163, Josephus quotes a decree of Augustus exempting Jews from court attendance on the Sabbath and 'on the preparation before it, from the ninth hour' (ἐν σάββασιν ἢ τῇ πρὸ αὐτῆς παρασκευῇ ἀπὸ ὥρας ἐνάτης). This is perhaps the earliest attestation of Friday as παρασκευή ('Preparation'), a term reflecting Jewish preparatory activity before the Sabbath; this term is specific to Greek language, and was soon to become standard for 'Friday'.

The same designation of Friday is famously used in the Gospels, in the context of the Passion. Here, mention of the days of the week comes much closer than in Josephus to the function of dating events within the narrative. Thus the Crucifixion thus took place on 'Preparation, which is the day before Sabbath' (παρασκευή, ὅ ἐστιν προσάββατον) (Mk 15:42); the Resurrection was on the 'first of the week' (τῇ μιᾷ τῶν σαββάτων) (Mk 16:2).[39] Similarly, a meeting of Paul and his companions to break bread is said, in Acts 20:7, to have been on 'the first of the week'; and in 1Cor. 16:2, Paul instructs his followers to make a collection every 'first of the week'.

A unique first-century CE Greek inscription provides further evidence of the use of the Sabbath week for dating purposes. It is a birth record from Tremithus (eastern Cyprus), dated 'year 7 of Domitian Caesar, birth of twins (?) ... month of Tybi 25, 1st hour of the day, *sambat* 6'. The latter means day 6 of the Sabbath,

---

38 On this passage see Stern, *Calendar and Community*, pp. 113–15.

39 Parallels in Mt 27:62, 28:1; Lk 23:54, 24:1; Jn 20:1. In John 19:31, mention of the Sabbath is given more explicit relevance: the Jews did not want crucified bodies to remain on the cross on Sabbath.

i.e. Friday. The reference to Sabbath suggests, in the context of the first century, that the author of the inscription was Jewish, although this cannot be proved.[40]

All these are indications that in Greek too, the days of the Jewish week were used for dating, at least by Jews, from the first century CE.

1.9 *The Significance of the Early First Century* CE

The evidence that happens to have survived is insufficient to prove that the early first century CE marked the beginning of a new usage of the Jewish seven-day week, as events [illegible] being dated according to the day of the week. The dating of the Aramaic ostraca, in particular, remains somewhat conjectural, and nothing precludes the existence of similarly dated ostraca from an earlier period that may be waiting to be discovered. Nevertheless, it is striking that the first appearance of the day of the week in Judaean dating formulas coincides with the emergence of a seven-day week, at about the same time, in a very different part of the Roman empire. The relationship between these two phenomena needs to be carefully explored.

2 The Roman Planetary Week

2.1 *The Origins of the Planetary Week*

The planetary week makes it first appearance in Rome and other parts of central Italy in the late first century BCE. This is also the first appearance, in Rome, of a seven-day week. As a planetary week, it presents itself very differently from the Jewish week. It is not structured around the Sabbath or on any specific day of the seven days, and it does not assume its seventh day to be a day of rest – a practice which was always associated specifically with the Jews, and from which Greeks and Romans until then always dissociated themselves.[41] The Roman planetary week seems to have served a mainly astrological function, by indicating which planets ruled over which days in rotation. Unlike the Jewish week, it did not determine work patterns or structure religious life. At first sight, the Roman planetary week was a separate tradition, that was formed independently of the Jewish week. However, as we shall see, the traditions are likely to have always been linked.

'Planets', in this context, are defined as the celestial bodies visible to the naked eye (and hence, to the ancient sky watchers) that, unlike the stars, have

40 For detailed analysis of the full date, which is given in the eastern Cypriot calendar and corresponds to 28 December 87 CE, see Stern, 'A "Jewish" birth record'.

41 To pick just one example, Agatharchides of Cnidus in the passage mentioned above, n. 19.

a course of their own and therefore appear not to be fixed to the firmament. This includes the five planets Mercury, Venus, Mars, Jupiter, and Saturn, as well as the Sun and Moon. The reason why the planetary week counts seven days is simply that these bodies are seven. The planetary names of the days of the week have survived in many modern languages, especially the romance languages, and are therefore not unfamiliar to the modern reader. As we shall see, in the Roman tradition the week starts on Saturday, the day of Saturn, followed by the days of the Sun (Sunday), Moon, Mars, Mercury, Jupiter, and Venus.

The planetary week should be regarded as specifically 'Roman' – alternatively 'Latin', or at the very most 'Italian' – with good reason. It is well attested in the city of Rome and other parts of central-southern Italy in the Augustan period and first century CE, but it did not spread to the eastern Mediterranean and other regions until much later; and even then, it remained a limited phenomenon. Contrary to what has often been asserted, there is no evidence of a planetary week in ancient Egypt, Mesopotamia, the Hellenistic world, or anywhere further east, before the second century CE (as we shall see below). The evidence suggests, therefore, that the tradition could only have originated in Rome or Italy.[42]

This is all the more plausible in that the planetary week is very different from the astrological traditions that are known to have existed in the East. The Egyptian and Mesopotamian astrological traditions (which were eventually absorbed, in various degrees, in the Hellenistic tradition) were based on empirical observation of the stars and planets and on the construction of theoretical models accounting for their celestial motions; in other words, the science of astronomy. On the basis of rigorous astronomical inquiry, the position of stars and planets could be accurately known and predicted, and their putative influence on the sub-lunar world could be determined. The planetary week, in contrast, was not founded on any scientific astronomy of this kind. The seven celestial bodies that it refers to do not correspond, in reality, to any seven-day cycle; the correspondence of days of the week with each of

42 See also Bultrighini, 'Thursday [*dies Iovis*]', pp. 62–65, and ead., '*Theōn hemerai*'. This conclusion differs radically from earlier scholarship, which conjectured, but without any evidence, that the planetary week originated from Babylonian and/or Egyptian, and hence Hellenistic, astrology: see, with references to earlier scholarship, Rüpke, *Kalender und Öffentlichkeit*, pp. 456–57; id., *Roman Calendar*, pp. 162, 176–77; and see also Smith, *Elegies of Albius Tibullus*, pp. 239–40; Salzman, 'Pagan and Christian notions', p. 188. Colson (*The Week*, pp. 55–59) is rightly sceptical about the ancient origins of the planetary week, but still falls back on the assumption that it 'spread from east to west' and not the reverse (ibid., p. 59). Similarly, Rüpke writes about the 'Greek/Hellenistic planetary week'; yet as Brind'Amour convincingly argues, there is no evidence of such a concept in any source of the Hellenistic period (Brind'Amour, *Le Calendrier romain*, p. 263).

these bodies is abstract and bears no relationship to astronomical reality. It is reasonable to assume that the planetary week was designed for astrological purposes, but this astrology was popular rather than scientific, and very different from what had been the astrological tradition in Babylonia, Egypt, and the Hellenistic world.[43] It is not surprising, therefore, that the planetary week cannot be found in the East, until it spread eastwards during the Roman imperial period. A western provenance, such as Rome or Italy, is for that reason quite plausible.

Nevertheless, if the provenance of the planetary week was Rome, an explanation is still needed as to why it emerged specifically in or around the late first century BCE, whereas there is no evidence of it at all in earlier Roman tradition. Several factors converged towards the creation of the planetary week at the end of the Republic and the early Augustan period. This was a period of heightened interest in astrology in Rome,[44] as well as one of intense calendrical activity, most importantly with the institution of the Julian calendar, its correction under Augustus, and its rapid diffusion, adoption, and adaptation in the Roman provinces,[45] alongside the promotion and diffusion of monumental Fasti (Roman calendar tables) at least throughout the Italian peninsula.[46] Astrology and calendar naturally converged in the creation of a new, seven-day planetary week.

In addition, the late Republican period seems to have been one of growth for the Diaspora Jewish community in Rome, perhaps as a result of Pompey's activities in Judaea in the 60s BCE.[47] The prominent presence of Jews in Rome heightened the awareness, among Romans, of the Jewish Sabbath and probably also of the seven-day week, as is attested in several passages of Augustan-period Latin poetry. As will be argued below, the Jewish seven-day week may have played a critical part in the creation, in Rome, of a planetary week. Whatever role the Jewish week exactly played in the formation of the planetary week, it

43 See very partial discussion in Colson, *The Week*, pp. 68–74.

44 Domenicucci, *Astra Caesarum*; Green, *Disclosure and Discretion*; Bultrighini, 'Thursday (*dies Iovis*)', p. 64; ead., '*Theōn hemerai*'. Manilius' *Astronomica* (on which see Hübner, *Manilius 'Astronomica'*) exemplifies Roman interest in astrology under Augustus, although Manilius chooses for some reason to pass the planets in silence (Green, *Disclosure and Discretion*, pp. 128, 191). See also Horace, *Odes* 2:17.

45 Feeney, *Caesar's Calendar*; Stern, *Calendars in Antiquity*, pp. 211–27, 259–94; and Chapter 2 in this volume.

46 Rüpke, *Kalender und Öffentlichkeit*.

47 This explanation has been generally assumed by historians: Leon, *Jews of Ancient Rome*, pp. 4–16; Schürer, *History of the Jewish People*, iii.1.75; Rutgers, 'Roman policy'; Williams, *Jews in a Graeco-Roman Environment*, pp. 34, 49, the latter also referring to Herod's capture of Jerusalem in 37 BCE.

seems likely that the two traditions were never completely separate or independent of one another.

The creation of the seven-day planetary week fitted well a Roman context for further reasons. The planetary week shares much in common with another cycle, the eight-day *nundinae*, which constituted a fundamental structure of the Roman calendar long before the introduction of the seven-day week. The eight nundinal days were named after the main market towns around Rome, and regulated the rotation of markets in the region; but they also regulated many other aspects of legal, political, social, and religious life.[48] Although this cycle of eight days ran independently from the rest of the calendar, it was inscribed in a column of its own in the Fasti, where the nundinal days were represented with the letters A to H in a recurring sequence; this can be found in the earliest surviving Fasti, the Fasti Antiates Maiores (from between 67 and 55 BCE), and in all subsequent Fasti.[49] The planetary week, a similarly short cycle of days running independently of the rest of the calendar, was thus a concept that fitted well the calendrical culture of Rome. In the early stages of its introduction, in the Augustan period, the planetary week did not take long to get inscribed in at least some of the Fasti, alongside the nundinal column and similar to it, but with the letters A to G (see further below); this demonstrates how easily it could be integrated in the existing structures of the Roman calendar.

2.2 *The Day of Saturn: Early Augustan Period*

The earliest evidence of a planetary day in Augustan Rome appears in an elegy of Tibullus, which may have been written as early as 30 BCE, and certainly no later than when the author died in 19 BCE. In this passage, he uses the day of Saturn as a pretext for remaining longer with his mistress Delia:

> Aut ego sum causatus aves aut omina dira | Saturni sacram me tenuisse diem (1.3.17–18)
>
> I pretended that either the birds, or bad omens, | or Saturn's sacred day were detaining me

The association of the 'sacred day of Saturn' with 'birds' and 'bad omens' suggests that it carries astrological significance, rather than representing some calendar day with religious, cultic significance. However, the passage only

48 Rüpke, *Kalender und Öffentlichkeit*, pp. 225–28, 453–55; id., *Roman* Calendar, pp. 32–34, 41–42, 160–62; Ker, 'Nundinae'.

49 Rüpke, *Kalender und Öffentlichkeit*, esp. 39–44; id., *Roman Calendar*, pp. 6–8.

proves the concept of a day dedicated to Saturn; it does not necessarily imply that the day of Saturn belonged to a seven-day week populated by the other planets, or even that this day of Saturn was reiterated every seven days. The concept of a day of Saturn could have been the first step towards the formation of a full planetary week, perhaps a few decades later.[50] But it is equally possible that the day of Saturn that Tibullus refers to belonged already, in his mind, to a fully-formed planetary week.

Another interpretation, however, is equally possible. A number of scholars have assumed, or even completely taken it for granted, that 'day of Saturn' in this passage is simply Tibullus' way of referring to the Jewish Sabbath.[51] This interpretation finds support from a contemporary parallel in a line of poetry of Horace, who describes how a friend excused himself from helping out on the grounds that it was the Sabbath (*sabbata*) – thus, similarly using this day, the Jewish Sabbath, as a pretext for inaction.[52] Accordingly, the naming of the Jewish Sabbath as 'day of Saturn', in Tibullus' poem, could thus be no more than an *interpretatio romana* of the Sabbath, in which the planetary or astrological element is present but only as a supplementary, perhaps explanatory, component. Indeed, the traditional, 'orientalist' assumption that the Near East was the cradle of astrology could easily have led Romans to believe that the Jewish scruples about the Sabbath were related to astrological concerns. But primarily, Tibullus would be referring in this passage to the Jewish Sabbath.[53]

50 Colson, *The Week*, p. 35.

51 So ibid., pp. 16–17 and 35; Leon, *Jews of Ancient Rome*, p. 13; Stern, *Greek and Latin Authors on Jews*, vol. 1, pp. 318–20; and Brind'Amour, *Le Calendrier romain*, pp. 258–60. See further Rüpke, *Kalender und Öffentlichkeit*, p. 458, and in much detail, Smith, *Elegies of Albius Tibullus*, pp. 238–40.

52 Horace, *Satires* 1:9:68–70. Further, contemporary references to the Jewish Sabbath as an unlucky day for travel or for business can be found in Ovid, *Remedia Amoris* 219–20 and *Ars Amatoria* 1:415–6.

53 The *interpretatio romana* of the Jewish Sabbath as the day of Saturn is explicit, over a century later, in Tacitus' account of the Jewish religion (*Histories* 5.4), where he writes that some people thought that the sabbatical rest was in honour of Saturn, 'because, of the seven planets that rule the mortals, the planet of Saturn moves in the highest orbit and with the greatest potency, and many of the celestial bodies traverse their paths and courses in multiples of seven'. Not only do we find here the Jewish week linked to its planetary counterpart, but also a clear statement of belief in the planetary control of human affairs (quod de septem sideribus, quis mortales reguntur). Other Latin sources, in this later period, call the Jewish Sabbath the 'day of Saturn': Frontinus, *Stratagemata* 2.1.17 (late first-century CE Rome, on Vespasian's assault on the Jews on the Sabbath). Later still, in Greek sources, it is rendered similarly as the 'day of Kronos': Cassius Dio (early third century) 37.16.2–4, 37.17.3, 66:7:2.

These alternative interpretations depend possibly on the exact meaning of Tibullus' phrase. We have translated *sacram diem*, above, as 'sacred day', but others have translated it as 'accursed day'. The difference between these translations is critical. 'Accursed day' agrees well with the astrological interpretation, since Saturn appears in other, contemporary Latin sources as astrologically unfavourable.[54] It does not agree with the Jewish or Biblical Sabbath, which is holy but certainly not accursed, although Romans may still have regarded this foreign practice as accursed. Less ambiguous is 'sacred day', which agrees well with the Jewish Sabbath, and far better than with a day believed to be under the influence of a pernicious planet.[55]

If the Jewish interpretation is correct, whereby Tibullus is primarily referring to the Jewish Sabbath, this short passage could have much to inform us about the origins of the Roman planetary week. It could suggest that the planetary week did not originate from some independent, home-grown Roman astrological tradition, but rather from the Jewish seven-day week, or at least from the Jewish Sabbath, and as an *interpretatio romana* of it. It may have been restricted, initially, to an interpretation of Sabbath as a day of Saturn; but in the course of Augustus' reign, the rest of the Jewish week would have been associated with the other planets – which conveniently was of the same number – and the planetary week would have been accordingly constructed.[56] Whether or not this scenario is correct, there is certainly room to argue that the Roman tradition of the planetary week was not independent from the Jewish week, even if it originated in Rome and not anywhere in the east.

2.3 *The Planetary Week: Development and Diffusion under Augustus and in the First Century* CE

The first explicit evidence of a fully constituted, Roman seven-day week appears a little later than Tibullus, although the sources are difficult to date precisely.

Two of the monumental Fasti that have been fragmentarily preserved from the Augustan period include a column dedicated to a seven-day cycle, with days

54 Horace, *Odes* 2:17:22–3 (impio ... Saturno, 'impious Saturn'); Propertius, *Elegies* 4:1:84 (grave Saturni sidus in omne caput, 'planet Saturn heavy on every head').

55 Brind'Amour (see n. 51 above), who favours the Jewish Sabbath interpretation, translates appropriately 'le jour consacré à Saturne'. Stern (n. 51 above) translates 'accursed day', which might work better with the astrological interpretation than with his own Jewish Sabbath interpretation, but this is consistent with his general policy, in *Greek and Latin Authors on Jews and Judaism*, of using existing translations rather than his own.

56 Colson (*The Week*, pp. 39–42) is doubtful that the planetary week was constructed around the day of Saturn, but his argument is intuitive and weak.

FIGURE 1.1
Fasti Sabini (*CIL* IX 4769)
IMAGE FROM THE ARTHUR E. GORDON COLLECTION, REPRODUCED WITH KIND PERMISSION OF THE CENTER FOR EPIGRAPHICAL AND PALAEOGRAPHICAL STUDIES, OHIO STATE UNIVERSITY

numbered from A to G: the Fasti Sabini and the Fasti Nolani.[57] In both cases, the seven-day column is located immediately to the left of the eight-day nundinal column (where days succeed each other in cycles running from A to H). An additional, very fragmentary example, are the Fasti Foronovani which preserve traces of the same columns.[58] These Fasti are not from Rome itself, and not from regions that are known to have been significantly populated by Jews, which suggests that already in this early period, the diffusion of the seven-day week was not necessarily dependent on direct Jewish influence. As their modern scholarly names indicate, the Fasti Sabini are from the territory of the Sabines, north of Latium, as are the Fasti Foronovani, whilst the Fasti Nolani are from Nola in Campania, at quite some distance to the south of Rome. Their date

57 Fasti Sabini: *CIL* IX 4769 = Inscr.It. XIII.2, no. 5 = *AE* 1953, 236; Rüpke, *Kalender und Öffentlichkeit*, pp. 95–97; also Brind'Amour, *Le Calendrier romain*, pp. 267–68. Fasti Nolani: *CIL* X 1233 = Inscr.It. XIII.2, no. 37 = *AE* 1959, 253: Rüpke, *Kalender und Öffentlichkeit*, pp. 100–4.

58 *CIL* IX 4770 = *Inscr.It.* XIII.2, no. 21, from Montebuono Sabino (Rieti) (ancient Forum Novum), which has also been dated to the Augustan period. The extant fragment suggests two parallel columns with a sequence A–G on the left and A–H on the right.

is Augustan; according to Rüpke and others, the Fasti Sabini are dated after 19 BCE but not much beyond the end of the century.

These Fasti tell us that the practice of counting days in continuous hebdomadal cycles was sufficiently well established, in various parts of Italy, to be formally included in public calendars. They do not specify what these cycles were about: just like the nundinal days, the days of the seven-day column are only assigned a letter, whose identity was flexible and could vary from one year to the next (although both columns would have begun, on 1 Januarius, with the letter A, the years would not always have started on the same nundinal or hebdomadal day; consequently, the letter A, and subsequent letters in the sequence, would have symbolised different days from one year to the next). We can only conjecture that these seven-day cycles were intended as planetary.[59]

More explicitly planetary are graffiti inscriptions from Pompeii. They are particularly difficult to date, and could be said to belong to anytime in the century leading up to the destruction of the city in 79 CE. However, they are undisputable evidence of a planetary week. The first is *CIL* IV 6778–6779, a graffito found on the *triclinium* wall of a private house in Insula IV, Regio V of Pompeii, which reads as follows:

V K(alendas) Idu(s) Qui(ntiles)
Saturni
Solis
Lunae
Martis
[Mercuri]
Iovis
Veneris

The list of planets follows the order of the planetary week, rather than a cosmological or any other order (the order of the planetary week will be discussed further below). This shows that the inscription cannot be a mere list of planets, but specifically a list of the days of the week. This is further confirmed by the genitive ('of Saturn, of Sun, etc.'), which presumably refers to an omitted *dies* ('day': 'day of Saturn', etc.).[60] One of the planets, Mercury, disappeared as a

59 So Rüpke (*Kalender und Öffentlichkeit*, p. 97), who conjectures further that their use must have been astrological. See also Bultrighini, 'Thursday (*dies Iovis*)', p. 62 n. 3.

60 Colson (*The Week*, p. 32) assumes that this graffito (along with the Greek one discussed next) could be interpreted as an attempt of a schoolboy to memorize the days of the week.

result of damage to the plaster. The top line, which seems to preserve two random calendar dates, is possibly unrelated to the planetary week that follows.[61]

A similar list of planetary days, but this one in Greek, is in CIL IV 5202. It is headed explicitly with Θεῶν ἡμέρας ('days of the gods'), followed by the names of the planets in the genitive, arranged vertically: Κρόνου Ἡλίου Σελήνης Ἄρεως Ἑρμοῦ Διός Ἀφροδείτης ('of Kronos, Helios, Selene, Ares, Hermes, Zeus, Aphrodite', the Greek equivalents of Saturn, Sun, Moon, Mars, Mercury, Jupiter, and Venus).[62] The use of the Greek language – and of the corresponding divine names for each of the planets – is a reflection of bilingualism in this part of Italy, and cannot be treated as evidence that the planetary week originated from the Greek East.

The integration of the planetary week in the Roman calendar is more evident from CIL IV 8863, an elaborate graffito that was found in a *taberna vasaria*, a shop selling wine and terracotta containers, in via dell'Abbondanza, and that lays out in separate, very uneven columns a set of different calendrical sequences.[63] In the middle of the inscription, the days of the Roman month are laid out on three columns (the days are counted down, in Roman style, and begin from the day after the Ides, which is day XIX before the Kalends). To the right of it, laid out in two columns and a bit, is a continuous sequence from I to XXX, which can be interpreted as the days of the lunar month. To the left, in one column, is a list of the eight nundinal days starting from *Pompeis* ('at Pompeii'), and explicitly headed with the title *Nundinae*. To the far left, finally, is a column headed with the title *Dies*, followed by the sequence of the planetary week (the names are abbreviated): *Sat Sol Lun Mar Mer Iov Ven*.

In addition to the graffiti, there are similarly undated, but early imperial, parapegmata which include the planetary week. A parapegma is a calendar with perforated holes, in which a peg is moved along on a daily basis to keep track of the days. CIL VI 32505, from a village in the Latium, is an elaborate parapegma of which enough has been preserved to show that it originally included the thirty days of the lunar month in a complex geometric design and with the heading, in both top corners, *Dies lunar*(*es*) (lunar days); along the right, a vertical list of the local *Nundinae*; and along the top, horizontally, the names of planets (written out in full, in the genitive) in the sequence of the planetary

61 Since the month Quintilis was renamed Julius in 44 BCE, this graffito should probably be dated not very long thereafter; *pace* Mau, 'Ausgrabungen von Pompeji', p. 361. Thus, if CIL IV 6778 and 6779 are not unrelated, the list of days of the planetary week could possibly be assigned a rather early date, perhaps in the Augustan period.

62 See Bultrighini, '*Theōn hemerai*'.

63 = Inscr.It. XIII.2, no. 53. See Brind'Amour, *Le Calendrier romain*, pp. 269–75; Bennett, 'Imperial Nundinal Cycle'.

week; every entry is accompanied with a peg hole.[64] It is presumed to be of the first century CE. Another parapegma from the same period, *IIt* XIII.2, no. 52, has two horizontal sequences, of planetary days and of nundinals; it is from Posillipo (Pausilipum, Campania).[65]

A passage of Petronius' *Satyricon* (30.3–4) has been interpreted as a description of a lunar and planetary parapegma. He describes how two tables (*tabulae*) were affixed to the door posts at the entrance of Trimalchio's dining room, one of which displayed images of the moon's phases and the seven planets, as well as lucky and unlucky days marked with distinctive knobs.[66]

Finally, a number of first-century CE inscriptions reveal that the days of the planetary week were often used for dating purposes, as part of the dating formulas. Most of these are graffiti from Pompeii: *CIL* IV 294, dated 'day of Saturn' (a birth announcement, close in time to Pompeii's destruction in 79 CE);[67] *CIL* IV 6838, dated 'day of Sun' (mid first-century CE); and *CIL* IV 8820, dated 'day of Jupiter'.

One first-century CE inscription is the earliest evidence of the planetary week from outside Italy, suggesting its incipient diffusion in other parts of the Roman Empire: *AE* 1993, 1217, a wall graffito dated 'fourth day before the Nones of April, day of Mars', from Aventicum, Germania Superior (Avenches, Switzerland), dating from between 35 and 80 CE, thus approximately coeval with the graffiti from Pompeii.[68]

2.4 *The Planetary Week in the Greek East*

In the Greek-speaking, eastern half of the Roman Empire, the diffusion of the planetary week was much slower, far less extensive, and did not begin before the second century CE.[69] The earliest evidence of planetary days in the Greek

64 = Inscr.It. XIII.2, no. 49 = *AE* 2010, 117. See Lehoux, *Astronomy*, pp. 32–34, 171–72. Outside the four corners of the geometric design, the dates and lengths of the seasons are also provided. The marble fragment is from an unknown place in Latium; it was once held by Fulvio Orsini in Rome, and is now in Naples (Museo Archeologico Nazionale, inv.no. 2635).

65 Lehoux, *Astronomy*, pp. 12–14, 18, 173–74. Now in the Johns Hopkins University Archaeological Collection, Baltimore, inv. no. 5384 a/b.

66 duae tabulae in utroque poste defixae (…) altera lunae cursum stellarumque septem imagines pictas; et qui dies boni quique incommodi essent, distinguente bulla notabantur. Brind'Amour (*Le Calendrier romain*, pp. 260–62) supports the view that a parapegma is intended. Colson's (*The Week*, pp. 32–33) scepticism reflects the fact that less epigraphic evidence was available to him in his period.

67 Varone, *Erotica Pompeiana*, pp. 168–69.

68 Frei-Stolba, '"Dienstag, den 2. April … n. Chr."'.

69 Philo does not ever mention the planetary week, and as Colson (*The Week*, p. 54 n. 1) remarks, this silence is significant, particularly in the context of Philo's treatise on the number seven (*de septenario*) within his *Special Laws*.

East is a lost passage of Plutarch, of which all that has survived is the title. In his *Quaestiones Conviviales*, book 4, 672c, question 7 (a number which may be significant?) is listed as follows: 'Why the days named by the planets are not counted according to their own order, but in a different order'.[70] The meaning of this question will be discussed further on, but it is at least clear from the question that Plutarch, in the early second century, knew the planetary week. His question, moreover, follows two questions about the Jews, which suggests an association of the planetary week with Jewish tradition.

But the sources of Plutarch's knowledge of the planetary week are more likely to have been astronomical or astrological than Jewish. Indeed, the next attestation of the planetary week in the Greek East is *P.Harris* 1.60, an astronomical ephemeris from Egypt for 140 CE, where every seventh day is marked with the letter κ for (ἡμέρα) Κρόνου, '(day) of Kronos'.[71]

In the third quarter of the second century, Vettius Valens, an eminent astrologer in Alexandria, devotes a substantial section of his work to the planetary week and the explanation of its structure (answering, effectively, Plutarch's question). He also explains how to calculate the day of the week of any given date, using as a paradigm the Alexandrian date of 13 Mechir, year 4 of Hadrian, day of Mercury, which corresponds to Wednesday 8 February 120 CE. This date, which he frequently uses and charts in his work, is widely believed to be his own birthday.[72] We shall return to this important passage below.

Non-astrological references to the planetary week are very rare in Greek sources of the second century. In the mid 150s, Justin Martyr identifies the Christian day of worship, Sunday, as 'the day of the Sun', which commemorates the first day of the Creation as well as Jesus Christ's resurrection, following his crucifixion on the day 'before that of Kronos' (τῇ πρὸ τῆς κρονικῆς).[73] This phrase, for Friday, is odd (it should have been the 'day of Aphrodite'), and suggests that Justin, a Greek native of Flavia Neapolis (Nablus) in Samaria (the central region of ancient Palestine), was not entirely familiar with the days of the planetary week[74] – even though he resided, at this time, in Rome. He

70 Διὰ τί τὰς ὁμωνύμους τοῖς πλάνησιν ἡμέρας οὐ κατὰ τὴν ἐκείνων τάξιν ἀλλ' ἐνηλλαγμένως ἀριθμοῦσιν.

71 Jones, 'An astronomical ephemeris'; and id., *Astronomical Papyri*, vol. 1, pp. 40–41, 175–76, 304; also Lehoux, *Astronomy*, p. 209. An ephemeris is a table showing the position of heavenly bodies through a sequence of days.

72 Vettius Valens, *Anthology* 1.10; see Riley, 'Survey of Vettius Valens'. http://www.csus.edu/indiv/r/rileymt/PDF_folder/VettiusValens.PDF, accessed 26 May 2019. Colson (*The Week*, p. 47) gives incorrectly 7 February 119, which cannot be year 4 of Hadrian.

73 Justin Martyr, *First Apology*, 67.3 and 8.

74 Colson, *The Week*, pp. 28–29: Justin was 'hazy' about planetary names.

may have been aware of the planetary names for Saturday and Sunday only, as the two most significant days of the seven-day week for Jews and Christians. His decision to use planetary designations, in this passage, can be explained on the basis of the Roman context in which he was writing, as well as of his explicit appeal, in the *Apology*, to the Roman emperor Antoninus Pius. If anything, therefore, this passage could suggest that in the mid second century CE, the planetary week was still poorly known in the eastern part of the Roman Empire. This is likely if its use was largely restricted to astrology, or to a popular form of astrology.[75]

By the end of the second century, however, the planetary week seems to have become more widespread in the Greek East, at least in Alexandria; for Clement of Alexandria (*Stromata* 7.12.75) has no trouble identifying the Christian fast days on the fourth day and the day of preparation (Wednesday and Friday) with the days of Hermes and Aphrodite respectively. Within his narrative of Pompey's invasion of Judaea, the historian Cassius Dio (early third century) provides an explanation of the meaning and structure of the planetary week, similar to that of Vettius Valens (Dio 37.18–19). In this passage, he claims that the seven-day week originated from the Egyptians, but he concedes nevertheless that its institution was only recent (οὐ πάλαι ... ἀρξάμενον) and that the ancient Greeks had actually never known it. Still, in his narrative of the capture of Jerusalem in 37 BCE, Dio states that the event occurred on what was 'already then' called Saturn's day (ἐν τῇ τοῦ Κρόνου καὶ τότε ἡμέρᾳ ὠνομασμένῃ: Dio 49.22.4–5).

Except for the astronomical ephemeris, all these sources are literary. Planetary days do not appear in any inscription from the Greek East in the first and second centuries CE.[76] This indicates that the use of the planetary week was still very limited in the Greek East; it may have been still confined largely to Italy and the Roman West.

Further to the east, the Hebrew esoteric work *Sefer Yetzirah* (the Book of Creation), which is most probably of Palestinian origin and dating from between the third and seventh centuries, associates the planets with the days of the week, but not in the normal sequence of the planetary week: it begins with Saturn on the Sabbath, but is followed by Jupiter on the first day of the week (instead of Sun), then Mars (instead of Moon), etc., following the standard cosmological order of the planets (on which see further below). This suggests that

75 See Bultrighini, '*Theōn hemerai*'.

76 *IG* XIV 2184 is a late second-century epitaph in Greek of a girl who was born and died on the 'day of Helios', but significantly, its provenance is Rome. On this inscription, see Bultrighini, '*Theōn hemerai*'.

the author of this work was actually unaware of the Roman planetary week; he only noted that the planets and the Jewish week were of the same number, and this basis, created a planetary week of his own making.[77]

In sum, all the evidence indicates that after the first appearance of the 'day of Saturn' in a poem of Tibullus of c. 30 BCE, the planetary week was rapidly developed and integrated into the Roman calendar, and was used as a dating method already during the Augustan period and through the first century CE. In this period, however, its diffusion remains largely limited to Italy. The origins of the planetary week were Roman or Italian, and could not have been from the Hellenistic East, where it is only first attested, in literary sources, in the second century CE.

2.5 *Planetary Hours and the Structure of the Planetary Week*

The order of the planets within the week, which appears to have been standard from the outset, appears problematic. In the Hellenistic astronomical and cosmological traditions, the planets were ordered according to their perceived distance from the earth and/or the length of their revolution. Both criteria coalesced into a single sequence which in earlier, Hellenistic sources appears as: Saturn, Jupiter, Mars, Mercury, Venus, Sun, Moon,[78] and in later, Roman sources as Saturn, Jupiter, Mars, Sun, Venus, Mercury, Moon.[79] The order of planets is discussed by Ptolemy in the *Almagest* (9.1); he favours the latter order, which he considers more ancient, and which appears anyway to have been dominant in the Roman period.

The question that Plutarch raised, as we have seen, is why the order of the planetary week differs from this cosmological sequence. The order of the planetary week is indeed very different, but it is certainly not random. The weekly sequence can be described as a reiteration of the cosmological sequence (that of Ptolemy) but skipping each time two planets: thus Saturn is followed by Sun (skipping Jupiter and Mars), then by Moon (skipping Venus and Mercury), then by Mars (skipping Saturn and Jupiter), and so on. This describes the structure of the planetary week, but it does not explain it.

---

77 *Sefer Yetzirah*, ch. 4 (in the short and long recensions): Hayman, *Sefer Yeṣira*, pp. 136–45 (paragraphs 41–44 in his edition). This cosmological order of the planets is standard and the same as Ptolemy's (see below).

78 Ps-Aristotle, *De Mundo*, 2 (392a).

79 This is the so-called Chaldean order, which according to Cicero was introduced by the Stoic Diogenes of Babylon (second century BCE): Cicero, *De Divinatione* 2.91; also in Pliny, *Natural History* 2.6; Vettius Valens, *Anthology* 1.10. In *De Natura Deorum* 2.52–3, Cicero switches round Mercury and Venus, and likewise Philo (*Quis Rerum Divinarum Heres* 45 (224), *Quaestiones* 2.75) places Mercury before Venus.

Plutarch's answer, if ever there was one, is not known; but a century later, Cassius Dio suggested two explanations. The first was based on a musical theory of tetrachords, which explained why two planets (of the cosmological sequence) had to be skipped at each step (Dio 37.18–19). The second was already formulated by Vettius Valens a half a century earlier; it is based on an astrological model that gained much credence in Late Antiquity and the Middle Ages, and as an explanation of the structure of the planetary week, it has been widely accepted by scholars right until today.

Vettius Valens argues, in his *Anthology* (1.10), that the planetary week is not structured by its days but rather by its hours (there are 168 hours in the week). Every hour of the week is deemed to be ruled by a planet in succession, following the standard cosmological sequence. Assuming the first hour of the week is ruled by Saturn – the first in the cosmological sequence – followed by Jupiter in the second hour, Mars in the third, etc., then after twenty-four hours, at the beginning of the second day, the ruling planet will be the Sun; twenty-four hours after that, the first hour of the day will be Moon, and so on. Each day of the week is thus named after the planet ruling its first hour.

This theory was almost certainly not Vettius Valens' invention, nor should we assume it originated from Alexandria. It is attested, indeed, in an inscription on a small, fragmentary marble slab from the area of Potenza Picena, ancient Potentia, in central Italy (near the Adriatic coast), which has been dated mainly palaeographically to around 100 CE or possibly even earlier.[80] The fragment preserves part of a reiterative list of planets in the cosmological sequence; each planet is given a number and a letter (B, N, or C) designating it as good (*bona*), harmful (*noxia*), or indifferent (*communis*); Saturn, notably, is 'harmful'. The only fully preserved line reads: *VII Sol*(*is*) *C*(*ommunis*).[81] There can be little doubt that this inscription presents the sequence of planetary hours as described by Vettius Valens, and even more similarly in the fourth century (or later) by the so-called Chronograph or Calendar Codex of Filocalus (or 'of 354'), where the planetary days are graphically represented and each personified

80 CIL IX 5808 = *Suppl.It.* XXIII 2007, 171. For an extensive discussion of the inscription (including its dating – which some would place, on palaeographical grounds, in the Augustan period), together with a first-ever published image of the fragment and inscription, see Heilen, 'Short Time', pp. 244–46.

81 On this basis the fragment could be identified as part of the night hours of Tuesday or the day hours of Friday, assuming a scheme, as in the Calendar Codex of 354, where days of the week are named after the first day hour. In a scheme like that of Vettius Valens, where days of the week are named after the first night hour (with night preceding day), the fragment would belong to the day hours of Monday or the night hours of Friday. Whatever the day or night of the week, it is clear that the original inscription must have included the full listing of hours for all seven days of the week.

planetary day is flanked with lists of the night-time and day-time planetary hours which are similarly numbered and characterized by the letters B, N, and C.[82] The Potentia inscription is likely to have been on public display, and to have served the purpose of displaying practical astrological guidance on a daily and hourly basis.

The practical, astrological use of the planetary hours is only attested after this in late antique sources from the fourth century and later, and only rather sporadically.[83] If the Potentia inscription has been correctly dated to a much earlier period, which seems very likely, its historical implications are considerable. Besides suggesting that the planetary hours were known much earlier and were already in popular, astrological use in the first century CE, it also demonstrates that the scheme of planetary hours was most likely of Italian origin, and not the invention of, for example, Alexandrian astrologers such as Vettius Valens. This would tie in well with the Italian origins of the week of planetary days which have been discussed above.

Furthermore, the early dating of the Potentia inscription leads to the conclusion that the scheme of planetary hours was not superimposed on the seven-day week at some later stage (e.g. in the late second century, with its first literary reference in Vettius Valens), but was more likely contemporaneous with the institution of the Roman planetary week, and closely intertwined with it.[84] It would only take a small step to proceed to the further conclusion, based

82 Text edition and facsimiles with introduction and discussion in Divjak and Wischmeyer, *Das Kalenderhandbuch von 354*, pp. 111–36; also Heilen, 'Short Time', pp. 247–48.

83 Such as in the Chronograph of 354, and in Paul of Alexandria's astrological treatise, dating from the late fourth century (*Eisagogika* 21, in Boer, *Paulou Alexandreōs eisagōgika*, pp. 42–45 and Greenbaum, *Late Classical Astrology*, pp. 39–40); see further references in Heilen, 'Short Time', pp. 248–49. In the Babylonian Talmud (*bShabbat* 156a), the theory of the influence of the planets over the hours is attributed to the early third-century R. Ḥanina, but the notion that the week is structured by the planetary hours is not explicit (besides, the reliability of attributions in the Babylonian Talmud is notoriously problematic). In *bBerakhot* 59b, a statement that the night of Wednesday begins at the hour of Saturn is attributed to the fourth-century sage Abaye (this would work in a planetary week where the days are named after the first hour of daytime); however, ms Oxford Bodl. Heb. d.46.88, a Cairo Genizah fragment of R. Ḥananel's commentary, suggests that this passage may be a later interpolation. Planetary hours are also associated with specific days of the week in *bShabbat* 129b, but in a passage that is part of the editorial layer of the Talmud, thus potentially quite late (c. sixth century?). The *Sefer Yetzirah* (ch. 4) makes a fleeting reference to the planetary hours, but it has been argued that this is a later interpolation (Hayman, *Sefer Yeṣira*, pp. 36–37, 143–45; the passage is in paragraph 42 of Hayman's edition, pp. 139–40, and appears in all the recensions).

84 It is important to note, in this context, that the 24-hour division of the day, which the scheme of planetary hours assumes, was known in Rome at the time when the planetary week was instituted, and reasonably well used. The 24 hours are first mentioned in Varro,

FIGURE 1.2 Calendar Codex of Filocalus, 354 CE: day of Saturn (Saturday). Ms Vatican Lat. 9135 (1620/1 CE), fol. 228r

again on the early dating of this small marble fragment, that it is the scheme of planetary hours that originally determined the naming of the planetary days within the seven-day week and hence shaped the structure of the Roman planetary week, exactly as Vettius Valens (and later Dio) assumed.[85]

2.6 *The Significance of the First Century* CE

The creation and diffusion of the planetary week in Rome and in other parts of Italy, in the late first century BCE and first century CE, must be evaluated in the context of the major changes that Roman time-reckoning underwent under Julius Caesar and Augustus – not least, the institution of the Julian calendar – which itself reflected, in a variety of ways, the institution of a new political order in the Roman state. Although the seven-day week did not become, in this period, an official part of the Roman calendar (in the same way as the Julian calendar did), it was still sufficiently important in public life to be included as an additional column in some of the monumental Fasti. The institution of the Principate cannot serve as an explanation, on its own, for the creation of the planetary week, but it certainly provides a significant context to it.

This period coincides with the earliest evidence we have of the seven-day week coming into practical use, for purposes of dating, in Judaea. It is tempting to regard this coincidence as significant, but much caution needs to be exercised. Although Judaea was under direct Roman rule through most of the first century CE, it seems unlikely that direct lines of influence, in one direction or the other, or some other direct relationship, existed between the incipient Judaean, Aramaic-language use of the seven-day week and the parallel rise of the Roman planetary week in the Italian peninsula (for which there is hardly any evidence, in this period, elsewhere in the Roman Empire). Moreover, the apparent coincidence of these developments in Judaea and in the Italian peninsula should be treated with caution, because although the origins of the Roman planetary week can be confidently dated to the Augustan period, for the beginnings of the use of the seven-day week in Judaea we rely on a handful

Lingua Latina 9.26 (referring however to twenty-four 'lunar hours'); and a little later, in the mid first century CE, in Seneca, *Epistles* 12.7: *dies est tempus viginti et quattuor horarum*. The twelve hours of the day are used much earlier, for example in Caesar, *Bellum Gallicum* 1.26.2; Livy, 23.44.6, 27.2.7, 28.15.4, 42.57.6. Although the night was usually divided into four watches (*vigilae*: Caesar, *Bellum Gallicum* 7.3.3; Livy, 25.38.16; Pliny, *Historia Naturalis* 10.46), twelve hours are also attested: e.g. Cicero, *Letters to Atticus* 4.3.5 (*haec ego scribebam hora noctis nona*), *Pro Sexto Roscio* 19; and *AE* 1920, 83. See Wolkenhauer, *Sonne und Mond*.

85 This is what most modern scholars have assumed (e.g. Rüpke, *Kalender und Öffentlichkeit*, pp. 456–57, and id., *Roman Calendar*, p. 162), but solely on the strength on the reports of Dio.

of recently discovered, first-century BCE ostraca that only happen to have survived. The discovery of similar ostraca but from an earlier, e.g. Hasmonean period – which could easily be made in the near future – would be sufficient to disrupt this apparent coincidence.

Nevertheless, the relationship between the rise of these two seven-day week traditions, Judaean and Roman, needs to be further explored, partly because – as has been suggested above – the Roman planetary week may have begun as an *interpretatio romana* of the Jewish week, and more importantly, because in the following centuries both weeks became increasingly identified and confused.

3 The Standardization of the Seven-Day Week: From Ptolemaic Egypt to the Early Roman Empire

3.1 *The Equation of the Jewish and the Planetary Weeks*

The equivalence of the Jewish and planetary weeks, whereby Sabbath was equated with the day of Saturn, appears never to have been questioned. Its origin is uncertain, but it is likely to go back to the very institution of the Roman planetary week. As we have seen, Tibullus already might be calling the Jewish Sabbath the 'day of Saturn'. But the first explicit evidence of equivalence of the Jewish Sabbath with the day of Saturn comes from Frontinus, in the late first century CE, followed closely by Tacitus.[86] It is confirmed in the mid-second century by Justin Martyr, who refers to the first day of the Jewish week, when God created the world and when Jesus was resurrected, as 'the day of the Sun';[87] this implies that the seventh day or Sabbath corresponds to the day of Saturn. In around 197 CE, Tertullian (in Carthage) similarly writes that Christians rejoice on the day of the Sun, and the Jews on the day of Saturn; whilst his contemporary Clement of Alexandria explicitly identifies the two weekly Christian fast days, the fourth day and the day of preparation, with the days of Hermes and Aphrodite (Wednesday and Friday).[88]

Until the end of the second century, all the evidence is literary and from Rome, whether written in Latin (Frontinus and Tacitus) or in Greek (Justin).

---

86 See above, n. 53.

87 Above, p. 32.

88 Tertullian, *Ad Nationes* 13 and *Apologeticum* 16.11; Clement, *Stromata* 7.12.75. The identification of Saturn with the Sabbath is reciprocated on the Jewish side with the Hebrew naming of the planet as *Shabbetai*, which implies of itself a connection with Sabbath. This name is attested in *Sefer Yetzirah* (see above, pp. 33–34), where, moreover, Saturn is identified with the day of Sabbath. It also appears in the Babylonian Talmud (*bShabbat* 156a), in a saying attributed to the early third-century R. Ḥanina.

This is hardly surprising, given that as we have seen, the planetary week is barely attested in the East before the third century. But what this might indicate is that the identification of the planetary with the Jewish week was not the result of a later merger of two originally distinct traditions, along the lines of 'East meets West', but rather goes back to the very origins of the planetary week, in Augustan Rome. It has been argued, indeed, that the equivalence of the Sabbath, focal point of the Jewish week, specifically with Saturn, the first of the planets in cosmological sequences, was not coincidental and demonstrates a deliberate intention, from the outset, to model the new, planetary week on the existing Jewish scheme.[89] Alternatively, as argued above, the planetary week could have been gradually built around an astrological interpretation of, initially, the Jewish Sabbath alone. Either way, the apparently stable equivalence between Sabbath and Saturday suggests an early link between the planetary and the Jewish weeks, even if explicit evidence of this equivalence only emerges in the late first century CE.

3.2 *The Standard Reckoning of the Week*

Far less certain, however, is whether the week – planetary or Jewish – was always reckoned in phase with what was later to become its standard reckoning; in other words, whether it always conformed to the seven-day week that has been reckoned continuously since Late Antiquity (when, as we shall see, the week became very widely used) until today. This can be determined when a day of the week is provided together with a full date, either in the Julian calendar or in a calendar that can be reliably converted to the Julian calendar; the correlation makes it possible to determine whether the day of the week is the same as what would be expected according to the standard reckoning.

The little evidence that we have before the third century CE suggests, indeed, that in some places at some times the week was diversely reckoned, with a planetary day or a Jewish weekday occurring on a different date than would be expected. Diversity of practice between different Jewish communities could be explained as due to poor communications in the ancient world, and the difficulty of keeping synchronic time frames across relatively long distances. This explanation, however, will not be satisfactory in the context of Egypt, which boasted throughout Antiquity, from Upper to Lower Egypt, effective communications as well as a unified system of time reckoning (the Egyptian 'civil' calendar), which should have enabled Jewish communities to synchronise their observance of the Sabbath without too much difficulty. Diversity in the reckoning of the week would still be possible, nevertheless, if synchronicity and

89 Colson, *The Week*, pp. 51–52.

unity of Sabbath observance was not considered necessary for the fulfilment of the Biblical commandment to 'work six days and rest on the seventh'.[90]

3.3 *Ptolemaic Egypt: Explicit Days of the Week*

The earliest evidence that correlates a day of the Jewish week with a firm calendar date is, very sporadically, from Ptolemaic Egypt. An account of deliveries from the Arsinoite nome, mentioned above (*PSI.Congr.XVII* 22), equates a day 1 of Hathyr with 'Sabbath' (recto col. 3, l. 2); further on, an entry is listed for 'year 3, Tybi 1' (verso col. 3, ll. 1–2). Although Tybi (month 5) comes after Hathyr (month 3), the proximity of the entries makes it most likely that both belong to the same year, and hence, that Hathyr 1 = Sabbath occurred in year 3 of an unspecified ruler. On palaeographic grounds, the papyrus should be dated to the late second century or early first century BCE, which yields two options for the date of Hathyr 1: year 3 of Ptolemy IX, 20 November 115 BCE, which in our reckoning was not a Saturday/Sabbath but a Wednesday; or year 3 of Ptolemy XII, 11 November 79 BCE, which in our reckoning was a Thursday.[91]

In contrast, *BGU* 20 2847 (also mentioned above), from the Herakleopolite nome, equates year 3, Epeiph 27 (= 27 July 49 BCE), with the day 'before Sabbath'; in our reckoning of the week, this date was indeed a Friday. The inconsistency between the two documents, and between the first document and our reckoning of the week, does not have a certain explanation. It could be due to error, or to a discrepancy of how the week was counted – and hence, of when Sabbath was observed by local Jewish communities – between the not-so-far-apart nomes of Arsinoe and Herakleopolis.

3.4 *Ptolemaic and Roman Egypt: Inference from Sabbath Observance*

The possibility of diversity of practice, in the reckoning of the week, among Jewish communities of Ptolemaic and early Roman Egypt, can be explored through the analysis of a much wider body of papyrological sources, on the basis of a hypothesis and methodology that will be briefly explained here.

The day of the week never became part of the dating formula in Roman Egypt, and is therefore almost completely absent from Egyptian papyri. However, Egyptian papyri are precisely dated throughout the history of ancient

---

90 This possibility is not sufficiently taken account of in Bloch's otherwise excellent survey of Sabbath observance in the neo-Babylonian and Achaemenid Empires; Bloch, 'Judean Identity'.

91 Further, less likely options are year 3 of Berenice IV, which yields Friday 5 November 57 BCE; or year 3 of Cleopatra VII, Friday 4 November 50 BCE. No reign affords a match with Saturday. For regnal years, see Chris Bennett in http://www.instonebrewer.com/TyndaleSites/Egypt/ptolemies/chron/chronology.htm, accessed 26 May 2019.

Egypt, with a calendar – the Egyptian calendar, adjusted to become the 'Alexandrian' calendar in the early Roman period – which was stable and well known. It is therefore possible to convert Egyptian dates into equivalent Julian dates, and to establish what would have been their days of the week, assuming the standard reckoning of the week.

Our hypothesis is that if the Jews of Egypt observed the Sabbath, a significant reduction (or even a complete cessation) of activity on one day per week should be reflected in the dated documents pertaining to Jews; and on this basis, it should also be possible to determine which day was treated as Sabbath and whether it was compatible to the later, standard week. As stated, Egyptian papyri and ostraca are generally precisely dated, and the conversion of their dates to Julian dates is methodologically non-problematic. But the ability to infer, from the dates of these documents, whether the Jews observed the Sabbath and on what day this Sabbath fell is fraught with methodological difficulties.[92]

The first is the difficulty of identifying a person in a document, e.g. through his name, as Jewish or as having at least sufficient Jewish identity to be expected to be Sabbath observing.[93] The second is the difficulty of determining what kind of activity, reflected in the documents, would have been considered forbidden on the Sabbath, and hence would not have been recorded in a document as having been carried on the Sabbath. It is not even clear whether the activity of writing was considered forbidden; no verse in the Bible explicitly prohibits it.[94]

---

92 Our starting point, in sourcing the relevant materials, was Tcherikover, Fuks, and Stern, *Corpus Papyrorum Judaicarum* (*CPJ*), but this work is now very outdated within the discipline of papyrology, as well as being fraught with many of the methodological problems outlined here. For more recent discoveries, we were helped by Tal Ilan whose team is preparing a supplement to the *CPJ*. We are also grateful to Nikolaos Gonis for his advice.

93 The editors of *CPJ* were well aware of this issue (Tcherikover et. al., *CPJ*, vol. 1, pp. xvii–xix), but they did not satisfactorily resolve it, mainly because they took 'Jew' for granted as a definite category, and only saw the problem as one of identifying this presumed category of 'Jew' in the papyri. For our purposes, the definition of 'Jew' is less important than that of the 'Sabbath-observant', although this is equally difficult to pinpoint. For example, *CPJ* 453 is a papyrus dated Year 17, Phaoph 22 (= 19 October 132 CE), which was Saturday, and signed by Phibion son of O[n]ias, whom the editors assume to be Jewish on the basis of the father's name; but even if his father identified as Jewish (which a name can hardly tell – even less when it is only a patronym), we still do not know if Phibion was committed to observance of Sabbath.

94 For example, *BGU* 14 2381 is a deed relating to a loan between 'Ptolemaios son of Sabbataios, Jew of the Epigone', and other Jews; it is dated year 5, Epeiph 1 = 2 August 176 BCE, which according the standard reckoning was a Saturday. However, it is unclear whether writing up this document (which could have been by a gentile scribe) and the loan itself would have been regarded a desecration of the Sabbath. Philo gives as examples of work

The third is the uncertainty, in many cases, of the regnal year referred to in the date, and hence of the dating of the document. The fourth is the possibility of scribal errors.[95] The fifth is whether the corpus is large enough for results to be significant. It could be argued that the many sources extant are too widely scattered across Egypt, and too thinly over a period spanning the Ptolemaic, Roman, and late Roman periods, for any significant conclusions to be drawn from an analysis of their dates.

Notwithstanding these limitations, two sub-corpora can be subjected to a useful analysis, as they preserve a relatively large sample within restricted geographical areas and chronological periods: the Jewish ostraca of Diospolis Magna (Thebes, Upper Egypt) of the second century BCE, of which 41 fully dated samples are extant, and the Jewish tax receipts from Apollinopolis Magna (Edfu, Upper Egypt) from the reigns of Vespasian to Trajan, of which 57 ostraca are extant (all in CPJ), dated from 72 to 116 CE.

In Diospolis, the dates are distributed as follows (assuming the standard reckoning of the week): Sunday 19.5%, Monday 19.5%, Tuesday 4.9%, Wednesday 9.8%, Thursday 19.5%, Friday 19.5%, Saturday 7.3%. The very low percentage on Tuesday suggests that perhaps this was the day of Sabbath, although Saturday is also quite low.

The analysis of the Edfu ostraca from 72 to 116 CE was already carried out by Clarysse, Remijsen and Depau. They found that the corpus as a whole did not reflect any pattern of Sabbath observance, as receipts were distributed among all days of the week. However, when restricting the corpus to receipts of the Jewish tax (imposed by Vespasian on all Jews of the Empire after the destruction of Jerusalem in 70 CE), of which 57 are extant from Edfu, nearly no receipts were dated to a Saturday, in contrast to the other days of the week that were well represented. They interpreted this result as evidence that the Jewish tax was not collected on Saturdays, perhaps because in contrast with other taxes, the Jewish tax was collected and administered by Jews who avoided work on the Sabbath.[96]

---

forbidden on the Sabbath to 'light fires or till the ground or carry loads or institute proceedings in court or act as jurors or demand the restoration of deposits or recover loans' (*Migration of Abraham* 91); and in the *Legatio ad Gaium* (158), he states more generally 'to receive anything, or to give anything, or in short, to perform any of the ordinary duties of life'. It is unknown to what extent his understanding of the Sabbath was widely shared in Ptolemaic and early Roman Egypt.

95 In *CPJ* there are also errors in some of the conversions from Egyptian to Julian dates; all these had to be reviewed and corrected.

96 Their interest was only in establishing whether (or to what extent) Jews in Egypt observed the Sabbath. They did not question which day of the week would have been treated as Sabbath, as they assumed that 'the seven-day cycle of the Roman period can be determined

We can infer from this, additionally, that in late first-century Edfu Saturday was observed as Sabbath, and thus that the seven-day week was in phase with its later, standard reckoning. According to our analysis, the dates of Jewish tax receipt ostraca are distributed as follows: Sunday 10.5%, Monday 17.5%, Tuesday 24.6%, Wednesday 14%, Thursday 19.3%, Friday 8.8%, Saturday 5.3%.[97] Here, the lowest percentage on Saturday, with low percentages on either side (Friday and Sunday), support the view that Sabbath was observed on the standard Saturday.

3.5 *The Early Roman Empire*

Moving over to mid-first-century CE Italy, we find a graffito from Pompeii with the following text (*CIL* IV.2 4182):

> Nerone Caesare Augusto
> Cosso Lentulo Cossi fil(io) co(n)s(ulibus)
> VIII Idus Febr(u)arias
> dies Solis luna XIIIIX nun(dinae) Cumis V (Idus Februarias) nun(dinae) Pompeis

> Under the consulship of Nero Caesar Augustus and Cossus Lentulus, son of Cossus, on the 8th day before the ides of February, day of the Sun, 16th lunar day, nundinae at Cumae; 5th day before the ides of February, nundinae at Pompeii.

---

by simply counting back from modern times' (Clarysse, Remijsen, and Depau, 'Observing the Sabbath', p. 53). Although this assumption, for the late first century CE, should not have been made, in the event this did not matter, as their analysis had the subsidiary result of confirming that in late first-century CE Edfu, the Sabbath was reckoned in phase with its later, standard reckoning. A similar analysis of a sample of dated Judaean Desert legal documents of the late first and early second centuries CE was carried out by Katzoff and Schreiber, 'Week and Sabbath', pp. 106–11, yielding similar results.

97 According to Clarysse et al. ('Observing the Sabbath'), no more than two Jewish tax receipts are dated to a Saturday; of these two ostraca, they argue that one, *CPJ* 200, is actually uncertain, as its fragmentary date of (ἔτους) η [Φα]ῶφ[ι] η could have read originally Phaoph 18 rather than 8; they also point out that they were not able to locate and verify either of the originals, and intimate that it may be in fact that no ostraca are dated on Saturday at all (ibid., pp. 54–55). However, they overlooked a third ostracon, *CPJ* 183, which is firmly dated year 4 Domitian, Payni 24; the date is wrongly converted by the *CPJ* editors to 18 May 85 CE (Tcherikover et al., *CPJ*, vol. 2, p. 125), but it should be 18 June, which was a Saturday. This pushes up the statistics for Saturday, although Saturday remains significantly under-represented. Our figure of 5.3% assumes all three ostraca, and can be taken as a maximum within the extant corpus.

There are two dates in this inscription, but the second, very partial, does not concern us now.[98] The first date corresponds to 6 February 60 CE, which in our reckoning was a Wednesday; the inscription, however, equates it with Sunday. A much quoted explanation has been that of Brind'Amour, who argues that the author of this inscription was basing the planetary week on the same hourly scheme as Vettius Valens, but named the days of the week after the first hour of the day rather than after the first hour of the (preceding) night; consequently, what was for Vettius Valens (and us) a Wednesday was for him a Sunday.[99] This explanation is possible but speculative and, in our view, over-scholarly. Given that the seven-day week is arbitrary and does not correspond to any astronomical (or other) reality, it is just as plausible to explain that the author of this inscription began the week on a different date. On any interpretation, it is evident that the week was reckoned differently.

The Pompeii graffito joins evidence from Ptolemaic Egypt, albeit sporadic, of diversity in the reckoning of the week: the papyrus from the Arsinoite nome from 115 or 79 BCE, and the inference from the distribution of weekdays in second-century BCE Diospolis.

But equally sporadic, and generally somewhat later, is evidence of consistency with the later, standard reckoning. After the Herakleopolite nome papyrus from 49 BCE, which is compatible to the standard reckoning, we have the above-mentioned inscription from Tremithus, eastern Cyprus, which is dated 'year 7 of Domitian Caesar ... month of Tybi 25, 1st hour of the day, *sambat* 6', i.e. the sixth day of the Sabbath week (Friday). As has been demonstrated elsewhere, this date is in the eastern Cypriot calendar and corresponds to 28 December 87 CE, which was indeed, in our reckoning, a Friday. Then, the Edfu ostraca from 72 to 116 CE, and Judaean Desert documents from the late first and early second centuries, suggest observance of the Sabbath in accordance with the standard reckoning.[100]

We then have *P.Harris* 1.60 and Vettius Valens, both from mid-second century CE Egypt, and both compatible to the standard reckoning (see above, p. 32). Vettius Valens' exposition of a method to calculate the day of the week, for any date in the calendar and in any year, implies in fact the assumption that the planetary week was a standard, fixed scheme, on a par with the Egyptian

98 According to the Pompeii calendar graffito, *CIL* IV 8863 (above mentioned), the nundinal day of Pompeii should come four days after Cumae, thus on IV Idus; this inconsistency is outside our scope. On the various aspects of this inscription, see Deman, 'Notes de chronologie romaine'; Brind'Amour, *Le Calendrier romain*, pp. 268–75.

99 Brind'Amour, ibid., pp. 268–69.

100 Tremithus inscription: Stern, 'A "Jewish" birth record'. Edfu ostraca and Judaean Desert documents: above, n. 96.

calendar, which could not be open to any diversity. This announces, better than any other source, the increasing standardization of the seven-day week in the Roman Empire. But there is no good evidence of this standard emerging before the late first century CE.

3.6 *The Week in the Christian Roman Empire: Politics and Ideology*

More substantial evidence becomes available in the third century, when the seven-day week became more frequently used for dating inscriptions, and when the first Christian Easter tables were designed. The earliest of these tables, attributed to Hippolytus, was redacted in Rome around 222 CE.[101] The Easter tables, which went on to develop mainly in Rome and in Alexandria in the course of the third-fifth centuries, provide for many years in advance the dates when Easter, always on a Sunday, is to be celebrated; effectively, they also indicate how the week is counted, and on what dates Sunday occurs. For this reason, it is likely that the production of these tables and their diffusion in the Christian world played an important part in the standardization of the seven-day week in the later Roman Empire. Indeed, at this stage in history, the wide diffusion of the seven-day week itself – more specifically, of a Christianized version of the Jewish Sabbath week – must be attributed to the general process of Christianization of the Roman Empire, which was further to gain pace after Constantine's conversion in the early fourth century.

The standardization of the seven-day week was also, even more generally, the product of the increasing cultural unification of the Roman Empire. This trend was exemplified in the very innovative decrees of the Councils of Arles (314) and Nicaea (325), convened by Constantine, that Easter be observed by all Christians on the same date; this also meant implicitly that Sunday should be reckoned by all on the same day. The promotion of a standard seven-day week furthered the development of an ideology that valorised social cohesion through temporal synchronicity, and that no longer made it possible for Jewish (and Sabbath-observing Christian) communities to work six days and rest on the seventh in their own time. These decrees were as much about setting an orthodox standard in Christianity as about politically unifying the Roman Empire under the sole rule of the Christian emperor.[102] Constantine's further

101 *ICUR* VII 19933–19935. See Mosshammer, *Easter Computus*, pp. 116–25; Stern, *Calendars in Antiquity*, pp. 388–409.

102 Stern, *Calendars in Antiquity*, pp. 395–402, 422–24, emphasizing the role of Constantine in forming these decrees, but noting that 'it would be over-simplistic to attribute this change entirely to the emperor, and not to appreciate the role of leading bishops – especially perhaps at the Council of Arles, where the role of Constantine appears to have been

decree on the observance of Sunday, which will be considered below, fulfilled similar agendas, and had the similar effect of turning the seven-day week, increasingly, into an official and standard component of the Roman calendar.

4 The Seven-Day Week in the Later Roman Empire and Late Antiquity (Third–Sixth Century CE): Standardization and Diffusion

After the merger of the traditions of the Jewish, Biblical week and the planetary week, the seven-day week became increasingly widespread from the third century and especially in the fourth century CE, when a variety of literary, papyrological, epigraphic, and further textual and material evidence point to the use of the hebdomadal cycle in a wide geographical area comprising the entire territory of the Roman Empire.[103] During the imperial period, early Christians adopted and adapted the nomenclature of the Jewish week; they made Sunday, the Christian Lord's day, the first and most important day of the week instead of the Sabbath. We shall return to the Christian week further below. For now, we note that the gradual Christianization of the Roman Empire resulted in the increasing use of the Christian nomenclature for the days of the seven-day week alongside the planetary names.

The use of the Christian and the planetary designations follows a quite distinct distribution pattern.[104] In the eastern side of the Roman Empire, the seven-day week occurs mostly in the Christian form. It appears in sources originating from Greece (mainly Athens, Corinth, and Crete), Thrace, Macedonia, and Asia Minor, as well as from Egypt and the Roman and Late Antique Near East. By contrast, the planetary week designations are widespread in the West (especially in Italy, Sicily, and Gaul), where the Christian nomenclature is, instead, poorly attested, even in manifestly Christian contexts.

---

more limited – in seizing the opportunity of Constantine's conversion to promote the long-standing but somewhat dormant notions of catholicism and orthodoxy' (p. 395).

103 As part of her work in the ERC project 'Calendars in Late Antiquity and the Middle Ages: Standardization and Fixation', Ilaria Bultrighini compiled a comprehensive database of references both to individual days of the week and to the seven-day week as a whole in the entire corpus of epigraphic, papyrological, and literary sources in Greek and Latin from throughout the Roman Empire. This material also includes artefactual and visual evidence and covers a time span approximately from the first century BCE to the sixth century CE.

104 Bultrighini, 'Thursday (*dies Iovis*)', p. 65.

4.1 *The Planetary Week in the Later Roman Empire and Late Antiquity: Epigraphic and Documentary Evidence*

As we have seen above, during the Augustan period and throughout the first century CE the diffusion of the planetary week remained largely limited to the Italian peninsula, with only one attestation elsewhere in the Roman West. In the eastern Mediterranean, the planetary week is first attested, in literary sources, in the second century CE. The distribution of the evidence from the later imperial and late antique periods confirms that the planetary designations for the days of the week continued to spread mainly in the western part of the Roman Empire, with a limited number of testimonies having been found in its eastern half.

Days of the planetary week occur over 300 times in Greek and Latin epigraphic texts, with Latin testimonies being more than double the Greek ones.[105] Most of the inscriptions that can be more or less precisely dated belong to the fourth and fifth centuries, with a particular concentration from the mid-fourth century onwards. The epigraphic evidence consists predominantly of epitaphs, whose greatest part can be identified as Christian. As a way of example, *ICUR* I 3978 is a marble plaque inscribed with a Greek epitaph commemorating a little girl who belonged to a Christian family and died sometime in the fourth century in Rome: 'Achillia passed away, having been baptised, at the age of 1 year and 5 months, on the 7th day before the calends of March, on the day of the Moon (Monday)'.[106] A Latin example is the sepulchral inscription of Vitalissima and Benenatus or Benenata, who were seemingly twins as they both 'lived more or less four years', and were 'buried on the day before the ides of August, on the day of Mercury (Wednesday)' in the year 425, in Milan.[107] Two orant figures drawn below the text, alluding to the deceased twins, point unequivocally to a Christian milieu.[108]

105 The total number of occurrences is 312, of which 98 are Greek and 214 are Latin; not unexpectedly, all Latin inscriptions originate from the West; the Greek dossier includes 44 inscriptions from the West and 54 inscriptions from the East. Of these latter, 45 are from a single site, Zoar/Ghor es-Safi (south of the Dead Sea), on which see further below.

106 ἐκοιμήθη Ἀχιλλία/νεοφώτιστος ἐνι/αυτοῦ μηνῶν ε′ πρ(ὸ) ζ′/καλανδῶν μαρτίων ἡμέρᾳ/ Σελήνης.

107 hic requiescit Vit[a]/lessema qui vixit an(nos)/pl(us) m(inus) IIII et Benenat[us]/qui vixit an(nos) pl(us) m(inus) IIII/d(epositi) pridie idus augustas/die mercuris dd(ominis) nn(ostris)/Teudosio Aug(usto) XI et Valent(ini)a/no puero florentissim/o Caesare. *CIL* V 6278 = *ICUR* I 3228 = *ILCV* 4394B = *ICI* XVI 194. The consular year is the eleventh of Theodosius II and the first of Valentinian III.

108 On the motif of the orant or praying figure in Christian funerary contexts, see Jensen, 'Ritual', pp. 589–93.

The sparser epigraphic evidence of the use of planetary names for the days of the week in the eastern half of the Roman Empire consists of three Christian epitaphs from Egypt,[109] a building inscription found near Heliopolis/Baalbek in Roman Syria,[110] a horoscope and a related text, both wall graffiti, from Dura-Europos,[111] an epitaph from Asia Minor,[112] and a Christian epitaph from Macedonia.[113] A special case is the site of ancient Zoar or Zoora, identified with the area of modern Ghor es-Safi at the southeastern end of the Dead Sea in Jordan. In the Roman period, Zoar was part of Provincia Arabia (starting from 106 CE, when the province was established). In the early fourth century the province was reorganised and renamed as Palaestina Tertia. During the fourth to sixth centuries, Zoar is known to have been the seat of a bishop and to have had a vibrant Christian community at least until the seventh century. The site produced an exceptionally large number of Christian tombstones in Greek,[114] mostly from the fourth and fifth century, whose dating formulae include days of the week both in the planetary and in the Christian form. The planetary designations appear in as many as 45 epitaphs from Zoar. It remains unclear how this apparently entirely isolated case should be interpreted in the context of the diffusion of the planetary week in the East.

As for other types of sources, planetary days of the week appear in a rather limited number of papyri, ostraca, and wooden tablets from Egypt, dating from the third, fourth, and later centuries. This material includes various astronomical, astrological, and magical texts. We have previously mentioned *P.Harris* I.60, an ephemeris for 140 CE in which Saturdays are marked by a recurring K, for Kronos. In four further ephemerides from the fourth and fifth

109 Lefebvre, *Inscriptiones Christianae Aegypti*, no. 391 (from Hermonthis, undated; 'day of Aphrodite'); ibid., no. 150 (from Akoris, fifth or sixth century CE; 'day of Aphrodite'); Priesigke (ed.), *Sammelbuch*, no. 1564 (from Antinoopolis, undated; 'day of Hermes').

110 *IGLS* VI 2915 (539 CE; 'day of Selene'). The last line of this inscription apparently preserves the beginning of the formula οἱ θεοὶ ἀθάνατοι, 'immortal gods'; on this basis, the document is generally regarded as originating from a pagan milieu.

111 No. 232 (horoscope, 218 CE; 'day of Kronos') in Welles, 'Graffiti'; ibid., no. 220 (inscription concerning a horoscope, variously dated between 219 and 242 CE). See Neugebauer and Van Hoesen, *Greek Horoscopes*, p. 54 nos. 219 I–Ib, and p. 167. Both texts, whose context is clearly astrological, were scratched (probably by the same individual) into the plaster of a wall of a house near the centre of Dura known as the House of the Archives.

112 *TAM* V.3.1851 (from Philadelphia, Lydia; 342/3 CE; 'day of Aphrodite'). The text was inscribed on a white marble column, which was found in a field near the Christian cemetery of the ancient city. Apart from the provenance, there are no textual or visual elements which suggest that this may be a Christian epitaph.

113 *IG* X.2.1 784 = *RICM* 159 = *ICG* 3180 (Thessaloniki, fifth century CE; 'day of Kronos').

114 Meimaris and Kritikakou-Nikolaropoulou, *Inscriptions from Palaestina Tertia, Ia and Ib*. On the Jewish tombstones from Zoar, in Aramaic, see above pp. 19–20.

centuries CE, hooks are used to mark every seventh day, i.e. Saturday.[115] The only known instance of a day of the planetary week in a papyrus horoscope is *P.Oxy.* 61.4274, from 503 CE. The text is fragmentary but apparently the dating formula specified that the geniture, which was nocturnal, occurred at the turn of a Saturday and a Sunday, the two days being represented by the symbols for Saturn and the Sun.[116] Days of the planetary week appear in three of the so-called Greek Magical Papyri. These are a body of papyri from Graeco-Roman Egypt (second century BCE to fifth century CE) containing a variety of magical spells, formulae, hymns and rituals. *PGM* XIII is a third/fourth-century papyrus that, inter alia, describes the proper invocations to appeal to the deities of the weeks, hours, and days.[117] It gives directions, in the form of a table, for discovering which deity ruled any given day of the week; the table includes a column listing the planets in week order. Quite unexpectedly, however, the list begins with Helios (Sunday) instead of Kronos (Saturday). This discrepancy is likely to have resulted from a Christian interpolation detectable in the text, which occurred in the first half of the fourth century CE. It follows that the original text was probably produced in the third century.[118] *PGM* XXXVI 320–332 is a contraceptive spell which has been dated to the fourth century CE on palaeographical grounds. Among other things, it prescribes to perform certain actions 'during the waning of the moon which is in a female sign of the zodiac on the day of Kronos (Saturday) or Hermes (Wednesday)' (ἐν ἡμέρᾳ Κρόνου ἢ Ἑρμοῦ).[119] Finally, *PGM* IV 1–25 is a fourth-century papyrus titled 'The spell for revelation'. Here, a certain action must be performed 'on the day of Zeus (Thursday) in the first hour' (ἡμέρᾳ Διὸς ὥρᾳ α΄). Another fourth-century document whose context is clearly astrological/magic is *P.Kellis* I 82. This is a wooden board, once part of a codex, preserving a calendar of good and bad days which was to be consulted in order to compute the right time to practice magic.[120] With the sole exception of Sunday, which is referred to as the Christian Lord's day, all days are designated by their planetary names.

A few further papyri, ostraca, and wooden tablets from Egypt preserve texts of different nature involving planetary names for the days of the week. These

---

115 *P.Oxy.* 61.4179 (348 CE: Jones, *Astronomical Papyri*, vol. 1, pp. 186–90); *P.Mich.* Inv. 1454 (Curtis and Robbins, 'Ephemeris of 467 A.D.'); *P.Vind.* G. 29370b (471 CE) and *P.Vind.* G. 29370 (489 CE: Jones, 'Two Astronomical Papyri Revisited').

116 Ibid., pp. 279–81.

117 See Smith, 'Eighth Book of Moses'.

118 Betz (ed.), *Greek Magical Papyri*, p. 180 n. 68.

119 For the astrological implications of these prescriptions, see Barton, *Ancient Astrology*, p. 193.

120 See Hoogendijk, 'A note on P.Kellis I 82'.

include four school exercises,[121] a schedule of workdays,[122] a private letter,[123] and a small number of official documents of various types.[124] In these cases the days of the week are either part of dating formulae or otherwise provide temporal indications. Although the relevant epigraphic and documentary sources from the eastern half of the Roman Empire are less numerous and more patchy than in the West, they nevertheless demonstrate that the planetary week was not entirely unknown in the Roman Near East during the imperial and late antique periods.

4.2 *The Planetary Week in the Later Roman Empire and Late Antiquity: Parapegmata and Other Material Evidence*

As mentioned above, parapegmata are instruments of different shapes and materials, which were used to keep track of various temporal and cyclical phenomena by the use of a moveable peg.[125] We have already considered two early imperial parapegmata, one from a village in Latium and the other one from Posillipo (ancient Pausilipum) in Campania, both including the planetary week. We have also discussed a passage from Petronius' *Satyricon* (30.3–4), which has been interpreted as a description of a lunar and planetary parapegma. All other extant parapegmata including the planetary week belong to later periods (roughly, second to fourth century CE). As is the case with the early imperial parapegmata and other types of evidence of the planetary week, the overwhelming majority of these later parapegmata were found in Italy and in the western provinces of the Roman Empire. A noticeable difference between the early imperial and the later imperial parapegmata is the absence in almost

121 *P.Oxy.*44.3174v.17 (papyrus; precisely dated to Wednesday, 'day of Hermes', 8 March, 243 CE); Cribiore, *Writing*, no. 146 (wooden tablet; 'day of Helios'; variously dated between 294 and 327 CE); Seldeslachts and Wouters, 'Christian word-list' (papyrus; 'day of Helios'; sixth century CE); *P.Bagnall* 15–16 (ostracon; the text is in copticising Greek; a list of Christian days and their planetary equivalents, except for the days of Kronos and Helios, i.e. Saturday and Sunday; sixth-eighth century CE).

122 *P.Bagnall* 2 (papyrus; 'day of Zeus' and 'day of Selene' along with Christian names of Sunday and Saturday; possibly fourth century CE).

123 *PSI* 7.843.13 (papyrus; ἄχρις τῆς Ἑρμοῦ 'until (the day) of Hermes'; fifth/sixth century CE).

124 *P.Oxy.*42.3026 (papyrus; 'day of Hermes' is restored; after 166 CE); *P.Oxy.*54.3741 (papyrus; '(day) of Zeus'; 313 CE); *P.Oxy.*22.2343 (papyrus; '(day) of Zeus'; 287 CE); *P.Oxy.*60.4075 (papyrus; '(day) of Zeus'; 318 CE). On the last three see Bultrighini, 'Thursday (*dies Iovis*)', pp. 72–75 and further below.

125 See above, pp. 30–31. A catalogue of Greek and Roman parapegmata can be found in Lehoux, *Astronomy*, pp. 147–491. This, however, does not include a few epigraphic examples that were discovered after 2007: see Stern's review of Lehoux, *Astronomy*, in *Scripta Classica Israelica* 27 (2008), p. 141; also Bultrighini, 'Die griechischen Parapegmata'; and ead., 'Das Parapegma von den Trajansthermen'.

all of the latter of the nundinal days. In these later exemplars, the days of the planetary week are consistently associated to the days of the lunar month, and occasionally to the signs of the zodiac or the four seasons.

A remarkable example of such a parapegma from the later imperial period was scratched in the wall plaster of a Roman house near the Baths of Trajan on the Oppian Hill in Rome, where it was preserved even after the building was repurposed as a Christian chapel.[126] The graffito was already damaged when it was unearthed in 1812. After its discovery, it was apparently left exposed to the elements, which quickly led to its complete erosion. Nevertheless, its original appearance has been preserved thanks to the illustrations that were produced by Piale in 1812 and de Romanis in 1822. Afterwards, two copies of the graffito emerged: a terracotta copy made from the original, which ended up in the Kunstgeschichtliches Museum of the University of Würzburg, and a plaster cast of this copy, now stored in the Museo della Civiltà Romana in Rome. Variously dated between the second and the fourth century CE, the graffito included a top row with the seven days of the week depicted as busts of the planetary gods and goddesses presiding over each of the days, with their typical attributes. The series originally began on the left hand side with an image of Saturn (Saturday), but this had already vanished when the graffito was discovered. Following were the Sun wearing a solar crown (Sunday), Luna with a crescent-shaped crown (Monday), Mars armed with a helmet and spear (Tuesday), Mercury with his winged hat (Wednesday), another blank in place of Jupiter (Thursday), and Venus sporting a tiara on her forehead (Friday). Below each of these images was a peg hole. In the centre of the parapegma was a zodiac wheel bearing representations of the twelve zodiac signs, also equipped with peg holes. A sequence of numbers from I to XXX, all with peg holes, was arranged in two columns to the left and right of the zodiac wheel. By manually moving a peg along the holes on its top, centre, and sides, the parapegma would have enabled its users to track the days of the planetary week, the zodiac signs, and the days of the lunar month (the sequence I–XXX must refer to the lunar month, as some Julian months have 31 days, and in Roman practice the days of calendar months are not numbered consecutively).[127]

---

126 *IIt.* XIII 2, 308–9, no. 56. See Lehoux, ibid., pp. 16–17, 168–70; Bultrighini, 'Das Parapegma von den Trajansthermen'.

127 Further parapegmata in which the days of the planetary week appear as images: a marble slab from the ancient town of Veleia near modern Piacenza in northern Italy (*CIL* XI 1194 = *IIt.* XIII 2, 313, no. 59; see Lehoux, ibid., p. 172; undated; with lunar days). A stone fragment of unknown date and provenance now held in Arlon, Belgium (ibid., p. 177). A limestone relief of unknown date and provenance, now held in Soulosse, France (ibid.; Moitrieux, *Recueil général des sculptures*, no. 959). A clay parapegma from

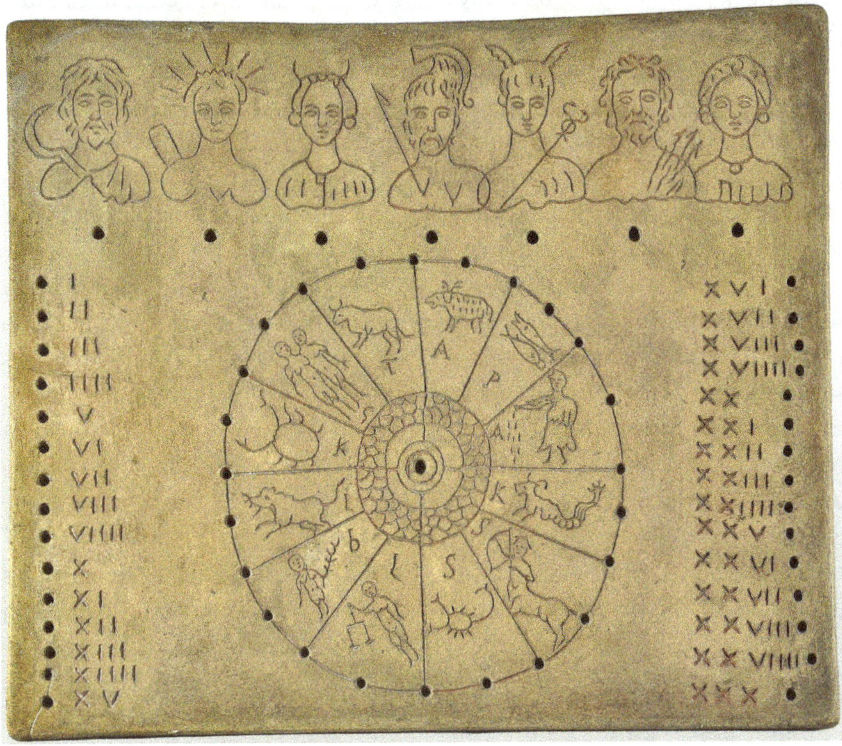

FIGURE 1.3 Trajan's Baths parapegma (*IIt* XIII 2, 308–9, no. 56), terracotta copy. With kind permission of the Martin von Wagner Museum of Wuerzburg University
PHOTOGRAPH: P. NECKERMANN

In the eastern Mediterranean, the only parapegma tracking the days of the planetary week to have been found is a graffito that was scratched on the wall of a Roman military barrack at Dura-Europos, during the period of Roman occupation of the city (late second–mid third centuries CE).[128] The graffito is known through a drawing that was made after much of it broke apart and was lost during excavation, which is why some caution is required in its interpretation. What is certain is that similarly to the Trajan's Baths parapegma, the graffito from Dura displayed the busts of the seven planetary gods and goddesses

Rottweil, Germany (Lehoux, ibid., pp. 178–79; mid-second century; with zodiac signs and lunar days). A fragmentary sandstone relief from Bad Rappenau, Germany (ibid., p. 179; late second or early third century). A clay fragment and a clay mould from Trier, Germany (Lehoux, 'Days', pp. 107–9; third/fourth century; with lunar days and personified images of the four seasons).

128 See Lehoux, *Astronomy*, pp. 170–71, with previous bibliography. Lehoux could not locate the remaining fragments of the graffito at Yale, where they are supposedly preserved.

of the week positioned horizontally across the top and arranged in week order (Saturn to Venus), each provided with a peg hole. The name of the relevant deity/planet was inscribed in Latin above each image (only 'Luna' is preserved). The graffito also included the days of the lunar month as well as, apparently, the nundinal days. Although found in Dura-Europos, this Latin-language parapegma with nundinal days clearly reflects a Roman cultural context; it seems reasonable to assume that the military official who scratched it originated from the western part of the Roman Empire.[129]

Although these late imperial parapegmata including the planetary week were used for tracking temporal cycles, they also possessed an astrological connotation. The absence of basic calendrical data such as Julian months and days suggests that the main intention behind these parapegmata was not calendrical, but rather astrological. This is apparent in the case of the Trajan's Baths parapegma (as well as the Rottweil parapegma, see n. 127), which enabled its users to track the passage of the Sun or Moon through the twelve signs of the zodiac.

In addition to parapegmata, a variety of further artefactual evidence bears depictions of the planetary week deities, representing the seven planets in week order – that is, in a sequence starting with Saturn (Saturday) and ending with Venus (Friday). This material includes mosaics, frescoes, stone reliefs, architectural members, bronze and silver statuettes, vessels, and other objects, as well as a homogeneous series of relief images on altars and statue bases from the Germanic and Gallic provinces. Symbolic representations occasionally occur, yet the seven heavenly bodies of the planetary week are most frequently depicted anthropomorphically, in the shape of their associated deities, typically as busts. These images are first attested roughly at the same time as the planetary week emerged, and became increasingly widespread during the imperial period. Although they have been found throughout the Roman Empire, the earliest examples and the majority of these representations originate from Italy and the western provinces. And it is seemingly upon western

129 Another parapegma in the form of a wall graffito was found in a third-century private house in Via dei dipinti at Ostia (*CIL* XIV 2037 = *IIt.* XIII 2, 312, no. 58; see Lehoux, ibid., p. 173; now lost). The first line showed the names of the planetary days in the genitive (only 'Lunae' being preserved), followed by numerals indicating the days of the lunar month. The only other parapegma which remains to be mentioned is an inscribed marble fragment of uncertain origin and date, currently held in Naples (*CIL* X 1605 = *IIt.* XIII 2, 307, no. 55; see Lehoux, ibid., p. 173; with lunar days). Here, each day of the planetary week was indicated by the name of the relevant planet in the genitive, as in the Ostia parapegma.

models that the limited number of depictions of the seven planetary week deities from the eastern Mediterranean is directly dependent.[130]

4.3 *The Planetary Week in the Later Roman Empire and Late Antiquity: Patristic Sources*

A substantial part of the literary evidence for the use of the seven-day week in its planetary form from the third century onwards is provided by patristic and other early Christian sources. The earliest Christian mention of the planetary week is the passage from Justin Martyr's *Apology* (c. 150 CE) discussed earlier, in which the Christian day of worship, Sunday, is identified as 'the day of the Sun', while the day of Jesus's crucifixion (Friday) is called the day 'before that of Kronos'. Slightly later, both Clement of Alexandria and Tertullian make reference to planetary days of the week, as we have already seen.[131]

In Tertullian we find an early instance of the Christian reproach for the use of the planetary week, which will later become considerably less veiled: while addressing the pagans about their cult of the Sun, Tertullian mentions 'the day of the Sun' and 'the day of Saturn', and affirms: 'it was certainly you who admitted the Sun within the list of the seven days'.[132] From this point onwards, Church Fathers and early Christian leaders regarded the habit of referring to the days of the week by their planetary designations as a sign of paganism and superstition, as these days were named after pagan deities and were connected to the pervasive belief that the planets influenced people's lives.

130 On visual representations of the seven planets as week deities, see De Witte, 'Les divinités des sept jours'; Haug, 'Die Wochengöttersteine'; Maass, *Die Tagesgötter*; Duval, 'Les dieux de la semaine'; Gundel, 'Planeten'; id., 'Pianeti'. See also Bultrighini, 'Notes on days of the week', pp. 187–89 and ead., '*Theōn hemerai*'. An article exploring the ways in which the planetary week deities were depicted on a wide range of material sources is currently in preparation by the same author. The monuments from Roman Gaul and Germany mentioned above are known as 'Jupiter columns' and were widespread from the second to the mid-third centuries CE. For the Gauls, see Picard, 'Imperator Caelestium'; for the Germanic provinces, Bauchhenß and Noelke, *Die Iupitersäulen*. See also Woolf, 'Representation as cult'; and Van Andringa, *La religion en Gaule romaine*, pp. 190–91.

131 In a passage of his *Protrepticus* dealing with the pagan worship of the Sun, the Moon, and the planets (4.63.1), Clement declares that these 'are not gods, but instruments for measuring time' (τὰ ὄργανα τοῦ χρόνου). This idea is echoed by the African mathematician Hilarius in the fourth or fifth century CE (*De Solstitiis*, *PLS* 1.567). In addition to generally hinting at the role of celestial bodies in time reckoning (cf. Gen. 1:14), Clement may also be referring more specifically to the planetary deities as rulers of the seven days of the planetary week (Bultrighini, '*Theōn hemerai*').

132 *Ad nat.* 1.13: vos certe estis, qui etiam in laterculum septem dierum solem recepistis.

The condemnation of the pagan and astrological implications of the seven-day planetary week escalated during the fourth century.[133] Philaster, bishop of Brescia in northern Italy, includes the use of the planetary nomenclature for the days of the week in his catalogue of heresies compiled around 384 CE: Habenda est et haeresis, quae dicit nomina dierum, Solis, Lunae, Martis, Mercurii, Jovis, Veneris, Saturni.[134] Other Christian leaders not only criticised the planetary designations, but also recommended the use of a distinctively Christian vocabulary to refer to the days of the week, which by then had been fully developed (see further below). In the late fourth or early fifth century, Augustine urges his fellow Christians to refrain from calling the days of the week by their planetary names, and encourages them to use the Christian nomenclature: 'The first of the week is the Lord's day; the second of the week is the second day, which people of the world call day of the Moon; the third of the week is the third day, which they term day of Mars. Therefore, the fourth of the week is the fourth day, which is called day of Mercury by pagans, and also by many Christians; but we do not want that; and we wish they would change for the better, and cease to do so; for they have their own language, which they should use'.[135]

In the fifth and sixth century Christian preachers attacked even more vehemently this habit, which was apparently still widespread at the time. Both Caesarius, bishop of Arles in Gallia Narbonensis (south-eastern France), and the Galician Martin, bishop of Bracara Augusta (present Braga in north-western Portugal) are known for their violent opposition to the custom of naming the days of the week after Greek and Roman gods and goddesses. In the first half of the sixth century, Caesarius harshly condemns the deities or 'demons' that gave their names to the seven days, and insists on the need to replace such appellations with the Christian designations of Biblical origin: 'But these

133 Salzman, 'Pagan and Christian notions', pp. 192–94.

134 *Haer.* 113: Haeresis de septem planetis. Cf. Irenaeus of Lyon, *Adversus haereses* 1.30.9 (c. 180 CE): 'They want the sacred Hebdomad to consist of the seven stars called planets' (Sanctam autem hebdomadem septem stellas, quas dicunt planetas, esse volunt). Irenaeus is talking about the Ophite Gnostics, and this passage seems to show that these used to call the planetary week 'sancta Hebdomas'. As part of his attacks against heresies based on the principles of astrology, Philaster also criticises the idea that people's fate depends upon the twelve signs of the zodiac (*Haer.* 123: Alia est heresis quae dicit secundum duodecim zodia nasci homines).

135 *Commentary on the Psalms* 93.3: Una sabbati, dies dominicus est; secunda sabbati, secunda feria, quem saeculares diem Lunae vocant; tertia sabbati, tertia feria, quem diem illi Martis vocant. Quarta ergo sabbatorum, quarta feria, qui Mercurii dies dicitur a paganis, et a multis christianis; sed nollemus; atque utinam corrigant, et non dicant sic. Habent enim linguam suam, qua utantur.

unhappy and ignorant men, who worship these most wretched and impious men we mentioned above, by fear of them rather than by love of them, for their sacrilegious cult, almost in their honour, named each day of the week after each of them. (…) We judge no day worthy of the name of demons (…). But also let us disdain to speak those most sordid names and never let us say 'Mars' day', 'Mercury's day', or 'Jupiter's day', but rather 'first day', 'second day', or 'third day'; according to what is written, let us call the days. Warn your families too about these names.'[136] Caesarius also rises up against superstitious and pagan attitudes associated with specific days of the week,[137] and the practice of opting for specific days regarded as auspicious to begin a journey or return home,[138] as well as other 'pagan' observances related to particular times of the seasons and the calendar year, such as the summer solstice,[139] and the calends of January.[140] A few decades later, Martin of Braga, certainly drawing on Caesarius, addresses similar issues in his sermon focused on the issue of rural paganism (*De correctione rusticorum*). In particular, he condemns the practice of observing propitious days of the week to perform specific activities and the habit of naming the days of the week after the planetary gods and goddesses of the Graeco-Roman pantheon. Like his predecessors, Martin recommends, instead, the use of the ecclesiastical designations.[141]

From the fourth century onwards, generations of Christian preachers struggled against the use of the planetary designations for the days of the week,

136 *Sermo* 193.4: sed miseri homines et imperiti, qui istos sordidissimos et impiissimos homines, ut supra diximus, timendo potius quam amando colebant, pro illorum sacrilege cultu, quasi in honore ipsorum, totos septimanae dies singulis eorum nominibus consecrarunt. (…) nullum diem daemonum appellation dignus est iudicemus. (…) sed etiam ipsa sordidissima nomina dedignemur et ore proferre, et numquam dicamus diem Martis, diem Mercurii, diem Iovis; sed primam et secundam vel tertiam feriam, secundum quod scriptum est, nominemus. De his etiam nominibus et vestras familias admonete.

137 In particular with Thursday (*Sermones* 13.5, 19.4, and 52.2), on which see Bultrighini, 'Thursday (*dies Iovis*)', pp. 66–67 and further below.

138 *Sermo* 1.12 (qua die in itinere egrediatur, vel qua die ad domum propriam revertatur). See also *Sermo* 54.1.

139 Here referred to as St John's day (*Sermo* 33.4: ne ullus in festivitate sancti Iohannis aut in fontibus aut in paludibus aut in fluminibus nocturnis aut matutinis horis se lavare praesumat).

140 *Sermones* 192 and 193 (De kalendis Ianuariis).

141 C. 8: 'They give the names of these demons to each day of the week, calling these days by the names of Mars, Mercury, Jupiter, Venus and Saturn, where these created no days, but were wicked and evil men among the Greeks' (sed sunt dubii in tantum ut nomina ipsa daemoniorum in singulos dies nominent, et appellent diem Martis et Mercurii et Iovis et Veneris et Saturni, qui nullum diem fecerunt, sed fuerunt homines pessimi et scelerati in gente Graecorum). See also c. 9, 16, and 18. See Bultrighini, 'Thursday (*dies Iovis*)', pp. 67–68, with full bibliography.

which they considered as a symptom of paganism. While it is doubtful that the habit of referring to the days of the week by pagan names retained much religious significance in Merovingian Gaul,[142] it seems that at least until the fourth century the seven-day week continued to possess an astrological connotation, and people still commonly believed to be constantly subject to the planets throughout their life.[143]

Despite the continued efforts of some Christian preachers to supplant the planetary names with the Christian nomenclature, in Western Europe the vast majority of people, Christians included, persisted in calling the days of the week after the seven planets into Late Antiquity and well beyond, as we are about to consider more in detail in the next section.

4.4 *The Christian Week*

As stated earlier, during the Roman imperial period the early Christians adopted the nomenclature of the Jewish, Biblical week and adapted it to produce what we may refer to as the Christian week. Most importantly from a Christian perspective is that the Jewish Sabbath (Saturday) was replaced by the Christian Lord's day (Sunday) as the focal day of the week, a day of worship and gatherings, as early as the second century CE.[144]

In accordance with the Jewish week, which has the Sabbath as its seventh day, Sunday was identified as the first day of the Christian week. This differed from the planetary hebdomadal cycle whose day one was Saturday.[145] By the fourth century, Sunday was systematically referred to as the first day of the seven-day week in all types of sources, including non-explicitly Christian ones.[146] In all likelihood, this order of the days of the week emerged as a result of the influence of the Jewish and Christian traditions.[147]

---

142 W.E. Klingshirn, *Caesarius of Arles: The Making of a Christian Community in Late Antique Gaul* (Cambridge: Cambridge University Press, 1994), pp. 215–16.

143 Salzman, 'Pagan and Christian notions of the week', p. 194.

144 *Didache* 14.1; Ignatius, *Ad Magn.* 9; Justin, *Apolog.* 1.67. Bacchiocchi, *From Sabbath to Sunday*; Llewelyn and Nobbs, 'Earliest dated reference', pp. 110, 112; Di Berardino, 'Liturgical celebrations', p. 211; Salzman, 'Pagan and Christian notions', pp. 198–99; Bultrighini, 'Thursday (*dies Iovis*)', p. 78.

145 Early evidence of Saturday being the first day of the planetary week includes some of the Pompeii graffiti discussed above (*CIL* IV 6779; *CIL* IV 5202; *CIL* IV 8863) as well as the parapegmata and the visual representations of the planetary week dealt with in one of the previous sections.

146 Ausonius, *Ecl.* 8; Paul of Alexandria, ch. 20; *PGM* XIII 218–224; *P.Kellis* I 82.

147 Llewelyn and Nobbs, 'Earliest dated reference', p. 117; Salzman, 'Pagan and Christian notions'; Bultrighini, 'Kaiser Konstantins Edikt', pp. 589–90. *Contra* Di Berardino ('Liturgical celebrations', p. 217), who perplexingly posits that this order 'comes from a

In its purely numbered form, the Greek version of the Jewish, Biblical week in Late Antiquity had its days expressed with adjectives or with alphabetic numerals: ἡμέρα α' or ἡμέρα πρώτη ('first day') for Sunday, ἡμέρα β' or ἡμέρα δευτέρα ('second day') for Monday, ἡμέρα γ' or ἡμέρα τρίτη ('third day') for Tuesday, ἡμέρα δ' or ἡμέρα τετάρτη ('fourth day') for Wednesday, ἡμέρα ε' or ἡμέρα πέμπτη ('fifth day') for Thursday, ἡμέρα ϛ' or ἡμέρα ἕκτη ('sixth day') for Friday, and ἡμέρα ζ' or ἡμέρα ἑβδόμη ('seventh day') for Saturday. As noted previously, Friday was also referred to as (ἡμέρα) παρασκευή, '(day of) preparation', and προσάββατον, 'the (day) before Sabbath'.[148] Saturday may also be termed ἡμέρα σαββάτου or σάββατον (also plural, σάββατα), 'day of Sabbath' or simply 'Sabbath'.

The essential Christian modification of the Biblical week pertained to Sunday. Greek-speaking Christians began to call Sunday ἡμέρα κυριακή or simply κυριακή ('the Lord's day') presumably after the unique mention of this term in the New Testament (Rev. 1:10).[149] The first day of the week was primarily intended as a weekly commemoration of Christ's resurrection.[150] Because of the importance the Christians attached to that event, Sunday was early on established as a day of worship, apparently in the second century CE.[151]

As for the Latin vocabulary of the Christian week, Western Christianity rendered the Greek numbered weekdays attested in the East as feria secunda (or feria II), feria tertia (or feria III), feria quarta (or feria IV), feria quinta (or

---

pagan source'. There is no evidence that the cult of Mithras or Sol Invictus helped in any way to favour dies Solis as the beginning of the week; indeed, no religious rite is known to have been regularly performed for any solar deity on dies Solis.

148 For example, a third or fourth-century CE lead curse tablet against an athlete from Corinth, Greece, which proclaims: 'may they not prevail on Friday (προσάββατον)' (*SEG* 44 308; Jordan, 'Magia nilotica', p. 699). On earlier attestations in the Septuagint and other sources, see above p. 15.

149 Kubitschek suggested that the Christians might have coined the term 'the Lord's day' on the model of the Egyptian practice of naming the first day of the month or its most important one as 'Σεβαστή' (day of the emperor); Kubitschek, *Grundriss der antiken Zeitrechnung*, p. 33. Early references to the Lord's day in Christian literature include Ignatius of Antioch (*ad Magn.* 9.1) who around 100 CE urged the Magnesians to 'no longer keep the Sabbath, but live in the observance of the Lord's day' (μηκέτι σαββατίζοντες, ἀλλὰ κατὰ κυριακὴν ζῶντες). Around the same period, the Didache (14.1) instructed the Christians to assemble on the Lord's day to worship (κατὰ κυριακὴν δὲ κυρίου συναχθέντες ...).

150 Just. *Apol.* 1.67; Tert. *de Orat.* 23. In Justin's passage, Sunday is referred to by its planetary name: 'On the day of the Sun (τὴν δὲ τοῦ ἡλίου ἡμέραν), we assemble, because it is the first day, in which God created the world and because on this day Jesus Christ, our Saviour, came back from the dead'. On Justin's use of the planetary designations, see above, pp. 32–33 and 39.

151 See references above, n. 144.

feria V), feria sexta (or feria VI), for Monday to Friday.[152] For Saturday, the Greek σάββατον was transliterated into the Latin alphabet as sabbatum. For Sunday, ἡμέρα κυριακή or κυριακή was translated as dies dominica (also dies dominicus) or simply dominica, 'the Lord's day'. The term 'feria', which is drawn from the Roman calendrical tradition, can be rendered in this context simply as 'day'. Interestingly, 'feriae' were originally holidays, festival days, as opposed to working days.[153] They were distinguished by the cessation of all profane activities and by the performance of religious practices. Patristic sources seem to suggest that the term 'feriae' indicating weekdays in the Christian week nomenclature conveyed the idea that every single day of the week belongs to and is a celebration of the Lord. This idea is apparently related to the Christian criticism towards the Jewish Sabbath: according to some Church Fathers, there is no day which is more sacred than the others, every day is 'a Sabbath day', every day belongs to and must be consecrated to the Lord.[154] Yet, paradoxically, the notion that 'all days belong to God' was explained in relation to the seven days of Creation,[155] and was therefore closely related to the Jewish Sabbath week. On a more practical level, it has been suggested that the use of the term 'feria' for weekdays may have originated from the custom of referring to the days of the Holy Week as 'secunda feria', 'tertia feria', and so on. This was a week of rest and the first of the ecclesiastical year. Consequently the days of any other week of the year may have begun to be called in the same way.[156] In fact, ancient sources are largely quiet on the origins of the Christian habit of referring to weekdays as 'feriae': the only testimony we have is provided by Bede

152 These formulae are occasionally preceded by 'die' ('on the day'). The 'feria' terminology is treated in Bultrighini, 'New light', pp. 422–23.

153 This meaning of the Latin term 'feriae' has persisted in Romance languages. For instance, 'férias' in Portuguese and 'ferie' in Italian are holidays, vacation from work. At the same time, the ecclesiastical sense of the Latin word 'feria' is reflected in the adjectives 'ferial' in Portuguese and 'feriale' in Italian, which refer to workdays, weekdays. In this context, it is worth noting that Portuguese is the only major Western European language to preserve traces of the 'feria' nomenclature (segunda-feira for Monday, terça-feira for Tuesday, etc.). The use of 'feira' in Portugal as well as in Galicia suggests that here the influence of the Church in Late Antiquity was stronger than elsewhere.

154 A passage from the *Didascalia Apostolorum*, an early third-century treatise addressing Christian communities in Syria (known through a Latin translation), states that omnes dies domini sunt, 'all days belong to the Lord' (6.18.16). See also *Ad.Diognet.* 4; Origen, *C.Celsum* 8, 21–2; Tertullian, *de baptismo* 19. See Pietri, 'Le temps de la semaine', p. 63; Eriksson, *Wochentagsgötter*, p. 32; Markus, *End of Ancient Christianity*, p. 100.

155 Martin of Braga, *de corr. rust.* 9.

156 'feriae'; Du Cange et al., *Glossarium*.

in the early eighth century, who reports that the use of this terminology was enacted by Pope Sylvester in the first half of the fourth century (314–335 CE).[157]

The earliest attestation of the use of the Christian names of the days of the week in Latin is in Tertullian, in the late second or early third century CE. In his *de Ieiunio* (On Fasting), he refers to Wednesday and Friday first as 'quarta et sexta feria' (2, 3) and subsequently as 'quarta et sexta sabbati' (14, 2–3); he also mentions sabbatum and dominica (14, 2–3 and 15, 2), and refers to Friday as parasceve (14, 2–3), a transliteration of the Greek παρασκευή (Day of Preparation).[158] However, these are isolated occurrences, after which one has to wait until the fourth century for the Latin nomenclature to reappear in literary sources.

4.5 *The Christian Week: Epigraphic and Documentary Evidence*

As seen earlier, apart from a number of examples from the early Empire, most of the inscriptions with days of the planetary week date from the fourth and fifth centuries and come primarily from the western area of the Roman Empire. The epigraphic evidence relating to the Christian week, on the other hand, dates predominantly from the fifth and sixth (as well as later) centuries and originates chiefly from the eastern side of the Roman Empire, in particular from Greece, Thrace, Macedonia, Asia Minor, and the Roman Near East.[159] Similarly to the case of the planetary week, the inscriptional evidence for the Christian week consists largely of epitaphs, which – needless to say – originate eminently from Christian milieus.

157 *de temp. rat.* 8; *de rat. comp.* 5. In the early seventh century CE, Isidore of Seville proposed a singular etymology for 'feriae' as weekdays, which related the names of the days to the act of speaking (*etym.* 5.30.12: a fando autem feriae nuncupatae sunt, quod sit in eis nobis tempus dictionis, id est in divino vel humano officio fari).

158 The term is used in the third-century Apostolic Tradition and reappears in the late fourth-century Vulgate. Parasceve is also attested as a female personal name in a small number of sepulchral inscriptions: *CIL* III 8935 (from Dalmatia, third-fourth century CE); *CIL* XII 5764 = *CIL* XII 5765 = *CAG*-13-3, 67 (from Massilia, undated); *CIL* XIV 1449a = Thylander, 121 (from Portus, undated); *CIL* X 1101 (from Nuceria, second–third century CE); *CIL* VI 16429 (from Rome, undated); *CIL* VI 19917 (from Rome, undated); *CIL* VI 38355 (from Rome, 150–250 CE).

159 Inscriptions dated to later than the sixth century are not taken into consideration in this survey. The total number of occurrences is 495, of which 456 are Greek and 39 are Latin. The Greek dossier comprises a couple dozen inscriptions from the West (mostly from Italy and Sicily) and well over 400 inscriptions from the East. Of these latter, 290 are from the site of Zoar/Ghor es-Safi. With one exception, all Latin inscriptions originate from the West. The unique Latin document from the East is *INikaia* 574, from Nicaea, in the Roman province of Bithynia and Pontus; it is a bilingual (Latin & Greek) Christian epitaph attributed to the fourth or fifth century, whose Latin text commences with a monogrammatic cross followed by the words 'in die dom[inica ...]' ('on the Lord's day ...').

Much of the epigraphic record in Greek displays essentially the same nomenclature which is attested in the literary sources, as illustrated in the previous section. An interesting example is the sixth-century marble tombstone of Zacharias, son of Erasinos, from the South Church at Avdat in the Negev desert (southern Israel), an area once belonging to the Roman province of Palaestina Tertia (*SEG* 28, 1396). The epitaph describes in minute details the dates of death and burial of the deceased: 'Zacharias, son of Erasinos, passed away on the tenth of the month Panemos, in the fourteenth indiction, on Sunday, at the third hour of the night, he was buried here on Tuesday, at the eighth hour, on the twelfth of Panemos, in the fourteenth indiction, in the year 476 according to (the era of) Elousa.'[160] The days of the week are expressed as ἡμήρᾳ (for ἡμέρᾳ) κυριακῇ, 'on the Lord's day' (Sunday) and τρίτῃ τοῦ σάμβατος, 'on the third of the week' (Tuesday).

A fourth/fifth-century epitaph inscribed on a round marble plaque from Catania, Sicily, mentions παρασκευή, the '(day of) preparation' (date of death), and σάββατον, 'Sabbath' (date of burial): 'Agathe died at the age of sixty years on Friday, the ninth day after the calends of September; she was buried on Saturday.'[161] The latter is expressed as δαὶ Σαββάτοις, 'δαί' apparently being an attempt to transcribe into Greek the Latin 'die' ('on the day'); Σαββάτοις is the dative of the plural σάββατα, which was widely used alongside the singular σάββατον (see above).

The epigraphic record occasionally attests to the combination of two different designations, such as in the case of a fourth-century (?) epitaph from Nea Anchialos in Thessaly (Greece), whose two final lines refer to ἡμέρα κυριακὴ πρώτη, which may be rendered as 'the first day (of the week), the Lord's day'.[162]

---

160 Negev, *Greek Inscriptions*, no. 18; *SEG* 28, 1396: + ἀνεπάη ὁ μακά/ριος Ζαχαρίας / Ἐρασίνου ἐν / μηνὶ Πανέμου / δεκάτῃ ἰνδ(ικτιῶνος) ιδ' ἡ/μήρᾳ Κυριακῇ ὥραν / τρίτῃ τῆς νυκτὸς· κα/τετέθη δὲ ἐνταῦθα / τῇ τρίτῃ τοῦ Σάμβα/τος ὥραν ὀγδόην / Πανέμῳ δωδεκά/τῃ ἰνδ(ικτιῶνος) ιδ' ἔτους κα/τὰ Ἐλούσ(αν) υος'(…). The dates are expressed according to the so-called calendar of the Province of Arabia, a solar calendar based on the Egypto-Macedonian model, and the era of the Provincia Arabia (here referred to as era of Elousa); the dates correspond to 29 June and 1 July of the year 581. See Meimaris, with Kritikakou and Bougia, *Chronological Systems*, pp. 149–51, 159, 258 no. 368.

161 *ICatania* 173: Ἀγάθη / ⳩c ἐτελεύτησεν / ἐτῶν ξ Παρα- ⳩c /⟨σ⟩κευῇ τε̂ς θ ἀπὸ κα/λανδῶν Σεπτεμ/βρίω⟨ν⟩· ἐξεκομίσθη/ δαὶ Σαββάτοις· (…) Korhonen observes that the month day expressed as 'ἀπὸ καλανδῶν', 'after the calends', is common in Syracuse and south-eastern Sicily in this period, though the formula is normally μηνί + month name + ἀπὸ καλανδῶν + figure. Agathe's epitaph combines this formula with the traditional Roman one (πρό + figure + καλανδῶν + month name).

162 On this inscription see Bultrighini, 'Notes', pp. 190–93. See p. 193 n. 24 for details about the epigraphic documentation for ἡμέρα κυριακή and ἡμέρα πρώτη.

FIGURE 1.4 Tombstone of Zacharias, Avdat, Israel (Negev, *Greek Inscriptions from the Negev*, no. 18; SEG 28, 1396)
PHOTOGRAPH I. BULTRIGHINI

As mentioned, the site of Byzantine Zoar/modern Ghor es-Safi (Jordan) produced an exceptionally large number of Christian tombstones inscribed in Greek, mostly dating from the fourth to the sixth century, where the days of the week are expressed according to both the planetary and the Christian nomenclature. However, while the planetary names appear in 45 inscriptions, the Christian designations are mentioned in as many as 290 epitaphs. Interestingly, several of these latter differ from the 'standard' terminology we have thus far considered. About 125 of the 290 inscriptions including Christian days of the week from this site display the formula (ἐν) ἡμέρᾳ Κυρίου + numeral, that is, 'on the X day of the Lord'. As an example, the epitaph of Thopse, daughter of Ellios (*IPalaestina Tertia* Ia.73), specifies that she passed away on 31 July 395 CE, ἡμέρᾳ Κυρίου πέντῃ (for πέμπτῃ), 'on the fifth day of the Lord' (Thursday).

At Zoar are also attested a few 'hybrid' formulae, such as *IPalaestina Tertia* Ia.93: ἡμ(έρᾳ) Κ(υρίο)υ Κρόνου, 'on the Lord's day of Kronos', where the Christian and the planetary nomenclature are freely mixed; *IPalaestina Tertia* Ia.98: ἡμέρᾳ Κυρίου τετάρτῃ δ', 'on the fourth day of the Lord', the numeral being expressed both as adjective and according to the alphabetic numeral system; *IPalaestina Tertia* Ia.29[illegible]: [ἡμέ]ρᾳ Κ(υρίο)υ κυρ[ιακῇ], 'on the Lord's day', essentially with duplication of 'the Lord's';[163] *IPalaestina Tertia* Ib.49: ἡμ(έρᾳ) Κ(υρίο)υ παρασκ(ευῇ), 'on the Lord's preparation day', where the locution ἡμέρᾳ Κυρίου is associated with the 'Preparation' denomination for Friday instead of a numeral.

Outside the numerous examples from Zoar, only three epitaphs of different origin seem to display the formula ἡμέρα Κυρίου plus a numeral: one from Trier, Germany (Augusta Treverorum, in the province of Gallia Belgica) and two from Sicily.[164] The phrase ἡμέρα Κυρίου seems to express the notion that every single day of the week belongs to and is a celebration of the Lord.[165] In light of this, it may be speculated whether this peculiar and not much diffused formula emerged as an attempt to offer a close translation into Greek of the 'feria' nomenclature, the official Latin terminology of the Christian Church for the days of week.

Whereas the use of the Christian designations is first attested in Latin literary sources in Tertullian and then, after a hiatus, in the fourth century, the earliest epigraphic evidence of the 'feria' designations belongs to as late as the sixth century.[166] However, dominica and sabbatum appear earlier than weekdays in inscriptional sources (in the fifth and fourth century, respectively). The number of late antique inscriptions displaying the Christian nomenclature in Latin is limited (see note 159); the evidence becomes somewhat richer after 600 CE. As one might expect, this is an entirely western phenomenon: the majority of these inscriptions (27) originate from the Italian peninsula, while fewer were found in North Africa, Gaul, and the Germanic provinces.[167]

---

163 See also *IPalaestina Tertia* Ia.225.

164 See Bultrighini, 'Notes', pp. 193–95 for details.

165 Meimaris and Kritikakou-Nikolaropoulou, *Inscriptions from Palaestina Tertia, Ia*, p. 50.

166 On 'feria' inscriptions, see Bultrighini, 'New light', pp. 421–24.

167 North Africa: *CIL* VIII 8630 = *ILCV* 2104 = *AE* 1967, 640 (from a Church in Setif, Province of Mauretania Sitifensis; precisely dated to Sunday, 'dies dominica', 3 August 452 CE); *CIL* VIII 16662 (p. 2732) = *ILAlg* 1 3452 = *ILCV* 3149 (from ancient Tebessa, Province of Africa Proconsularis; 'd(ie) s(an)c(t)a' interpreted as 'Sunday'; possibly fifth century);

Once again, epigraphic evidence of the term 'feria' consists mainly of epitaphs. Let us consider an example from ancient Brundisium (Brindisi, Italy), which has been attributed to the sixth century: 'Pretiosus, the bishop of the Catholic Church of the holy Brundisium, was buried on Friday (sexta feria), on the 15th day before the calends of September (…)'.[168] As few as five further (certain) examples of 'feria inscriptions' dating from no later than the sixth century CE are preserved.[169] It must be observed that at least half of these 'feria inscriptions' belong to ecclesiastical settings, that is, they are Church regulations or epitaphs of priests and other members of the clergy – as in the case of Pretiosus's tombstone.

The epigraphic evidence thus suggests that the Christian nomenclature in Latin – and especially the 'feria' denominations for weekdays – did not take hold in the Latin West beyond ecclesiastical circles. Despite reiterated attempts on the part of Church Fathers and early Christian preachers to eradicate this habit, the lay population (including Christians) continued to refer to the days of the week by their planetary designations well into Late Antiquity. This phenomenon reflects the longer tradition of the planetary week in this area of

---

IRT 839 = *AE* 2011, 95 (from Leptis Magna, Province of Africa Proconsularis; 'dies dominica', fifth century); *CIL* VIII 2013 (p. 2731) = *ILCV* 1385 = *CIL* VIII 16516 = *ILAlg* I 3424 = *AE* 2012, 1905 (from ancient Tebessa, Province of Africa Proconsularis; 'dies sabbatorum', 484–513 CE); *CIL* VIII 25045 = *ILTun* 1008 = *ILCV* 1003 = *ICKarth* 3, 381 (found near ancient Chartago, Province of Africa Proconsularis; 'quarta feria', possibly fifth century). Gaul & Germanic provinces: *CIL* XII 1045 (p. 821) = *ILCV* 1689 = *CAG*-30-3, 743 (from ancient Avennio, Province of Gallia Narbonensis; 'dies dominicus', 586 CE); *CIL* XIII 5463 = *ILCV* 3129 = *CAG*-21-2, 333 (from ancient Dibio, Province of Germania Superior; 'sabbatum', sixth century); *CIL* XIII 7526 = *FIM* 108 = Terrien-2007, 38 (from ancient Cambodunum, Province of Germania Superior; 'sabbatum', 580–620 CE); *RICG* 142A = *FITrier* 11 = Binsfeld, 'Kirchliche Würdenträger', p. 44 (from Augusta Treverorum, Province of Gallia Belgica; 'quarta feria', 395–423 CE).

168 *CIL* IX 6150 = *ILCV* 1026 = Rugo, *Le iscrizioni*, no. 44 = *ICI* XIII 48: Pretiosus, aepescopus/a[e]cletiae catolicae sanc/te brundisine, depositus/sexta feria, quod est/xv [k]al(endas) septembris, requiebit/in [s]omno pacis.

169 *ILCV* 3495 = *ICUR* II 4289 (metrical epitaph of a girl from the catacomb of San Pancrazio, Rome; 'feria III', 542–575 CE); *CIL* VIII 25045 = *ILTun* 1008 = *ILCV* 1003 = *ICKarth* 3, 381 (variously interpreted as the text of a judicial canon of the Montanist church, a Donatist regulation, or a conciliar or episcopal decree; found near ancient Chartago, Province of Africa Proconsularis; 'quarta feria', possibly fifth century); *RICG* 142A = *FITrier* 11 = Binsfeld, 'Kirchliche Würdenträger', p. 44 (epitaph of a priest from Augusta Treverorum, Province of Gallia Belgica; 'quarta feria', 395–423 CE); *AE* 2012, 389 = Bultrighini, 'New light', pp. 421–24 (epitaph of an elite woman; from near Avellino, ancient Hirpinia; 'die IIII f(eria);' 23 January 547 CE); *CIL* X 4630 = *ILCV* 218 = *ITrebula* 110 (epitaph of a woman of high social rank; from a church near Caserta; 'die sexta feria'; 24 October 559 CE).

the Roman Empire.[170] In this context, it should be observed that while in the East the Christian nomenclature has generally persisted to this day (both in modern Greek and in other languages of the Eastern Church) most Romance languages spoken in Western Europe still employ the planetary nomenclature, albeit with some variations; as a general rule, in Romance languages Monday to Friday bear names that refer to the planets, while Saturday and Sunday are derived from sabbatum and dominica, respectively; the only exceptions are Portuguese and Galician, both of which retained the ecclesiastical denominations for weekdays as well (see note [illegible]).

The documentary evidence from Egypt relating to the Christian week is relatively meagre, or rather, unevenly distributed. It comprises about 30 occurrences of κυριακή, the Lord's day (Sunday), most of which date from the sixth century,[171] and about 10 documents that refer to σάββατον, the Sabbath (Saturday), which are variously dated and also include references to the Sabbath as the Jewish sacred day.[172] Prior to the sixth century, κυριακή appears in only three documents from the fourth and fifth century.[173] Both these latter and the more numerous sixth-century documents which mention κυριακή do so in the context of the prohibition of working on Sunday,[174] which leads to our next and final section.

4.6 *Constantine's Law of 321 CE and the Rise of Sunday as the Focal Day of the Week*

On 3 March 321 CE, the emperor Constantine issued a law by which judges, the urban population, and craftsmen were to abstain from work on 'dies Solis', the day of the Sun (Sunday). The instruction applied to towns and cities; in the countryside, farmers were permitted to work in the fields. To quote: 'all judges

170 Bultrighini, 'New light', p. 423; ead., 'Thursday (*dies Iovis*)', p. 65.

171 For a list of the sixth-century documents see Ast, 'Schedule of Work days', p. 13 n. 23.

172 σάββατον as day of the week appears in the following papyri: *P.Genova* I 38.6, 9 (private letter; sixth-seventh century); *P.Oxy.*6.903 = CPJ III 457d (accusation by a wife against her husband; fourth century); SB 14.11541.4 (private letter; sixth–seventh century); SB 5.7872 = Naldini, *Il Cristianesimo in Egitto*, no. 75 (private letter; 306–337?).

173 *P.Oxy.* 48.3407 (private letter; fourth century); SB.20.15134 (contract; 483 CE); *P.Oxy.* 54.3759 (proceedings before the *logistes*; 2 October 325 CE). This third document represents the earliest papyrological evidence of Sunday as a non-working day in Roman Egypt. See also *POxy.* 54.3758, 119–20.

174 The only exception is *P.Oxy.* 48.3407 (cf. previous note), which on the contrary attests to people working ἐν τῇ/κυριακῇ{ν} ἡμέρᾳ, τουτ/έστιν αὔριον ια, 'on the Lord's day (Sunday), that is tomorrow the 11th' (ll. 15–17). On the documentary evidence of the use of κυριακή, see Llewelyn and Nobbs, 'Earliest dated reference', pp. 106–13; Ast, 'Schedule of Work days', pp. 12–13.

and the people in the city shall rest, and the work in all crafts shall cease, on the holy day of the Sun. But the people in the country may freely and lawfully apply themselves to cultivating their fields, so that the benefit conferred by the providence of God may not perish in an instant, since it often happens that grain can be sown in the furrows and vines planted in the trenches on no better day'.[175] Later in the same year (3 July) the emperor issued another regulation that sanctioned further exceptions to the ruling previously issued, by authorising the manumission of slaves and the emancipation of children on Sundays: 'Just as it appears to us most unseemly that the day of the Sun, which is celebrated on account of its own veneration, should be occupied with legal altercations and with noxious controversies of the litigation of contending parties, so it is pleasant and fitting that those acts which are especially desired shall be accomplished on that day. Therefore all men shall have the right to emancipate and to manumit on this festive day, and the legal formalities thereof are not forbidden'.[176] Copies of these two rulings were addressed to the 'governors of every province', first in the West, and after 324 CE, following Constantine's victory over Licinius and his capture of the East, in the East as well.[177]

These are the earliest extant imperial laws to recognise Sunday as a legal holiday and the first evidence of Sunday being regarded as a day of rest. These instructions, defining Sundays as free from court activity and labour but encouraging agricultural labour when needed, were in line with Roman tradition: in the Roman calendar, festival days (feriae publicae) were characterised by the interruption of judicial activities and rest from labour (applying to all men, including slaves), except for agricultural work where rest was permitted only at specified times of the seasonal year when its interruption was not likely to hinder produce.[178] In this sense, Constantine's regulations essentially extended to Sunday the rules that applied to traditional Roman

175 *Justinian Code* 3.12.2: omnes iudices urbanaeque plebes et artium officia cunctarum venerabili die solis quiescant. Ruri tamen positi agrorum culturae libere licenterque inserviant, quoniam frequenter evenit, ut non alio aptius die frumenta sulcis aut vineae scrobibus commendentur, ne occasione momenti pereat commoditas caelesti provisione concessa.

176 *Theodosian Code* 2.8.1: Sicut indignissimum videbatur, diem solis, veneratione sui celebrem, altercantibus iurgiis et noxiis partium contentionibus occupari, ita gratum ac iucundum est, eo die, quae sunt maxime votiva, compleri. Atque ideo emancipandi et manumittendi die festo cuncti licentiam habeant, et super his rebus acta non prohibeantur.

177 Eusebius, *Vit. Const.* 4.23. A detailed discussion of Constantine's rulings on Sunday can be found in Bultrighini, 'Kaiser Konstantins Edikt', with reference to further bibliography.

178 Cicero, *De legibus* 2.29. Cf. Servius, *Georg.* 1.268.

holidays, thereby giving 'dies Solis' the legal character of the civic rest day.[179] Nevertheless, Constantine's decree was also the logical development of a Christian tradition that began with Ignatius of Antioch (above-mentioned), whereby Sunday was to become the substitute of the Jewish Sabbath.

Constantine's regulations refer to Sunday by its planetary designation (dies Solis). The use of the planetary name for Sunday has led some to assume that these laws refer to a weekly pagan holiday dedicated to the cult of the Sun god.[180] This interpretation, however, is most unlikely. The first law of 321 (JC 3.12.2) bans legal and similar business on the venerable day [illegible] Solis). In TC 2.8.1 the day of the Sun is declared a dies festus (festive day, holiday) and described as a day 'celebrated on account of its own veneration' (veneratione sui celebrem). This manner of referring to Sunday seems to imply that this was a day on which some significant religious rite was regularly performed. Indeed, it was apparently out of respect for this supposed act of worship that the rule of abstention from judicial activity issued in March 321 CE (JC 3.12.2) was reiterated four months later.[181] There is, however, no evidence that during the Roman imperial period dies Solis was marked by any cultic observance as part of a solar cult such as of Sol Invictus or Mithras, nor that dies Solis was singled out as the most important and sacred day of the planetary week; nor, finally, that the day was recognised as a pagan feast day.[182]

In fact, was there ever such a thing as a weekly pagan holiday? Two scanty but complex groups of sources suggest that for a relatively limited period of time in the late Empire, Thursday (rather than Sunday) may have been regarded as the focal day of the week. These two groups of sources are, on the one hand, a number of ecclesiastical texts from late antique and early medieval Gaul and Galicia, and on the other, three documentary papyri from Oxyrhynchus (Egypt).[183] The former group implies that an unofficial observance of Jupiter's day (dies Iovis) persisted among the populace despite

179 Dölger, 'Die Planetenwoche', pp. 235–36; Girardet, 'L'invention du dimanche', p. 348; Bultrighini, 'Kaiser Konstantins Edikt', p. 589.

180 Pharr, *Theodosian Code*, p. 44 n. 3; Moreno Resano, 'El dies Solis'.

181 Cf. TC 2.8.1, '… it appears to us most unseemly that the day of the Sun, which is celebrated on account of its own veneration, should be occupied with legal altercations and with noxious controversies of the litigation of contending parties …'. Agnati, 'Costantino', pp. 30–31.

182 Rordorf, *Sunday*, p. 37; Di Berardino, 'Liturgical celebrations', p. 217; Salzman, 'Pagan and Christian notions', p. 201; Girardet, 'L'invention du dimanche', p. 341; Agnati, 'Costantino', pp. 24, 42, 45; Bultrighini, 'Kaiser Konstantins Edikt', p. 589.

183 Caesarius of Arles, *sermons* 13.5, 19.4, 52.2; Martin of Braga, *De correctione rusticorum* 18; Council of Narbonne 13.6; *P.Oxy.*54.3741; *P.Oxy.*22.2343; *P.Oxy.*60.4075. For a detailed analysis of the evidence, see Bultrighini, 'Thursday (*dies Iovis*)'.

Church opposition to such deviant behaviour. The latter group of sources hints at Thursday (ἡμέρα Διός) being a non-working day for official bureaux during the third and early fourth centuries.

Thursday was not only the day dedicated to Zeus/Jupiter, the chief god of the Graeco-Roman pantheon, but also the day associated with the astrologically favourable planet that was named after him. Together, religious and astrological beliefs may have contributed towards Thursday temporarily becoming the most important and sacred day of the week. Although there is no evidence that it ever acquired the status of dies ferialis (holiday) in the Roman calendar, Jupiter's day may have been observed in a private or unofficial setting through sacrifices or other rites. Furthermore, astrological beliefs pertaining to the days of the planetary week may have led to the practice of performing certain activities on Thursdays.[184]

Given the Christian origin of Sunday as a day of worship (albeit re-interpreted by Constantine as a day of rest),[185] there remains to explain why in his laws Constantine referred to it as dies Solis – the day of the Sun – rather than by the then available Christian alternative dies dominica – the Lord's day. The use of dies Solis in Constantine's legislative texts of 321 CE along with his promotion of the cult of Sol Invictus[186] have led scholars to endlessly debate whether the emperor intended, through his legislation, to make Sunday a day of rest, to favour the solar cult or Christianity.[187]

In fact, the use of dies Solis in Constantine's laws attests to the popularity of the planetary designations in the Roman Empire, especially in its western half. In addition to the sources discussed above, imperial legislative texts consistently refer to Sunday as dies Solis until the end of the fourth

184 Sources from the Roman imperial and late antique periods give instructions on what activities should or should not be carried out on certain days of the week; a complete hemerology of the seven-day week appears in the Calendar Codex of 354 (see above, pp. 35–37). Specifically on Thursday, see for example Ausonius, *Ecl.* 23, on the idea that men should shave on Thursdays (barbam Iove); and a number of late antique sources on astrological botany indicate Jupiter's day as the appropriate time to collect a number of different plants: see *CCAG* VIII 3, 164, 2–21; *CCAG* XII, p. 119, l. 12; *CCAG* XII, p. 126 ff.; Marcellus Empiricus, *De medicamentis* 8.49, 11.32, 12.24, 14.68–9, 15.9, 15.109, 16.101, 23.78, 25.11, 25.13, 25.21, 25.46, 26.94–5, 26.134–135, 29.23–24, 34.67, 34.84.

185 Rordorf, *Sunday*, p. 220ff.; Di Berardino, 'Liturgical celebrations', p. 213; Salzman, 'Pagan and Christian notions', p. 187; Agnati, 'Costantino', p. 25; Bultrighini, 'Thursday (*dies Iovis*)', p. 78; and ead., 'Kaiser Konstantins Edikt', pp. 589–93.

186 For the scholarly discussion on Constantine's devotion towards the solar god and his subsequent support for Christianity, see Drake, 'Impact'.

187 See Agnati, 'Costantino', p. 31 n. 32 for the different views on the subject.

century CE.[188] Occasionally, Church Fathers and Christian preachers themselves refer to Sunday as 'the day of the Sun', especially when addressing non-Christian audiences.[189] The surviving evidence makes it clear that the use of the planetary week designations was common not only for pagans, but also for Christians. The expression dies Solis in Constantine's legislative texts may thus be regarded as neutral,[190] and was used to ensure that everyone, not only Christians, would understand which day was being referred to.

Constantine's legislation on the observance of Sunday as a day of rest is touched upon by [illegible]. Both consider it as indisputably referring to the Christian festivity of Sunday. Eusebius calls Sunday 'the Lord's day' (κυριακή), 'which also bears the names of day of light and day of the Sun' (ἣν καὶ φωτὸς εἶναι καὶ ἡλίου ἐπώνυμον συμβαίνει).[191] Christian theologians deliberately persisted in using the planetary designation to refer to the Christian Sunday in order to emphasise the symbolism of Christ as the Sun of Justice and the Sun of Resurrection who takes away darkness and sin (cf. John 1:4–5).[192] In fact, however, Eusebius's account remains ambiguous. His description of the emperor's obligation for non-Christian soldiers to join in a common prayer every Sunday marked by the lifting of the hands to the heavens and the address to a nameless god in terms of victories won (*Vit. Const.* 4.20) may be regarded as a reference to the cult of Sol Invictus. The ambivalence of the sources relating to Constantine's regulations on Sunday may suggest that in 321 CE the emperor was aiming to appeal at once to pagan worshippers of Sol and Mithras and to the Christian clergy in Rome and Italy.[193]

The rest of the fourth as well as the fifth century were marked by successive imperial legislations on the status of Sunday as a public holiday.[194] These later laws prohibited public entertainments such as circus games, which were

188 The first occurrence of the Christian designation for Sunday in imperial legislation dates from 386 CE: Solis die, quem dominicum rite dixere maiores, 'on the day of the Sun, which our ancestors rightly called the Lord's day' (TC 2.8.18).

189 See above, pp. 32–33. Further references can be found in Agnati, 'Costantino', pp. 36–37.

190 Dölger, 'Die Planetenwoche', p. 230; Agnati, ibid., p. 36; Bultrighini, 'Kaiser Konstantins Edikt', pp. 590–91.

191 *Life of Constantine* 14.19. See Girardet, 'L'invention du dimanche', p. 351 n. 57 and p. 358 n. 92 for further instances of Eusebius referring to Sunday as 'the day of light'.

192 Girardet, 'L'invention du dimanche', p. 349; Bultrighini, 'Kaiser Konstantins Edikt', pp. 591–92.

193 See further Salzman, 'Pagan and Christian notions', pp. 200–1. Also, Bultrighini, ibid., p. 592.

194 Pietri, 'Le temps de la semaine', pp. 75–76; Di Berardino, 'Liturgical celebrations', pp. 219–20; Girardet, 'L'invention du dimanche', pp. 368–69; Agnati, 'Costantino', pp. 32–35.

deemed incompatible with Sunday's celebration; thus, the Christian notion of Sunday became increasingly apparent.[195] The issuing of a series of laws on the status of Sunday as a day of rest over a long period of time suggests that the legislation and the new rhythm of social life which this enforced were met with some resistance: it appears that people were not generally ready to suspend certain activities on one day out of seven.[196] The gradualness with which the legislation on Sunday as a holiday was implemented implies that in the fourth century CE the seven-day week was not yet generally used to organise people's time across the Empire. It seems that at the time of Constantine's legislation the use of the seven-day cycle was still essentially restricted to the private sphere.[197]

Constantine's edict on Sunday of 321 CE represents the first major step toward the establishment of the seven-day week of Biblical and astrological origins as the standard time unit for the organisation of social activities across the territory of the Roman Empire.[198] The process of standardization of the week and calendar, thus initiated, was conceived to act as a medium for political unity and cultural cohesion.[199]

References

U. Agnati, 'Costantino e la scansione cristiana del tempo (Cod. Theod. II 8.1 e Cod. Iust. III 12.2)', in F. Raviola (ed.), *L'indagine e la rima. Scritti per Lorenzo Braccesi* (Hesperìa, 30; Rome: 'L'Erma' di Bretschneider, 2013), pp. 23–61.

R. Ast, 'Schedule of Work days', in R. Ast, H. Cuvigny, T.M. Hickey, and J. Lougovaya (eds), *Papyrological Texts in Honor of Roger S. Bagnall* (American Studies in Papyrology 53; Durham, NC: The American Society of Papyrologists, 2013), pp. 9–16.

S. Bacchiocchi, *From Sabbath to Sunday* (Rome: Pontifical Gregorian University, 1977).

M. Baillet, *Qumrân grotte 4 3 (4Q482–4Q520)* (Discoveries in the Judaean Desert 7; Oxford: Clarendon Press, 1982).

A. Barrows, *The Cosmic Time of Empire: Modern Britain and World Literature* (Berkeley: University of California Press, 2010).

195 Salzman, 'Pagan and Christian notions', pp. 201–2.

196 Llewelyn and Nobbs, 'Earliest dated reference', p. 118; Di Berardino, 'Liturgical celebrations', pp. 215, 219.

197 Salzman, 'Pagan and Christian notions', pp. 202–8; Bultrighini, 'Kaiser Konstantins Edikt', pp. 592–93.

198 Di Berardino, 'Liturgical celebrations', pp. 215, 218–19; Salzman, ibid.; Agnati, 'Costantino', p. 50; Bultrighini, ibid.

199 Agnati, ibid., p. 31; Bultrighini, ibid.

I.R. Bartky, *One Time Fits All: the Campaigns for Global Uniformity* (Stanford: Stanford University Press, 2007).

T. Barton, *Ancient Astrology* (London: Routledge, 1994).

G. Bauchhenß and P. Noelke, *Die Iupitersäulen in den germanischen Provinzen* (Cologne: Rheinland-Verlag; Bonn: Habelt, 1981).

J. Ben-Dov, *Head of All Years. Astronomy and Calendars at Qumran in their Ancient Context* (STDJ 78; Leiden: Brill, 2008).

C. Bennett, 'The Imperial Nundinal Cycle', *Zeitschrift für Papyrologie und Epigraphik* 147 (2004), pp. 175–77.

C. Bennett, 'The Ptolemaic Dynasty' (website), http://www.instonebrewer.com/TyndaleSites/Egypt/ptolemies/ptolemies.htm (accessed 22 September 2020).

H.D. Betz (ed.), *The Greek Magical Papyri in Translation* (Chicago: University of Chicago Press, 1986).

A. Binsfeld, 'Kirchliche Würdenträger in Trierer Inschriften', in L. Clemens, H. Merten, and C. Schäfer (eds), *Frühchristliche Grabinschriften im Westen des Römischen Reiches* (Trier: Kliomedia, 2015), pp. 37–60.

K. Birth, 'Standards in the shadows for everyone to see: the supranational regulation of time and the concern over temporal pluralism', in S. Benyon-Jones and E. Grabham (eds), *Law and Time* (London: Routledge, 2018), ch. 10.

J. Bitton, N. Dweck, and S. Fine, 'Yet another Jewish tombstone from late antique Zoar/Zoora: the funerary marker of Hannah daughter of Levi', in M.J. Lundberg, S. Fine, and W.T. Pitard (eds), *Puzzling out the Past* (Leiden: Brill, 2012), pp. 7–12.

Y. Bloch, 'Judean Identity during the Exile: Concluding Deals on a Sabbath in Babylonia and Egypt under the Neo-Babylonian and the Achaemenid Empires', in D. Rivlin Katz, N. Hacham, G. Herman, and L. Sagiv (eds), *A Question of Identity* (Berlin: De Gruyter, 2019), pp. 43–70.

A. Boer, *Paulou Alexandreōs eisagōgika = Pauli Alexandrini Elementa apotelesmatica* (with contributions of O. Neugebauer; Leipzig: B.G. Teubner, 1958).

P. Brind'Amour, *Le Calendrier romain. Recherches chronologiques* (Ottawa, Ontario: Editions Universitaires d'Ottawa, 1983).

I. Bultrighini, 'New light on five Latin inscriptions of the later imperial period, with special reference to their dating formulae', *Epigraphica. Periodico internazionale di epigrafia* 79 (2017), pp. 411–24.

I. Bultrighini, 'Notes on days of the week and other date-related aspects in three Greek inscriptions of the late Roman period', *Zeitschrift für Papyrologie und Epigraphik* 201 (2017), pp. 187–96.

I. Bultrighini, 'Thursday (*dies Iovis*) in the Later Roman Empire', *Papers of the British School at Rome* 86 (2018), pp. 61–84.

I. Bultrighini, 'Die griechischen Parapegmata aus Milet', in R. Färber and R. Gautschy (eds), *Zeit in den Kulturen des Altertums. Antike Chronologie im Spiegel der Quellen* (Cologne: Böhlau, 2020), pp. 405–18.

I. Bultrighini, 'Das Parapegma von den Trajansthermen in Rom', in R. Färber and R. Gautschy (eds), *Zeit in den Kulturen des Altertums. Antike Chronologie im Spiegel der Quellen* (Cologne: Böhlau, 2020), pp. 577–84.

I. Bultrighini, 'Kaiser Konstantins Edikt von 321 n. Chr. über den Sonntag', in R. Färber and R. Gautschy (eds), *Zeit in den Kulturen des Altertums. Antike Chronologie im Spiegel der Quellen* (Cologne: Böhlau, 2020), pp. 585–96.

I. Bultrighini, '*Theōn hemerai*: astrology, the planetary week, and the cult of the seven planets in the Graeco-Roman world', in I. Salvo and T. Scheer (eds), *Religion and Education in the Ancient Greek World* (Tübingen: Mohr Siebeck, forthcoming).

W. Clarysse, S. Remijsen, and M. Depau, 'Observing the Sabbath in the Roman Empire: a Case Study', *Scripta Classica Israelica* 29 (2010), pp. 51–57.

F.H. Colson, *The Week: An Essay on the Origin and Development of the Seven-Day Cycle* (Cambridge: The University Press, 1926).

R. Cribiore, *Writing, Teachers, and Students in Graeco-Roman Egypt* (Atlanta: Scholars Press, 1996).

H.D. Curtis and F.E. Robbins, 'An Ephemeris of 467 A.D.', *Publications of the Observatory of the University of Michigan* 6 (1935), pp. 77–100.

A. Deman, 'Notes de chronologie romaine', *Historia* 23 (1974), pp. 271–78.

J.G. Dercksen, 'Weeks, months, and years in Old Assyrian chronology', *Bibliotheca Orientalis* 68 (2011), pp. 233–43.

A. Di Berardino, 'Liturgical celebrations and imperial legislation in the fourth century', in B. Neil, G.D. Dunn, and L. Cross (eds), *Prayer and Spirituality in the Early Church. Vol. 3, Liturgy and Life* (Strathfield, NSW: St. Pauls, 2003), pp. 211–32.

J. Divjak and W. Wischmeyer, *Das Kalenderhandbuch von 354. Der Chronograph des Filocalus*. 2 vols (Vienna: Holzhausen, 2014).

L. Doering, *Schabbat. Sabbathalacha und -praxis im antiken Judentum und Urchristentum* (Texts and Studies in Ancient Judaism 78; Tübingen: Mohr Siebeck, 1999).

L. Doering, *Ancient Jewish letters and the Beginnings of Christian Epistolography* (Wissenschaftliche Untersuchungen zum Neuen Testament 298; Tübingen: Mohr Siebeck, 2012).

F.J. Dölger, 'Die Planetenwoche der griechisch-roemischen Antike und der christliche Sonntag', *Antike und Christentum* 6 (1940), pp. 202–38.

P. Domenicucci, *Astra Caesarum: astronomia, astrologia e catasterismo da Cesare a Domiziano* (Pisa: ETS, 1996).

H.A. Drake, 'The impact of Constantine on Christianity', in N. Lenski (ed.), *The Cambridge Companion to the Age of Constantine* (Cambridge: Cambridge University Press, 2012), pp. 111–36.

C. Du Cange et al. *Glossarium mediae et infimae latinitatis* (Niort: L. Favre, 1883–1887).

P.M. Duval, 'Les dieux de la semaine', *Gallia* 11 (1953), pp. 282–93.

S. Eriksson, *Wochentagsgötter, Mond und Tierkreis, Laienastrologie in der römischen Kaiserzeit* (Stockholm: Almqvist & Wiksell, 1956).

R. Färber and R. Gautschy (eds), *Zeit in den Kulturen des Altertums. Antike Chronologie im Spiegel der Quellen* (Cologne: Böhlau, 2020).

D. Feeney, *Caesar's Calendar: Ancient Time and the Beginnings of History* (Berkeley: University of California Press, 2007).

R. Frei-Stolba, '"Dienstag, den 2. April ... n. Chr.": zu einem ungewöhnlichen Datum aus Aventicum', *Archäologie der Schweiz: Mitteilungsblatt der Schweizerischen Gesellschaft für Ur- und Frühgeschichte = Archéologie suisse: bulletin de la Société suisse de préhistoire et d'archéologie* 16 (1993), pp. 128–33.

K.M. Girardet, 'L'invention du dimanche: du jour du soleil au dimanche. Les dies solis dans la législation et la politique de Constantin le Grand', in J.N. Guinot and F. Richard (eds), *Empire chrétien et église aux IVe et Ve siècles. Intégration ou concordat? Le témoignage du Code Théodosien. Actes du Colloque international de Lyon 2005* (Paris: Editions du Cerf, 2008), pp. 341–70.

S.J. Green, *Disclosure and Discretion in Roman Astrology. Manilius and his Augustan Contemporaries* (Oxford: Oxford University Press, 2014).

D.G. Greenbaum, *Late Classical Astrology: Paulus Alexandrinus and Olympiodorus* (Reston, Va.: Arhat, 2001).

H.G. Gundel, 'Planeten', *RE* 20 (1950), pp. 2017–185.

H.G. Gundel, 'Pianeti', *EAA. Supplemento 1970* (Rome: Istituto della Enciclopedia italiana, 1973), pp. 614–23.

F. Haug, 'Die Wochengöttersteine', *Westdeutsche Zeitschrift für Geschichte und Kunst* 9 (1890), pp. 17–53.

A.P. Hayman, *Sefer Yeṣira: Edition, Translation and Text-Critical Commentary* (Tübingen: Mohr Siebeck, 2004).

S. Heilen, 'Short Time in Greco-Roman Astrology', in K.J. Miller and S. Symons (eds), *Down to the Hour: Short Time in the Ancient Mediterranean and Near East* (Leiden: Brill, 2020), pp. 239–70.

F.A.J. Hoogendijk, 'A note on P.Kellis I 82', *Zeitschrift für Papyrologie und Epigraphik* 113 (1996), pp. 216–18.

W. Hübner, *Manilius 'Astronomica', Buch V*, 2 vols (Berlin, New York: De Gruyter, 2010).

R.M. Jensen, 'Ritual and early Christian art', in R. Uro, J.J. day, R. Roitto, and R.E. DeMaris (eds), *The Oxford Handbook of Early Christian Ritual* (Oxford: Oxford University Press, 2018), pp. 587–609.

A. Jones, 'An astronomical ephemeris for AD 140: P. Harris I.60', *Zeitschrift für Papyrologie und Epigraphik* 100 (1994), pp. 59–63.

A. Jones, 'Two Astronomical Papyri Revisited', *Analecta Papyrologica* 6 (1994), pp. 111–126.

A. Jones, *Astronomical Papyri from Oxyrhynchus*, 2 vols (Philadelphia: American Philosophical Society, 1999).

D.R. Jordan, 'Magia nilotica sulle rive del Tevere', *Mediterraneo Antico* 7 (2004), pp. 693–710.

R. Katzoff and B.M. Schreiber, 'Week and Sabbath in Judaean Desert Documents', *Scripta Classica Israelica* 17 (1998), pp. 102–113.

J. Ker, 'Nundinae: the culture of the Roman week', *Phoenix* 64 (2010), pp. 360–85.

W.E. Klingshirn, *Caesarius of Arles: The Making of a Christian Community in Late Antique Gaul* (Cambridge: Cambridge University Press, 1994).

W. Kubitschek, *Grundriss der antiken Zeitrechnung* (Handbuch der Altertumswissenschaft I. 7; Munich: Beck, 1928).

G. Lefebvre, *Inscriptiones Christianae Aegypti: recueil des inscriptions grecques-chrétiennes d'Égypt* (Chicago: Ares, 1978).

D. Lehoux, *Astronomy, Weather, and Calendars in the Ancient World: Parapegmata and Related Texts in Classical and Near Eastern Societies* (Cambridge: Cambridge University Press, 2007).

D. Lehoux, 'Days, months, years, and other time cycles', in A. Jones (ed.), *Time and Cosmos in Greco-Roman Antiquity* (Princeton: Princeton University Press, 2016), pp. 95–122.

H. Leon, *The Jews of Ancient Rome* (Philadelphia: Jewish Publication Society of America, 1960).

S.R. Llewelyn and A.M. Nobbs, 'The earliest dated reference to Sunday in the papyri', in S.R. Llewelyn (ed.), *New Documents Illustrating Early Christianity IX: A Review of the Greek Inscriptions and Papyri Published in 1986–1987* (Grand Rapids, MI: Eerdmans, 2002), pp. 106–18.

E. Maass, *Die Tagesgötter in Rom und den Provinzen: aus der Kultur des Niederganges der antiken Welt* (Berlin: Weidmann, 1902).

R. Markus, *The End of Ancient Christianity* (Cambridge: Cambridge University Press, 1990).

A. Mau, 'Ausgrabungen von Pompeji', *Mitteilungen des Deutschen Archäologischen Instituts, Römische Abteilung* 16 (1901), pp. 333–65.

Y.E. Meimaris, with K. Kritikakou and P. Bougia, *Chronological Systems in Roman-Byzantine Palestine and Arabia: The Evidence of Dated Greek Inscriptions* (Athens: National Hellenic Research Foundation, 1992).

Y.E. Meimaris and K.I. Kritikakou-Nikolaropoulou, *Inscriptions from Palaestina Tertia, Ia: The Greek Inscriptions from Ghor Es-Safi (Byzantine Zoora)* (Meletemata 41;

Athens: National Hellenic Research Foundation, Research Centre for Greek and Roman Antiquity, 2005).

Y.E. Meimaris and K.I. Kritikakou-Nikolaropoulou, *Inscriptions from Palaestina Tertia, Ib: The Greek Inscriptions from Ghor Es-Safi (Byzantine Zoora) (Supplement), Khirbet Qazone and Feinan* (Meletemata 57; Athens: National Hellenic Research Foundation, Research Centre for Greek and Roman Antiquity, 2008).

Y.E. Meimaris and K.I. Kritikakou-Nikolaropoulou, in collaboration with S.P. Brock, *Inscriptions from Palaestina Tertia*, IC: *The Jewish Aramaic Inscriptions from Ghor* [illegible] Foundation, 2016).

H. Misgav, 'Two Jewish tombstones from Zoar', *Israel Museum Studies in Archaeology* 5 (2006), pp. 35–46.

G. Moitrieux, *Recueil général des sculptures sur pierre de la Gaule: Toul et la cité des Leuques* (Nouvel Espérandieu III; Paris: Académie des inscriptions et belles-lettres, 2010).

E. Moreno Resano, 'El dies Solis en la legislación constantiniana', *Antiquité Tardive* 17 (2009), pp. 289–305.

A.A. Mosshammer, *The Easter Computus and the Origins of the Christian Era* (Oxford: Oxford University Press, 2008).

M. Naldini, *Il Cristianesimo in Egitto. Lettere private nei papiri dei secoli II–IV* (Florence: Le Monnier, 1968).

J. Naveh, 'Aramaic tombstones from Zoar', *Tarbiz* 64 (1995), pp. 477–97 (in Hebrew).

J. Naveh, 'More on the tombstones from Zoar', *Tarbiz* 68 (1999), pp. 581–86 (in Hebrew).

J. Naveh, 'Seven new epitaphs from Zoar', *Tarbiz* 69 (2000), pp. 619–35 (in Hebrew).

A. Negev, *The Greek Inscriptions from the Negev* (Jerusalem: Franciscan Printing Press, 1981).

O. Neugebauer and H.B. van Hoesen, *Greek Horoscopes* (Philadelphia: American Philosophical Society, 1959).

A. Pelletier, 'Σαββατα. Transcription grecque de l'araméen', *Vetus Testamentum* 22 (1972), pp. 436–77.

P. Perdrizet, 'Le jeu alexandrin de l'icosaèdre', *Bulletin de l'Institut français d'archéologie orientale du Caire* 30 (1931), pp. 1–16.

C. Pharr, *The Theodosian Code and Novels and the Sirmondian Constitutions* (Princeton: Princeton University Press, 1952).

G.C. Picard, 'Imperator Caelestium', *Gallia* 35 (1977), pp. 89–113.

C. Pietri, 'Le temps de la semaine à Rome et dans l'Italie chrétienne (IVe–VIe siècle.)', in J.-M. Leroux (ed.), *Le temps chrétien de la fin de l'Antiquité au Moyen Âge IIIe–XIIIe siècles. Actes du colloque Paris, 9–12 mars 1981* (Paris: Editions du Centre national de la recherche scientifique, 1984), pp, 63–97.

B. Porten, *Archives from Elephantine* (Berkeley: University of California Press, 1968).

F. Priesigke (ed.), *Sammelbuch griechischer Urkunden aus Ägypten*, I (Strassburg: K.J. Trübner, 1915).

F. Reiter (ed.), *Dokumentarische Texte der Berliner Papyrussammlung aus ptolemäischer und römischer Zeit: Zur Wiedereröffnung des Neuen Museums* (Berliner Griechische Urkunden 20; Berlin: De Gruyter, 2014).

M. Riley, 'A Survey of Vettius Valens', 1996. http://www.csus.edu/indiv/r/rileymt/PDF_folder/VettiusValens.PDF, accessed 26 May 2019.

W. Rordorf, *Sunday: The History of the day of Rest and Worship in the Earliest Centuries of the Christian Church* (trans. A.A.K. Graham; London: SCM Press, 1968).

P. Rugo, *Le iscrizioni dei sec. VI–VII–VIII esistenti in Italia, vol. 4: I ducati di Spoleto e Benevento* (Cittadella: Tip. Bertoncello, 1978).

J. Rüpke, *Kalender und Öffentlichkeit: Die Geschichte der Repräsentation und religiösen Qualifikation von Zeit in Rom* (Religionsgeschichtliche Versuche und Vorarbeiten 40; Berlin: de Gruyter, 1995).

J. Rüpke, *The Roman Calendar from Numa to Constantine* (trans. D.M.B. Richardson; Chichester: Wiley-Blackwell, 2011).

L.V. Rutgers, 'Roman policy towards the Jews: expulsions from the city of Rome during the first century C.E.', *Classical Antiquity* 13 (1994), pp. 56–74.

M.R. Salzman, 'Pagan and Christian notions of the week in the 4th century C.E. western Roman empire', in R.M. Rosen (ed.), *Time and Temporality in the Ancient World* (Philadelphia: University of Pennsylvania Museum of Archaeology and Anthropology, 2004), pp. 185–211.

J.L.W. Schaper, 'The Septuagint Psalter', in W.P. Brown (ed.), *The Oxford Handbook of the Psalms* (Oxford: Oxford University Press, 2014), pp. 173–84.

E. Schürer, *The History of the Jewish People in the Age of Jesus Christ (175 B.C.–A.D. 135)*, rev. and ed. by G. Vermes and F.G.B. Millar (with M. Black [vols. i–ii] and M. Goodman [vol. iii/1–2]), 3 vols. in 4 pts. (Edinburgh: T. & T. Clark, 1973–87).

H. Seldeslachts and A. Wouters, 'A Christian word-list on a papyrus of the Bibliothèque Royale at Brussels (MS. Brux. IV 590)', *Zeitschrift für Papyrologie und Epigraphik* 96 (1993), pp. 141–52.

C. Sirat, P. Cauderlier, M. Dukan, and M.A. Friedman, *La Ketouba de Cologne. Un contrat de mariage juif à Antinoopolis* (Papyrologica Coloniensia 12; Opladen: Westdeutscher Verlag, 1986).

K. Smith, *The Elegies of Albius Tibullus: The Corpus Tibullianum, edited with introduction and notes on Books I, II and IV, 2–14* (Darmstadt: Wissenschaftliche Buchgesellschaft, 1971).

M. Smith, 'The Eighth Book of Moses and how it grew (Pleid. J 395)', in *Atti del XVII Congresso Internazionale di Papirologia* (Napoli: Centro internazionale per lo studio dei papiri ercolanesi, 1984), pp. 683–93.

M. Stern, *Greek and Latin Authors on Jews and Judaism*, 3 vols (Jerusalem: Israel Academy of Sciences, 1974–84).

S. Stern, 'New Tombstones from Zoar (Moussaieff Collection)', *Tarbiz* 68 (1999), pp. 177–85 (in Hebrew).

S. Stern, *Calendar and Community. A History of the Jewish Calendar, 2nd century BCE–10th century CE* (Oxford: Oxford University Press, 2001).

S. Stern, Review of D. Lehoux, *Astronomy, Weather, and Calendars in the Ancient World: Parapegmata and Related Texts in Classical and Near Eastern Societies* (Cambridge: [illegible]

S. Stern, 'Qumran calendars and Sectarianism', in T. Lim and J.J. Collins (eds), *The Oxford Handbook of the Dead Sea Scrolls* (Oxford: Oxford University Press, 2010), pp. 232–53.

S. Stern, 'A "Jewish" birth record, *sambat-*, and the calendar of Salamis', *Zeitschrift für Papyrologie und Epigraphik* 172 (2010), pp. 105–14.

S. Stern, 'The "Sectarian" calendar of Qumran', in idem (ed.), *Sects and Sectarianism in Jewish History* (Leiden: Brill, 2011), pp. 39–62.

S. Stern, *Calendars in Antiquity: Empires, States, and Societies* (Oxford: Oxford University Press, 2012).

S. Stern, 'The problem of the date line in medieval Jewish sources', in I.-S. Nha, W. Orchiston, and F.R. Stephenson (eds), *The History of World Calendars and Calendar-making. Proceedings of the International Conference in Commemoration of the 600th Anniversary of the Birth of Kim Dam.* (Seoul: Yonsei University Press, 2017), pp. 113–22.

S. Stern, 'The Jewish Aramaic tombstones from Zoar', *Journal of Jewish Studies* 68 (2017), pp. 158–179.

S. Stern, *The Jewish Calendar Controversy of 921/2 CE* (Leiden: Brill, 2019).

S. Stern and H. Misgav, 'Four additional tombstones from Zoar', *Tarbiz* 74 (2005), pp. 137–51 (in Hebrew).

G.D. Stiebel, 'Meager bread and scant water' – food for thought at Masada', in A.I. Baumgarten, H. Eshel, R. Katzoff, and S. Tzoref (eds), *Halakhah in Light of Epigraphy, Journal of Ancient Judaism* 3 (2011), pp. 283–303.

S. Talmon, J. Ben-Dov, and U. Glessmer, *Qumran Cave 4 (XVI). Calendrical Texts* (Discoveries in the Judaean Desert 21; Oxford: Clarendon Press, 2001).

V. Tcherikover, A. Fuks, and M. Stern, *Corpus Papyrorum Judaicarum (CPJ)*, 3 vols (Cambridge: Harvard University Press, 1957–64).

W. Van Andringa, *La religion en Gaule romaine. Piété et politique (I–III. siécle apr. J-C.)* (Paris: Éditions Errance, 2002).

J. VanderKam, *Calendars in the Dead Sea scrolls: Measuring Time* (London: Routledge, 1998).

A. Varone, *Erotica Pompeiana. Love Inscriptions on the Walls of Pompeji* (Rome: 'L'Erma' di Bretschneider, 2002).

K.R. Veenhof, 'The Old Assyrian Hamuštum period: a seven-day week', *Jaarbericht 'Ex Oriente Lux'*, 34 (1995–6), pp. 5–26.

C.B. Welles, 'Graffiti', in P.V.C. Baur, M.I. Rostovtzeff and Alfred R. Bellinger (eds), *The Excavations at Dura-Europos conducted by Yale university and the French Academy of Inscriptions and Letters: preliminary report of fourth season of work, October 1930–March 1931* (New Haven: Yale University Press, 1933), pp. 79–178.

M.H. Williams, *Jews in a Graeco-Roman Environment* (Tübingen: Mohr Siebeck, 2013).

J. de Witte, 'Les divinités des sept jours de la semaine', *Gazette archéologique* 3 (1877), pp. 50–57, 77–85; 5 (1879), pp. 1–6.

A. Wolkenhauer, *Sonne und Mond, Kalender und Uhr. Studien zur Darstellung und poetischen Reflexion der Zeitordnung in der römischen Literatur* (Untersuchungen zur antiken Literatur und Geschichte 103; Berlin and New York: De Gruyter, 2011).

G. Woolf, 'Representation as cult: the case of the Jupiter columns', in H. Cancik, J. Rüpke, and W. Spickermann (eds), *Religion in den germanischen Provinzen Roms* (Tuebingen: Mohr Siebeck, 2001), pp. 117–34.

Y. Yadin, J.C. Greenfield, A. Yardeni, and B.A. Levine, *The Documents of the Bar Kokhba Period in the Cave of Letters. Hebrew, Aramaic and Nabatean-Aramaic Papyri* (Judean Desert Studies; Jerusalem: Israel Exploration Society, 2002).

A. Yardeni, 'New Jewish Aramaic ostraka', *Israel Exploration Journal* 40 (1990), pp. 130–52.

A. Yardeni, 'Twelve published and unpublished Jewish Aramaic ostraca written in the "Jewish" cursive script', in A.F. Botta (ed.), *In the Shadow of Bezalel. Aramaic, Biblical, and Ancient Near Eastern Studies in Honor of Bezalel Porten* (Culture and History of the Ancient Near East 60; Leiden: Brill, 2012).

E. Zerubavel, *The Seven day Circle: The History and Meaning of the Week* (New York: Free Press; London: Collier Macmillan, 1985).

CHAPTER 2

Calendars of the Greek East under Rome

A New Look at the Hemerologia *Tables*

Ilaria Bultrighini

1 Introduction

The term *Hemerologia* identifies a set of calendar conversion tables preserved in a few early medieval Greek manuscripts. They list in parallel columns the days of the month of the Roman (Julian) calendar and those of a number of local calendars of the provinces and cities of the eastern half of the Roman Empire. The manuscripts comprise separate tables for each of the twelve months of the Julian calendar. The Julian calendar thus serves as common reference point; its days are listed in the first column, while the other columns give the equivalent days in thirteen to fifteen different calendars of the Roman East. Different *Hemerologia* tables are known in four manuscripts:

1. Leiden, Bibliotheek der Rijksuniversiteit, BPG 78, ff. 145v–150v, 152 (http://hdl.handle.net/1887.1/item:1597397)
2. Città del Vaticano, Biblioteca Apostolica Vaticana, *Vat.gr.* 1291, ff. 10–15v (https://digi.vatlib.it/view/MSS_Vat.gr.1291)
3. Firenze, Biblioteca Medicea Laurenziana, *Plut.* 28.26, ff. 45r–50v (http://mss.bmlonline.it/catalogo.aspx?Collection=Plutei&Shelfmark=Plut.28.26)
4. Firenze, Biblioteca Medicea Laurenziana, *Plut.* 28.12, ff. 187r–192v (http://mss.bmlonline.it/catalogo.aspx?Collection=Plutei&Shelfmark=Plut.28.12)[1]

The earliest manuscripts are *Leidensis* BPG 78 and *Vaticanus gr.* 1291, which were both copied in the early ninth century; *Laurentianus* 28.26 has been attributed to the late ninth or early tenth century, while *Laurentianus* 28.12 is a fourteenth century copy of the Leiden manuscript.[2] The only full edition

1 While *Leidensis* BPG 78, *Laurentianus* 28.26, and *Laurentianus* 28.12 include thirteen calendars of the Roman East, the eastern calendars are fifteen in *Vaticanus gr.* 1291.

2 The *Hemerologia* tables included in *Leidensis* BPG 78 belong to the earliest portion of the codex, traditionally dated to 813–20; Tihon, *Les 'Tables Faciles' de Ptolémée*, pp. 24–25, 30–31. There is a general consensus that the *Hemerologia* in *Vaticanus gr.* 1291 should be assigned approximately to the same period; Janz, 'Scribes and the date of the Vat. gr. 1291'; Tihon, *Les 'Tables Faciles' de Ptolémée*, p. 34. Different dates were proposed, e.g., by Spatharakis, 'Some observations'; Cutler and Spieser, *Byzance médiévale: 700–1204*, p. 46. The *Hemerologia* in

 | DOI:10.1163/9789004459694_004

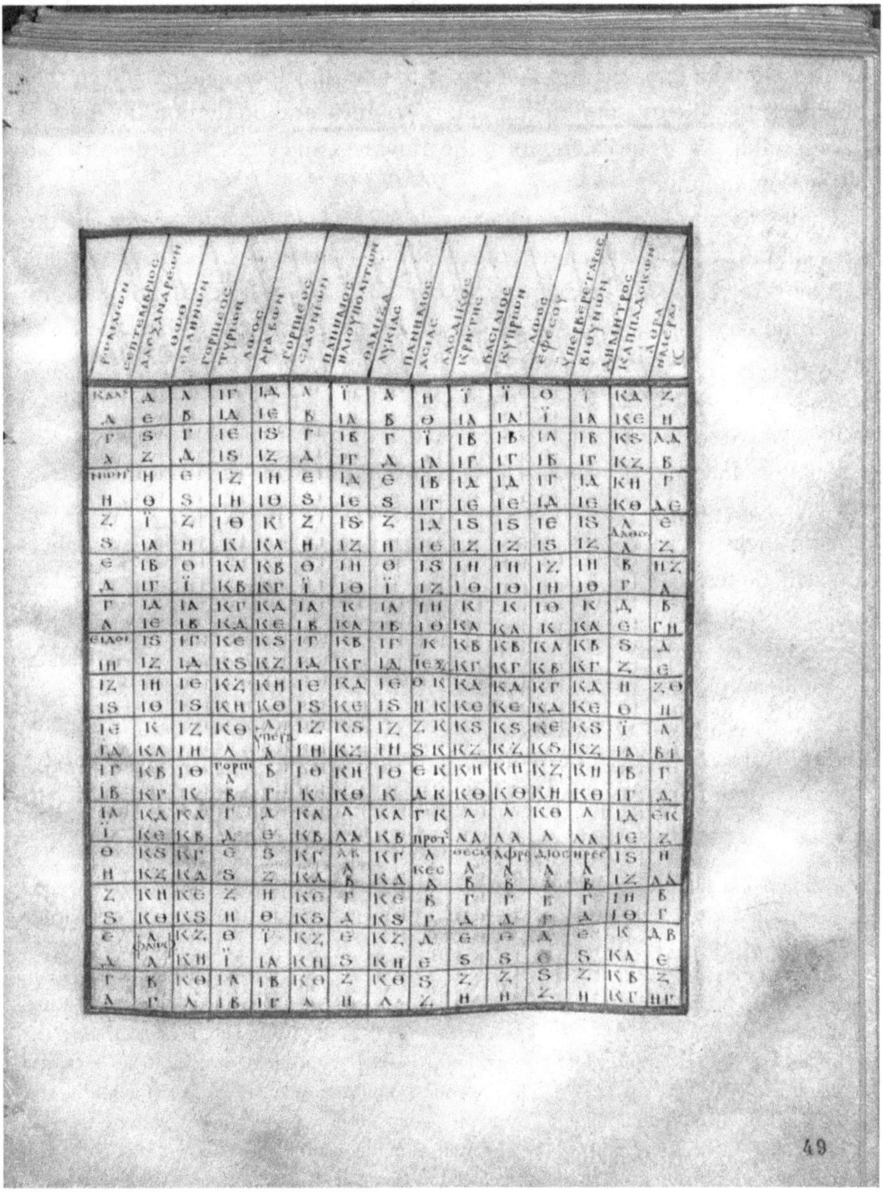

FIGURE 2.1 *Hemerologion*, month of September (Firenze, Biblioteca Medicea Laurenziana, Ms. Plut. 28.26, fol. 49r)

of the *Hemerologia* was produced by Wilhelm Kubitschek, who published the tables in 1915.[3] Kubitschek edited the manuscripts with transcriptions and critical commentary. He also compared the data of the *Hemerologia* to that of epigraphic, literary, and documentary sources, and offered on this basis a detailed analysis of the calendars of the provinces and cities of the eastern half of the Roman Empire.

Figure 2.1 shows the September table from *Laurentianus* 28.26. On the first column to the left, under the heading Ῥωμαίων / Σεπτέμβριος, which [illegible] listed the days of the month of September in the Julian calendar: the three fixed dates of the Roman month –calends, nones, ides– are labelled as καλς (for καλάνδαι), νωνς (for νῶναι), and εἰδοί, in the first, third, and seventh box, respectively (each box contains two days). For the other days of the month, Greek numerals and the Roman system are simultaneously used: for instance, the calends (καλς) are followed by the Greek letter delta, which indicates (by Roman inclusive reckoning) the fourth day before the nones of September; after the nones (νωνς) is an eta, marking the eighth day before the ides, and so on. The other columns refer to a number of calendars of the Roman East and give each day, in these local calendars, equivalent to each Roman month day. In the heading line, next to Ῥωμαίων / Σεπτέμβριος, are Ἀλεξανδρέων / Θωθ, (calendar) of the Alexandrians – (month of) Thoth,[4] Ἑλλήνων / Γορπιεῖος, (calendar) of the *Hellenes*, (month of) Gorpaios,[5] Τυρίων / Λῶος, (calendar) of the Tyrians,[6] (month of) Loos, Ἀράβων / Γορπιεῖος, (calendar) of the *Araboi*, (month

---

Laurentianus 28/26 were copied between 886 and 912; Bianconi, 'Il *Laur. Plut.* 28.26'; Orsini, 'Genesi e articolazioni della "maiuscola liturgica"', pp. 33–34; Tihon, *Les 'Tables Faciles' de Ptolémée*, p. 24. *Laurentianus* 28/12 belongs to the fourteenth century and is a copy of *Leidensis* BPG 78; see, most recently, Lempire, *Le commentaire astronomique*, pp. 13–14. Tihon, *Les 'Tables Faciles' de Ptolémée*, p. 48, assigns it more specifically to the first half of the century.

3 Kubitschek, *Kalenderbücher*. On the *Hemerologia* see also Lietzmann, *Zeitrechnung der römischen Kaiserzeit*, pp. 78–81, 106–17; Laffi, 'Le iscrizioni', pp. 72–75; Samuel, *Greek and Roman Chronology*, pp. 171–78; id., 'Calendars and time-telling', p. 394; Meimaris and Kritikakou-Nikolaropoulou, *Inscriptions*, pp. 35–36; Stern, *Calendars in Antiquity*, pp. 260–62.

4 The label Ἀλεξανδρέων refers to the Egyptian calendar after it was modified to adapt to the Julian in 22 BCE. On the Alexandrian calendar: Grumel, *Traité d'études byzantines*, pp. 166–68; Hannah, *Greek and Roman Calendars*, pp. 85–91; Bennett, 'Early Augustan calendars'; id., 'Two Egyptian birthdays of Augustus'; Stern, *Calendars in Antiquity*, pp. 263–69.

5 As already recognised by Kubitschek (*Kalenderbücher*, pp. 81, 101), the calendar of the *Hellenes* identifies with the so-called calendar of Antioch, on which see especially Stern, *Calendars in Antiquity*, Ch. 5. Cf. Grumel, *Traité d'études byzantines*, p. 174; Samuel, *Greek and Roman Chronology*, p. 174.

6 See Grumel, *Traité d'études byzantines*, pp. 173–74; Samuel, *Greek and Roman Chronology*, p. 176; Meimaris, in collaboration with Kritikakou and Bougia, *Chronological Systems*, p. 41; Stern, *Calendars in Antiquity*, p. 285.

of) Gorpaios,[7] etc. The last column on the right was designed to track the days of the seven-day week, which are marked by the series of Greek letters from alpha to eta (i.e. the alphabetical sequence α, β, γ, δ, ε, ζ, η, in place of the usual numeral system, according to which numbers six and seven are indicated by the letters ς and ζ, respectively), as well as the course of the lunar month, expressed by the succession of Greek letters from alpha to kappa. The heading displays the word ἡμέραι, 'days', along with the symbol of the moon. This particular column appears in the Florence *Hemerologion* only.[8] The month days of the thirteen calendars of the cities and provinces of the Roman East listed next to the Julian calendar are numbered consecutively by using Greek numerals. Several of these calendars shared the same Macedonian month-names: as an example, both Γορπιἑος (also spelt Γορπιαῖος) and Λῷος were originally month-names of the Macedonian calendar.[9] As several of these calendars had months whose beginning did not fall on the calends (first day) of the Julian months, the table also indicates the point where the ensuing months started: for instance, 22 September (the tenth day before the calends of October) in the Julian calendar coincided with the thirty-first day of the month *Basilios* in the Cretan calendar (listed under the heading Κρήτης / Βασίλιος), after which the month *Thesmophorion* started – marked by the abbreviation Θεσμ with a superscript omicron in smaller print. It should be observed that the four extant *Hemerologia* manuscripts preserve information on different calendars: as an example, both the Vatican manuscript (*Vat. gr.* 1291) and the Leiden manuscript (*Leid.* BPG 78) include the calendars of Ascalon (marked as Ἀσκαλωˢ) and Gaza (labelled as Γαζέων), which, however, do not appear in the earlier of the Florence manuscripts (*Laur.* 28.26).

7 By Ἀράβων is meant the so-called calendar of Provincia Arabia, which was apparently instituted at the moment of creation of the province of Arabia in 106 CE, and was also used in various areas of Roman Judaea/Palaestina: Grumel, *Traité d'études byzantines*, p. 173; Samuel, *Greek and Roman Chronology*, p. 177; Meimaris, *Chronological Systems*, pp. 40–41; Meimaris and Kritikakou-Nikolaropoulou, *Inscriptions*, p. 47; Stern, *Calendars in Antiquity*, pp. 291–22, 374.

8 Kubitschek, *Kalenderbücher*, pp. 79–81.

9 After the conquest of the Persian Empire by Alexander the Great, the Macedonian calendar became widespread throughout Asia Minor, the Near East, and Egypt, both in its original form and by fusing with local calendars across Alexander's new empire. As a result, the Macedonian month names continued to be widely used throughout the Roman East until late antiquity. See Grumel, *Traité d'études byzantines*, p. 169; Samuel, *Greek and Roman Chronology*, pp. 139–51, 177–82; Hannah, *Greek and Roman Calendars*, pp. 82–83; Stern, *Calendars in Antiquity*, pp. 72, 234–59.

In what follows I shall present the results of my research on the *Hemerologia*,[10] which involved a re-assessment of the information provided by these manuscripts by taking into account newly discovered evidence since Kubitschek's work over a century ago.

2 The Julian Calendar and Its Diffusion in the Roman East: The Calendar of Asia

The existence of calendar conversion tables of the type of the *Hemerologia* is directly dependent upon the institution of the Julian calendar and its subsequent diffusion in the eastern provinces of the Roman Empire. As is known, in 46 BCE Julius Caesar introduced a new calendar consisting of a fixed 365-day year, with the regular intercalation of one day every four years, the so-called leap or bissextile year.[11] The calendar of the Roman Republic, whose intercalation system had failed to keep the civil year in step with the seasons, was abandoned in favour of a purely solar one, which was 'essentially an improved version of the Egyptian calendar'.[12] Soon after its introduction, the Julian calendar started spreading throughout the Empire, although its diffusion differed considerably between the western and the eastern provinces: whereas it appears that in the West the new calendar spread massively and soon replaced completely all local calendars,[13] in the eastern provinces – specifically in Asia Minor, Egypt, and the Near East – it also spread quickly but did not supplant completely pre-existing calendars.[14] Before the eastern

10 The research was conducted from September 2013 to January 2018 within the framework of the ERC-funded project 'Calendars in Antiquity and the Middle Ages: Standardization and Fixation', based at UCL and led by Sacha Stern. My search for possible further *Hemerologia* manuscripts that may have been overlooked by Kubitschek or may have been inaccessible to him returned no results.

11 On the institution of the Julian calendar, see especially Grumel, *Traité d'études byzantines*, pp. 175–76; Samuel, *Greek and Roman Chronology*, pp. 155–58; Bickerman, *Chronology*, pp. 47–51; Samuel, 'Calendars and time-telling', pp. 392–93; Meimaris, *Chronological Systems*, pp. 41–45; Rüpke, *Kalender und Öffentlichkeit*, pp. 369–91 (= id., *Roman Calendar*, pp. 109–21); Hannah, *Greek and Roman Calendars*, pp. 112–24; Stern, *Calendars in Antiquity*, pp. 204–5, 211–27.

12 Stern, *Calendars in Antiquity*, p. 214.

13 With the notable exception of the so-called Coligny inscription, which represents a unique source for our knowledge of the Gallic calendar, evidence of the use of non-Roman calendars after 46 BCE in the western half of the Empire is next to non-existent. On the Coligny inscription and the Gallic calendar, see most recently Stern, *Calendars in Antiquity*, pp. 303–13; Stern, 'Calendars, politics, and power relations', pp. 44–48.

14 Stern, *Calendars in Antiquity*, pp. 259–60.

Mediterranean came under the control of Rome, the several Hellenistic kingdoms and cities operated under a multiplicity of calendrical systems, which were mostly lunar.[15] After the introduction of the new solar calendar by Julius Caesar in 46 BCE, the eastern cities and provinces did not simply adopt it but adapted their local calendars to its length. This process led to the creation of several different calendars, which, however, were arranged so as to follow the Julian year. The month-names of the local calendars were generally preserved. Some calendars had months equal in length but not coterminous with the Julian months; some began the year with Augustus' birthday, while others had different New Year's dates; some calendars had a fixed 365-day year with a leap year every four years, others had 30-day months plus intercalary days to bring alignment to the Julian year, while others had more complex methods of adjustment.

Apart from the Alexandrian calendar, which is very well documented, one of the better known calendars of the eastern half of the Roman Empire is that of the province of Asia. We owe such knowledge to a series of epigraphic fragments found in different cities of the province and preserving a decree of the provincial association (*koinon*) of the Greeks of Asia, which prescribed the introduction of the new calendar in the province in 8 BCE.[16] The decree adopts the suggestion of the proconsul (the Roman governor of the province) Paullus Fabius Maximus[17] to move the beginning of the civil year to Augustus' birthday, 23 September. A dossier comprising of the proconsul's edict – in both Greek and Latin – and two implementing decrees issued at the proposal of the high priest of the provincial *koinon* (the leading figure of the confederation)

15 On these, see Samuel, *Greek and Roman Chronology*, pp. 139–52; Hannah, *Greek and Roman Calendars*, pp. 71–97; Stern, *Calendars in Antiquity*, pp. 231–59.

16 On the introduction of the calendar of the province of Asia and the relevant decree: Ehrenberg and Jones, *Documents*, pp. 81–83, no. 98; Laffi, 'Le iscrizioni'; Sherk, *Roman Documents*, pp. 328–37, no. 65 (Greek and Latin texts); Samuel, *Greek and Roman Chronology*, pp. 174–76, 181–82; Sherk, *Rome and the Greek East*, pp. 124–27, no. 101 (English translation); Hannah, *Greek and Roman Calendars*, pp. 131–35; Dreyer and Engelmann, 'Augustus und Germanicus', pp. 175–82 (= *SEG* 56, 1233 = *AE* 2006, 1452: new Metropolis copy); Witulski, *Kaiserkult in Kleinasien*, pp. 25–32; Stern, *Calendars in Antiquity*, pp. 274–84; id., 'Calendars, politics, and power relations', pp. 34–38; Heller, 'Domination subie', pp. 222–32; *IPriene (2014)* 14 (new edition of Priene copy, on which cf. *SEG* 63, 982; *SEG* 64, 1122); Thonemann, 'Calendar'. The date of introduction of the new calendar has been the subject of some debate; for instance, Buxton and Hannah made a case for 5 BCE ('*OGIS* 458'), while others opted for 10/9 BCE, e.g. Hurlet and Suspène, 'Le proconsul et le prince'; Kirbihler, 'César, Auguste et l'Asie', p. 128. In an in-depth analysis of the issue, Stern (*Calendars in Antiquity*, pp. 274–76) has convincingly argued that the new calendar was established in 8 BCE (cf. Hannah, *Greek and Roman Calendars*, pp. 134–35).

17 Consul in 11 BCE: *PIR2*, F 47; W. Eck in *DNP* 4 (1998), p. 377.

was inscribed on stone and publicly exhibited in a number of cities of the province: copies of the text – in varying states of preservation – come from Priene, Apamea, Eumeneia, Dorylaion, Metropolis, and Meonia. In addition to prescribing that the New Year shall henceforth occur on the ninth day before the calends of October (23 September), the decree also reaffirmed a previous ruling that the first month of the year be named Kaisar – again in honour of Augustus. It also established that the beginning of each month of the year should fall on the ninth day, by Roman reckoning, before the calends of the following month. In leap years, the extra day was to be added to the month Xandikos (after day 1), which on those years would thus consist of 32 days. The decree includes a list of the months and their lengths: Kaisar, 31 days; Apellaios, 30 days; Audynaios, 31 days; Peritios, 31 days; Dystros, 28 days; Xandikos, 31 days; Artemision, 30 days; Daisios, 31 days; Panemos, 30 days; Loos, 31 days; Gorpiaios, 31 days; Hyperberetaios, 30 days. Apart from Kaisar, the month-names of the new calendar correspond to the old Macedonian month-names. The text specifies that the total number of days of the year is 365. Finally, details are provided on how to switch from the old to the new system.[18] The circumstances that led to the proconsul's proposal to reform the calendar are described in the decree of the *koinon*: around 30 BCE, in Smyrna, the province of Asia decreed that a golden wreath should be awarded to whoever could excogitate the greatest manner of honouring Augustus. Twenty years on, the decree informs us that the prize is to be given to the Roman proconsul Paullus Fabius Maximus for his proposal that Augustus' birthday be the first day of the Asian year, which, therefore, was also set to be the day on which 'all men should enter into their public office' in the province of Asia.[19] The story of the contest sponsored by the provincial assembly of Asia makes it unequivocal that the proconsul's suggestion to modify the local calendar, enthusiastically endorsed by the *koinon*, was mainly motivated by the desire of both the Roman governor and the province to demonstrate loyalty and devotion to the emperor Augustus – a fact already apparent from the contents, language, and tone of both the proconsul's edict and the decree of the *koinon*. It is therefore clear that the introduction of the calendar of Asia should ultimately be regarded as an expression of political allegiance.[20] In this context, it is important to underline the central role of the cult worship of Augustus as new facet of the interplay between

18 See Sherk, *Roman Documents*, p. 332, ll. 71–76 (Greek text) and id., *Rome and the Greek East*, p. 126 (English translation).

19 Sherk, *Roman Documents*, ll. 22–23; translation from id., *Rome and the Greek East*, pp. 124–25.

20 Stern, *Calendars in Antiquity*, p. 277; idem, 'Calendars, politics, and power relations', pp. 34–38.

Rome and her eastern provinces.[21] Differently from the West, where no tradition of devoting divine honours to mortals had ever existed before the introduction of the Roman imperial cult, the Greek East had the notable precedent of Hellenistic ruler cults, and it was presumably the familiarity with the concept of a human attaining divine status that led the East to embrace the imperial cult at a very early stage: in 30/29 BCE the provinces of Asia and Bithynia were granted permission by Octavian to worship *Dea Roma* and *Divus Iulius* (the deceased Caesar).[22] In the decree of the *koinon* of Asia, Augustus is addressed as 'saviour of mankind' and described as the greatest benefactors of all times, and his birthday is regarded as a new start, a day of salvation, joy and happiness for the whole world (ll. 34–41); moreover, the emperor is explicitly and repeatedly referred to as 'the god' (ll. 40–1: [ἡ γενέθλιος ἡμέ]ρα τοῦ θεοῦ – '[the birthday] of the god'; l. 43: τῶι μεγίστας γ' εἰς τὸν θεὸν καθευρόντι τειμὰς εἶναι στέφανον – 'that the person who found the greatest honours for the god should have a crown'). As effectively illustrated by the story of the province of Asia bestowing an award to the person who could invent the most creative way of celebrating Augustus, the imperial cult generated an intense competition in the worship of the emperor among cities of the Greek East. Moreover, the imperial cult allowed for a closer bond with Rome and Augustus, and created a new and stronger sense of belonging to the Roman Empire.[23] While it may be assumed that the Greeks of the province of Asia adopted the Julian calendar to demonstrate their loyalty to the emperor following the Roman governor's directive, it is unlikely that a major reform of the calendar of a large province was independently designed and set in motion by the proconsul Paullus Fabius Maximus alone.[24] Significantly, in the same year when the decree was

21 On the religious significance of the introduction of the new calendar as a key event in the history of the early imperial cult in Asia Minor, cf. Sherk, *Roman Documents*, pp. 334–37; Weinstock, *Divus Iulius*, pp. 206–10; Price, *Rituals and Power*, pp. 54–55; Zanker, *Power of Images*, pp. 302–6; Witulski, *Kaiserkult in Kleinasien*, pp. 25–32.

22 Dio Cassius 51.20.6–9. It has to be emphasised that the worship of Roman emperors emerged as an entirely new phenomenon and should not in any way be regarded as an evolution from the cults for Hellenistic kings; Millar, *Rome, the Greek World, and the East, Vol. 1*, pp. 308–9. On the idea that Octavian's measures of 30/29 BCE were not a new foundation but an alteration of an older cult, which would have been in place as early as c. 40/39 BCE, see Kirbihler and Zabrana, 'Zeugnisse'; Kirbihler, 'Ruler cults', pp. 196–201.

23 Zanker, *Power of Images*, pp. 302–6.

24 Cf. Buxton and Hannah, '*OGIS* 458', p. 300; Stern, *Calendars in Antiquity*, p. 278; Heller, 'Domination subie', pp. 222–26. The fact that the proconsul's edict is only concerned with the institution of the New Year on Augustus' birthday does not, in my view, demonstrate a lack of awareness or disregard of the complex consequences and technicalities involved in converting the local calendar to a 365-day year (*pace* Stern, *Calendars in Antiquity*, p. 277; id., 'Calendars, politics, and power relations', p. 36). Rather, the focus on the observance

issued (8 BCE), Augustus corrected an error in the intercalation of the Julian calendar *and* was honoured by having the month Sextilis renamed after himself (Augustus).[25] All this suggests that around that particular year 'calendar reform was clearly in the air'.[26] It has long been observed that Augustus made extensive use of the calendar as a tool to assert his hegemony: by creating new festivals related to events that revolved around the emperor and his family – victories, births, marriages, the assumption of civil and religious offices, etc. – Augustus appropriated civic time and incorporated himself and his own house into the Roman calendar.[27] The emperor's *dies natalis* (birthday), in particular, had been celebrated all over the empire well before it became the New Year in the province of Asia and is, in fact, the most frequently attested new imperial festivity.[28] In view of the massive use that Augustus made of the Roman calendar as an instrument of propaganda, it does not seem far-fetched to assume that the emperor was involved first-hand in the reform of the calendar of the Asiatic province.[29] While the fact that the Greeks of Asia did not fully replace their own calendar with the Julian has been seen by some primarily as the result of the Roman calendar being 'a distinctive marker of Romanness',[30] i.e. as a trait specific to Roman identity and culture which would not be suitable to

---

of the emperor's birthday as a New Year's day is consistent with the primary aim of the proconsul's move, that is, promoting the celebration of Augustus and instilling a greater sense of belonging to the Roman Empire in the province; after all, Paullus' edict is essentially an honorific document. Additionally, it appears unlikely that the proconsul or the local population should be informed of the technical details relating to the implementation of the new calendar.

25 Macrobius, *Saturnalia* 1.12.35, 1.14.13–5; Pliny, *Natural History* 18.211; Cassius Dio 55.6.6; Censorinus, *De die natali*, 22.16; Suetonius, *Augustus* 31.2; Solinus, 1.45–7. Cf. Rüpke, *Kalender und Öffentlichkeit*, p. 85; Stern, *Calendars in Antiquity*, pp. 214–15, 278.

26 Stern, *Calendars in Antiquity*, p. 278. On calendar changes in the Roman East as a 'direct result of Roman imperial rule', see further ibid., p. 263. Cf. Bickerman, *Chronology*, p. 48.

27 See especially Fraschetti, *Roma e il Principe*, pp. 9–39.

28 Almagno and Gregori, 'L'istituzione'; id., *Roman Calendars*, pp. 15–25. Augustus' birthday became a public festival in Rome in 30 BCE; subsequently, games were added which in 8 BCE were made annual: Dio 51.19.2; 54.8.5; 55.6.6. Cf. also *ILS* 112, 15 = *CIL* 12, 4333 (from Narbonne). There is no doubt that the public sacrifices instituted on Caesar's birthday in 45 BCE functioned as a model for the celebration of Augustus' birthday; in turn, the introduction of the commemoration of Caesar's *dies natalis* had been influenced by the habit of having public festivals on the birthdays of Hellenistic rulers: Weinstock, *Divus Iulius*, pp. 206–10; Price, *Rituals and Power*, p. 105; Heller, 'Domination subie', pp. 226–27; Almagno and Gregori, 'L'istituzione', p. 447; id., *Roman Calendars*, pp. 10, 135.

29 Cf. Heller, 'Domination subie', p. 225: '(...) la mesure adoptée par Paullus Fabius Maximus (...) pour la province d'Asie traduit la volonté impériale et correspond à un vaste projet destiné à inscrire le nouveau pouvoir dans le temps sacré et civil.'

30 Feeney, *Caesar's Calendar*, pp. 209–11. Cf. Stern, *Calendars in Antiquity*, p. 226 n. 196.

the Greeks of the East, it appears plausible that the Romans themselves did not attempt to impose their calendar on the Greek East, consistently with their laissez-faire policy in provincial administration. This approach on the part of the Romans would have offered the additional advantage of creating an impression of independence with their subjects. After all, the Romans never failed to recognise the strong and long-lived cultural traditions of the Greeks, including, I would argue, their calendars.[31] At the same time, some form of calendrical homogeneity across the Roman provinces would have arguably been deemed useful for managing such a vast empire. This suggests that the Roman imperial power opted for the 'compromise' of tolerating that the Greeks of Asia, and in fact the whole Roman East, retained several features of their calendars – thus preserving an important cultural trait – while at the same time having these calendars – for convenience – adapted to the structure and length of the Julian year. This strategy allowed the Romans, as it were, to kill two birds with one stone: it facilitated both conversions between the calendars and relations between the Romans and their subjects in the provinces of the East.

3 Equating Calendars in the Roman Empire

Although the four manuscripts preserving the *Hemerologia* were produced in medieval times, it is clear that the information included in the tables (i.e. the calendar equations themselves) originated during the Roman imperial period or in late antiquity, when the calendars that appear in the tables were still in use.[32] Equivalences among these calendars appear in literary, documentary, and epigraphic sources dating from the early imperial period to late antiquity. These, however, are not usually arranged in tabular format and compare dates of a limited number of different calendars. A notable exception in terms of the amount of calendars involved is a passage from Epiphanius'

31 The plural here is better suited to the calendrical fragmentation that characterised the ancient Greek world throughout its history, as is glaringly exemplified by the multiplicity of calendars used in the various Hellenistic kingdoms and cities before the Roman annexation of the Greek East (see references above in n. 15). As a way of example, see *IPriene* (*2014*) 132, an inscription dating from the early second century BCE which includes equivalences between the calendars of Rhodes and Priene (l. 42–44): ὡς μὲν Ῥόδιοι ἄγοντι μηνὸς Πανάμου ἐνάται [ἐπὶ δέκα?], ὡς δὲ Πριανεῖς [μην]ὸ[ς Θαργη]λ⟨ι⟩ῶνο[ς].

32 Cf. Kubitschek, *Kalenderbücher*, pp. 61, 75, 79; Samuel, *Greek and Roman Chronology*, p. 173; Meimaris, *Chronological Systems*, p. 36; Stern, *Calendars in Antiquity*, p. 260, all assuming that the tables were copied from earlier works.

Panarion (late fourth century CE), where the dates of Jesus' birth and baptism are given according to the Roman, Egyptian, Syrian, Cypriot, Paphian, Arabian, Cappadocian, Athenian, and Hebrew calendars.[33] Other literary sources displaying the ability to produce dates according to different calendars include Ptolemy's *Almagest* (mid-second century CE), which dates an observation according to the Old Egyptian (or pre-Alexandrian) and the Bithynian calendars,[34] the *Life of Porphyry of Gaza* (fifth century CE) in which a number of dates are expressed according to the Roman calendar and the calendar of Gaza,[35] Eusebius (early fourth century CE), who frequently dates events according to multiple calendars,[36] Theophilus of Alexandria (late fourth–early fifth century CE),[37] Athanasius of Alexandria (fourth century CE),[38] Evagrius Scholasticus (sixth century CE),[39] and a series of Christian Councils of the

---

33 *Panarion* 51.24. Cf. Kubitschek, *Kalenderbücher*, pp. 73–75. Stern, *Calendars in Antiquity*, p. 260, assumes that Epiphanius consulted a *hemerologion* to supply the dates in such a large variety of calendars.

34 *Almagest* 7.3. Cf. Kubitschek, *Kalenderbücher*, p. 97; Toomer, *Ptolemy's Almagest*, pp. 14, 334. While the Old Egyptian calendar is not represented in the *Hemerologia*, the Bithynian calendar appears in all the extant manuscripts and is marked as 'of the Bithynians' (Βιθυνῶν). The date mentioned by Ptolemy corresponds to 29 November 92 CE, the Bithynian date being 'the 7th of Metroos', which matches the information given in the *Hemerologia*.

35 Cf. Hill, *Life of Porphyry of Gaza*, pp. 122, 130; Kubitschek, *Kalenderbücher*, p. 99. The calendar of Gaza occurs in *Leid.* BPG 78 and *Vat. gr.* 1291 (marked as Γαζέων) but is absent from *Laur.* 28/26. All the equivalences between the Roman and the Gazean calendars in the *Life of Porphyry of Gaza* are in agreement with the *Hemerologia*, except one (c. 34): ἡμέρᾳ ὀγδόῃ καὶ εἰκάδι Γορπιαίου, κατὰ δὲ Ῥωμαίους Σεπτεμβρίῳ εἰκάδι τρίτῃ, that is, 'the twenty-eighth day of Gorpiaios, according to the Romans the twenty-third of September'; in the *Hemerologia*, 23 September corresponds to 26 Gorpiaios in the calendar of Gaza.

36 Cf. Kubitschek, *Kalenderbücher*, p. 101. Eusebius' dates are usually given according to the Julian calendar and the calendar of Antioch; occasionally (e.g. in *Historia ecclesiastica* 7.32.14) he also includes the equivalent date in the Egyptian (= Alexandrian) calendar.

37 Cf. Kubitschek, *Kalenderbücher*, p. 101; Mosshammer, *Prologues on Easter*, pp. 58–9. In his Easter letter, Theophilus sets the date of the spring equinox (21 March) on 'the 12th day before the calends of April, which is 25 Phamenoth, and, according to the Syrians of Antioch and the Macedonians, 21 Dystros'. This matches the equivalence between the Julian, Egyptian, and Antiochene calendars in the *Hemerologia*.

38 Athanasius was bishop of Alexandria in 328–73 CE. In *De synodis Arimini in Italia et Seleuciae in Isauria* 12.1, he provides a concordance between Julian, Egyptian, and Macedonian months (the latter presumably referring to the calendar of Antioch).

39 Cf. Kubitschek, *Kalenderbücher*, p. 101. In his *Historia ecclesiastica* (2.12), Evagrius gives the date of an earthquake that occurred in Antioch in 457 CE according to the local calendar and adds the equivalent Julian month: 'the 14th day of the month Gorpiaios, which the Romans call September'. Cf. Whitby, *Ecclesiastical History*, pp. 94–95.

fifth and sixth centuries CE.[40] These sources, spanning from the mid-second century CE to late antiquity, may imply that their authors had access to calendar conversion tables of the type of the *Hemerologia*.

In terms of both format and contents, the closest parallels for the *Hemerologia* tables are found in the epigraphic and documentary evidence. A notable and in fact unique case is that of an inscription preserved on three joining fragments of a white marble slab, which were discovered about twenty years ago in the ancient city of Metropolis (Lydia, Asia Minor).[41] The inscription has been attributed on paleographical grounds to the early first century CE, and more precisely to the reign of Tiberius (14–37 CE). Not only is the marble slab from Metropolis one of the most significant pieces of external evidence for the study of the *Hemerologia*, but it also counts among the earliest attestations of the practice of providing correspondences between dates of the (relatively) recently instituted Julian calendar and dates of other calendars that were in use in the eastern provinces of the Empire. The inscription comprises equivalences between the Julian calendar and the calendar in use at Metropolis (i.e. the calendar of the province of Asia), arranged in columns. The following is a transcription of the first column of text:

ιδʹ N[ω]. Ὀκ[τω-]
βρίαι
ιεʹ πρὸ ηʹ εἰ-
δῶν Ὀκτω-
βρίων
ιϛʹ πρὸ ζʹ
ιζʹ πρὸ ϛʹ
ιηʹ πρὸ εʹ

To the left are Greek numerals referring to the days of a month in the calendar of the province of Asia, followed, to the right, by their corresponding dates

40 For example, in the Acts of the Council of Chalcedon of the year 451 CE, the date of 25 February 449 CE is provided according to the calendar of Tyre and the Julian calendar: 'After the consulship of Flavius Zenon and Postumianus, on the fifth day before the calends of March, in the colony of Tyre (…), in the year 574, on the tenth of the month Peritios, according to the Romans on the twenty-fifth of February (…).' Cf. Kubitschek, *Kalenderbücher*, p. 109; Price and Gaddis, *Acts of the Council of Chalcedon: Volume 2*, p. 261. The calendar of Tyre appears in all the *Hemerologia* (marked as Τυρίων), where the fifth day before the calends of March of the Julian calendar does correspond to 10 Peritios in the calendar of Tyre.

41 Engelmann, 'Inschriften aus Metropolis', pp. 142–43 (*AE* 1999, 1538 = *SEG* 49, 1523). Cf. Stern, *Calendars in Antiquity*, pp. 260–61; Thonemann, 'Calendar', p. 126.

in the Julian calendar, expressed in Greek. The first two lines, for instance, equate the 14th day of a month in the local calendar with the nones of October (7 October) in the Julian calendar. The arrangement of the inscribed text is comparable to the layout of the *Hemerologia* tables; yet the date equivalences in the Metropolis inscription are considered from an opposite perspective vis-à-vis the *Hemerologia* manuscripts: in the latter, the days of the Roman month are consistently listed in the first column on the left, which is followed by a series of columns showing the corresponding days in numerous calendars of the provinces and cities of the Roman East; this arrangement clearly shows that the Roman calendar served as common reference point in the *Hemerologia*. Conversely, the Metropolis inscription prioritises the calendar of the province of Asia, whose dates appear before the Julian dates on the stone. The inscribed marble slab was most likely produced to be displayed in a public place – perhaps in the agora or in proximity of some major civic building –[42] and was meant to be used as an aid to convert dates of the calendar of the province of Asia into Julian dates, and vice versa. The inscription must have represented a helpful resource for locals and Romans alike, though the priority assigned to the local calendar on the stone suggests that the inscribed document was presumably designed with the Greek inhabitants in mind.[43] This particular arrangement of the text on the marble slab is also likely to reflect the relatively early stage of development of the Julian calendar at the time when the inscription was engraved: in the age of Tiberius the calendar first introduced by Julius Caesar and later restructured by Augustus had probably not yet fully developed into the standard, dominant dating system in use throughout the provinces of the Roman Empire. Among the various functions of the inscription from Metropolis was undoubtedly that of spreading knowledge about the Julian calendar in the Roman East, while at the same time making a political point about the relationship between the province of Asia and the Roman Empire. This two-column *hemerologion*[44] suggests that similar epigraphic monuments may have been displayed in central locations of other cities of the eastern provinces of the Roman Empire to facilitate date conversions between the Roman and local calendars. Although it can hardly be demonstrated that texts such as the Metropolis inscription acted as a direct model for the production of the *Hemerologia* that appear in the four early medieval manuscripts described earlier, based on the available evidence, the

---

42 Engelmann, 'Inschriften aus Metropolis', p. 143.

43 Engelmann (ibid.) assumes that the aim of the inscription was to facilitate date conversions primarily for those who were involved in provincial administration.

44 I am borrowing the expression used by Stern, *Calendars in Antiquity*, p. 261.

Metropolis inscription represents the earliest and closest parallel for the early medieval multi-column calendar conversion tables.

In addition to the inscription from Metropolis and the assorted textual sources including double dates or dates given according to multiple calendars,[45] it is worth considering a limited number of Greek documents on papyrus that establish approximate concordances among months of different calendars. As one would expect, these sources regularly include the Egyptian calendar: in three cases its months appear along with the Roman months,[46] while a sixth/seventh century CE scrap of parchment codex provides correspondences between Egyptian and Bithynian months.[47] In addition, there are two cases of month concordances among three different calendars: a late seventh/early eighth century CE papyrus, possibly from Fayum, provides equivalences among Roman, Cappadocian, and Egyptian months,[48] and a fifth century CE papyrus from Lycopolis displays a synoptic table of Egyptian, Roman, and Macedonian months.[49] While the former, however, compares Roman and Egyptian months and, on separate columns, Cappadocian and Egyptian months, in the papyrus from Lycopolis Egyptian, Roman, and Macedonian month-names are listed alongside each other, on three parallel columns, and are therefore simultaneously equated; in addition, a fourth column consisting of numerals accompanies the list of Roman months, detailing the number of days in each of them:[50]

45 A list of new sources updating the evidence presented by Kubitschek in *Kalenderbücher* is appended to Section VI.

46 *P.Fay.* 135 V = *C. Gloss. Biling.* 1.11 (Fayum, fourth century CE); *SB* 26 16521 (ostracon – provenance unknown, fourth/fifth century CE); *P.Iand.* inv. 654 (provenance unknown, sixth/seventh century CE).

47 Montserrat, 'Fragment' (provenance unknown).

48 *P.Rain. Cent.* 31 (provenance unknown). Grassien offers a comprehensive edition of the text and compares its calendrical data with the information provided by the *Hemerologia*; Grassien, 'Deux hymnes', pp. 67–70, 81–84. She points out that the Cappadocian language is very little known: this is the first and only Egyptian papyrus preserving Cappadocian words (transliterated into Greek); there are no inscriptions in Cappadocian from Asia Minor, and the very limited number of literary sources (such as the writings of the Church Father Gregory of Nazanzius) preserve merely a few month names (again, transliterated into Greek/Latin). In her study of *P.Rain.Cent.* 31 – which comprises a number of further texts in addition to the table of month correspondences – Grassien assumes that the papyrus had some sort of Christian amuletic function. Her interpretation is partly based on the occurrence of Cappadocian month names in the document: observing that Cappadocian is a mysterious language and that, perhaps significantly, everything we know about it are month names, Grassien assumes that Cappadocian was a purely liturgical language.

49 *P.Acad. inv.* 1 r° = Fournet and Gascou, 'Un lot d'archives'.

50 The names of the Roman months are similarly complemented by numerals indicating their lengths in a sixth century CE mosaic inscription from Scythopolis (*SEG* 8, 42).

	⳨			
⳨	Τυβι	[Ἰα]νουάριος	λα	Αὐδυναῖος
	Μεχειρ	[Φ]εβράριος	κη	Περίτιο[ς]
	Φαμεν[ω]θ	[Μ]άρτιος	λα̣	Δύστρο[ς]
	Φαρμου[θ]ι	[Ἀ]πρίλιος	λ̣	Ξανθι̣[κ]ο̣ς
	Παχων	Μάϊος	λα	Ἀρτεμίσιος
	Παυνι	Ἰούνιος	λ	Δέσιος
	[Ἐπειφ]	Ἰούλιος	λα	Πάνεμος
	[illegible]	[illegible]	[illegible]	[illegible]
	Θωθ	Σεπτέμβερ	λ	Γορπιαῖος
	Φαωφι	Ὀκτώβερ	λα	Ὑπερβερετε̂ος
	Ἀθυρ	Νοέμβερ	λ	Δῖος
	Χο[ι]ακ	Δεκέμβερ	λα	Ἀπιλλε̂ος
	κατ᾽Αἰγυπτίους	κατὰ Ῥωμαίους		κατὰ Ἀσιανούς

Although the Egyptian months occupy the first column on the left, the order followed by the three calendars is that of the Roman year, running from January to December. The Egyptian year started with Thoth and ended with Mesore, while our text lists Egyptian months from Tybi to Choiak.[51] Each of the three calendars is identified by a formula written at the bottom of each column: κατ᾽Αἰγυπτίους, 'according to the Egyptians', κατὰ Ῥωμαίους, 'according to the Romans', and κατὰ Ἀσιανούς, 'according to the Asians'. These formulae effectively function as headings – albeit in an upside down position – and parallel the labels that appear across the top on the *Hemerologia* manuscripts: there, above each column, the name of the calendar, along with that of the local month corresponding to the relevant Roman month, identify the specific calendar whose month-days are listed underneath. Κατὰ Ῥωμαίους clearly identifies the Julian calendar. Although the formula κατ᾽Αἰγυπτίους is occasionally used with reference to the Old Egyptian calendar,[52] in this case it is more likely

51 On the order of the months in the Egyptian year see, e.g., Samuel, *Greek and Roman Chronology*, p. 177. It appears that the writer of the comparative table was not particularly familiar with the Greek version of the Roman month-names, as is especially revealed by the transliterations Σεπτέμβερ, Ὀκτώβερ, Νοέμβερ, Δεκέμβερ in place of the canonical Greek forms Σεπτέμβριος, Ὀκτώβριος, Νοέμβριος, Δεκέμβριος.

52 This appears to be the case in sources that give a date according to both the Old Egyptian and the Alexandrian calendar, such as in some ancient horoscopes, where Old Egyptian dates are sometimes specified as either κατὰ Αἰγυπτίους (more rarely as παρὰ Αἰγυπτίοις) or κατ᾽ἀρχαίους, 'according to the Egyptians', or 'according to the ancients', whereas dates designated as καθ᾽ Ἕλληνας (or καθ᾽ Ἑλλήνων), 'according to the Greeks', refer to the Alexandrian calendar; Baccani, *Oroscopi greci*, pp. 60–61; Jones, *Astronomical Papyri*, pp. 12–33; Ross, 'Introduction', p. 151. The distinction between κατὰ Αἰγυπτίους and καθ᾽ Ἕλληνας dates as referring, respectively, to the Old Egyptian and the Alexandrian

to identify the 'new Egyptian' or Alexandrian calendar.[53] The calendar whose month-names are listed on the right hand column of the Lycopolis papyrus is labelled as 'according to the Asians': what calendar could be characterised in this manner? The obvious response would be the calendar of the province of Asia.[54] The *Hemerologia* list this calendar under various headings – Ἐφέσου, Ἀσίας, Παμφυλίας, Ἀσιανῶν – which identify a series of calendars that were in use in the provinces of western Asia Minor (corresponding to the area known as Diocese of Asia in the fifth century CE). All these calendars shared the same structure and yet presented differences in the names and order of their months, as well as minor one-day variations. It has to be observed that the order of the months on the κατὰ Ἀσιανούς column of the Lycopolis papyrus does not match the order of the months of any of the various 'Asian' calendars that used Macedonian month-names and that appear in the *Hemerologia*. On the other hand, the month order of the κατὰ Ἀσιανούς column does correspond to that of the calendar labelled as 'of the *Hellenes*' (Ἑλλήνων) on the *Hemerologia*, which identifies with the so-called calendar of Antioch. Arguably, this was the most widespread calendar in the Diocese of the East during the fifth century.[55] It is conceivable that there never was total uniformity and consistency in the denomination of the multitude of calendars of the provinces and cities of the eastern half of the Roman Empire; even more so, presumably, in the case of the various calendars that used the Macedonian month-names and that were not peculiar to a specific city, such as the calendar of Asia and the calendar of Antioch. Under these circumstances, one could perhaps expect that the calendar labelled as 'of the *Hellenes*' in the early medieval *Hemerologia* manuscripts could be referred to as 'of the Asians' in a fifth-century document on papyrus from Egypt.[56] After all, '*Hellenes*' is terribly vague. Moreover, it would have been easy for someone in Egypt to commit an error, either with the label

calendar, can also be observed in *P.Kellis* 41 (dated to 12 July 310 CE), on which see further no. 37 in the list of new sources below.

53 As observed earlier, in the *Hemerologia* the Alexandrian calendar is labelled as Ἀλεξανδρέων, 'of the Alexandrians', and not as Αἰγυπτίων, even though the Old Egyptian calendar is not represented there, a circumstance that would have left no room for confusion. Similarly, in a Greek inscription from Rome (*IGUR* I 77) the date (6 May 146 CE) is expressed according to the Roman as well as the Egyptian calendar, the latter being specified as κατὰ Ἀλεξανδρεῖς, 'according to the Alexandrians', rather than with the more common formula κατὰ Αἰγυπτίους, 'according to the Egyptians'.

54 As assumed by Fournet and Gascou, 'Un lot d'archives', pp. 1053–54.

55 *Pace* Fournet and Gascou, 'Un lot d'archives', p. 1054, who affirm that the calendar normally used in the Diocese of East during that period was the calendar of Asia.

56 In this context, it appears significant to note that in the Leiden manuscript (*Leid.* BPG 78) the *Hemerologia* are followed by a synoptic table (ff. 151–152^{v}) of the Egyptian, Roman, and Antiochene years. On this table, see Kubitschek, *Kalenderbücher*, pp. 70–71.

or with the month names, without anyone there to correct him. The editors of *P.Acad. inv.* 1 r° inform us that the person who transcribed the synoptic table of Egyptian, Roman, and Macedonian month-names also wrote a petition on the verso of the same papyrus, as well as two further documents in the same dossier, including a second petition.[57] Interestingly, it appears that both petitions were addressed to the praetorian prefect of the East. The nature of these texts leads the editors of *P.Acad. inv.* 1 r° to the very plausible conclusion that, differently from most documentary texts providing concordances, this is not a [illegible] rus presumably functioned as a reference tool for converting dates between the local calendar (the Egyptian) and two of the main systems in use at the time in the Roman East. The month concordance table may have been useful for both the sender and the addressee of the petitions. In fact, however, since the months in the three calendars were not coterminous, the table would have been of limited use, providing, as it did, only loose concordances between the months in the three systems. It is nevertheless worth pointing out that this is the only known text on papyrus that correlates the Egyptian (i.e. Alexandrian), Roman, and (if the hypothesis expressed here is correct) Antiochene calendars. Although the Lycopolis papyrus obviously differs from the *Hemerologia* in that it does not provide full-length calendar tables but only (approximate) equivalences among the months of the three calendars, nonetheless it does represent the best documentary counterpart of the *Hemerologia* manuscripts.

What precedes has shown that a variety of literary, documentary, and epigraphic sources suggest that calendar conversion tables of the type of the *Hemerologia* may have been in use already in the early Roman imperial period and continued to do so until late antiquity. Yet the question of when the *Hemerologia* were first composed and in what form (e.g. manuscripts, papyri, inscriptions), as well as the question of the mode of their transmission prior to their appearance in the early ninth century as part of the Greek manuscripts *Leidensis* BPG 78 and *Vaticanus gr.* 1291 (followed by *Laurentianus* 28/26 in the late ninth or early tenth century, and *Laurentianus* 28/12 in the fourteenth century) remain to be answered. One may argue that the dearth of inscriptions, papyri, ostraca, and other writing media displaying conversion tables among different calendars in use throughout the Roman Empire signals that such tables were not widespread – hence their chances of preservation have been

57 These fragmentary texts have not yet been published.

58 It cannot be ruled out that the letter on the verso is only a copy of the letter that was sent, which would explain why it stayed in Egypt. If that were the case, then the fact that these texts appear on both sides of the same papyrus would not be significant.

considerably lower than those of other, more prevalent, types of texts. However, the evidence – or rather the lack thereof – may also imply that the *Hemerologia* were not copied from an imperial or late antique model; it could be supposed that the tables were first compiled in the early Middle Ages by drawing from a miscellany of earlier sources. In other words, is it conceivable that the one and same 'Grundliste'[59] from which the preserved *Hemerologia* derive was composed not long before the early ninth century, when *Leidensis* BPG 78 and *Vaticanus gr.* 1291 were produced?

4 The *Hemerologia* Manuscripts: Times, Contexts, and Modes of Composition and Transmission

The question of whether the *Hemerologia* originated in the imperial, late antique, or early medieval period ought to be considered in the light of the context of production and transmission of the manuscripts to which the preserved texts belong. The four sets of *Hemerologia* published by Kubitschek in 1915 are all incorporated within medieval copies of Ptolemy's *Handy Tables* (Πρόχειροι Κανόνες).[60] This work consists of a series of astronomical tables designed to provide directions for using the tables included in Ptolemy's *Almagest*. The *Handy Tables* permit the calculation of the positions of the Sun, the Moon, and the planets, as well as the rising and setting of the stars, which in turn enable one to solve related astronomical and astrological problems, such as the determination of horoscopes and the character of lunar and solar eclipses. In addition to astronomical tables proper, as well as geographical tables listing 'renowned cities' (πόλεις ἐπίσημοι), both of which were certainly designed by Ptolemy, the manuscript tradition of the *Handy Tables* preserves a variety of additional tables, images, and diagrams of uncertain date and provenance. Anne Tihon has classified the supplementary material included in the *Handy Tables* manuscripts according to the following categories: tables specific to the climate of Byzantium, chronological tables, and ancillary tables of various types.[61] The *Hemerologia* belong to the chronological material, which include tables of regnal years – starting either with Nabonassar (747 BCE) or with Philip, Alexander's successor (324/3 BCE) –, lists of consuls, tables of bissexts,

59 The term is used by Kubitschek, *Kalenderbücher*.

60 A critical edition of the *Handy Tables* is offered in Tihon's *Les 'Tables Faciles' de Ptolémée* and in Mercier, Πτολεμαίου Πρόχειροι Κανόνες. See also Tihon, 'Les *Tables Faciles* de Ptolémée'; ead., 'Theon of Alexandria'; and ead., 'Une édition critique'.

61 Tihon, *Les 'Tables Faciles' de Ptolémée*, p. 8.

tables of epacts, tables for determining the day of the week, tables of concordance among Egyptian, Roman, and Greek months, tables of lunar dates, etc. The only tables explicitly mentioned by Ptolemy as complementary to the use of the astronomical tables are the lists of regnal years.[62] The rest of the chronological material should therefore be regarded as spurious, which implies that the *Hemerologia* were not originally part of Ptolemy's *Handy Tables*.[63] Thus, although we now have a *terminus post quem* for their production, we are still left with the question of the authorship and date of composition of the *Hemerologia*. The fact that these were consistently handed down together with Ptolemy's *Handy Tables* is certainly significant; yet, this circumstance does not necessarily imply that the tables were originally produced as part of a scholarly, scientific tradition during the Roman imperial period or in late antiquity. Indeed, there is no reason to exclude the possibility, for instance, that the *Hemerologia* were created – at some unspecified time – for one or more different purposes and subsequently 'reused' in association with Ptolemy's *Handy Tables* to assist medieval astronomers and astrologers in their calculations.

Whatever the time and context of their production, one may raise the objection that our tables for calendar conversion would have served little purpose in medieval times, as by the ninth century the calendars represented in the *Hemerologia* were no longer in use, with the sole exception of the Julian calendar. However, it should be observed that examples of dates and astronomical events from antiquity were commonly employed in medieval scientific practice.[64] Even though the extant astronomical-astrological sources may not feature dates in the calendars of Gaza, Cappadocia, or Ascalon (to name some of the 'unusual' calendars that appear in the *Hemerologia*), the general scarcity of evidence does not preclude the possibility that similar dates may have appeared in other material that might have been of use to a medieval astrologer: for instance, astronomical phenomena mentioned in chronicles, the foundation date of a city, a birthday or a dated horoscope. Indeed, as is witnessed by the early ninth century manuscripts *Leidensis* BPG 78 and *Vaticanus gr.* 1291, the late ninth or early tenth century *Laurentianus* 28/26, and the fourteenth century *Laurentianus* 28/12, the *Hemerologia* were deemed worth copying throughout the Middle Ages. On the basis of these premises, it appears equally plausible that the *Hemerologia* were first composed sometime in the

62 Ibid., 10.

63 *Pace* Kubitschek (*Kalenderbücher*, p. 79), who assumed that *similar* calendar conversion tables might have already been joined with the *Handy Tables* by Ptolemy.

64 Thanks are due to Philipp Nothaft for pointing out this important piece of information during the ERC conference *Calendars in Antiquity and the Middle Ages*, which took place at UCL on 3–5 July 2017.

early middle ages from a variety of earlier material, rather than being a product of the imperial or late antique period as it has been assumed so far.[65]

Ptolemy's *Handy Tables* experienced enduring popularity in the Byzantine Empire since its inception and until the fifteenth century.[66] They were particularly suited to making astrological forecasts, and it has been suggested that certain deluxe manuscripts produced between the eighth and the early tenth century and incorporating the *Handy Tables*, such as *Laurentianus* 28/26, were intended for use by court astrologers who were requested to cast horoscopes by Byzantine emperors.[67] More generally, apart from the specific use of the *Handy Tables* for the purpose of making astrological predictions, the works of Ptolemy and ancient commentaries on them constituted a major portion of the legacy of ancient astronomical writings preserved by the Byzantines. The interest in ancient texts and especially in scientific and astrological works from Greek and Roman antiquity reached a peak during two specific periods of the Byzantine Empire: with the advent of the Macedonian dynasty in the second half of the ninth century, especially during the reign of Constantine VII Porphyrogenitus (first half of the tenth century), and in the fourteenth and fifteenth centuries, with the re-conquest of Constantinople and the rise of the Paleologan dynasty.[68] Significantly, the *Handy Tables* are preserved in forty-five manuscripts that mostly date either to the ninth/tenth or to the fourteenth to sixteenth centuries, with the *Hemerologia* being included in four manuscripts dating to the ninth/tenth and fourteenth century, thus confirming this distribution pattern.

While all this may shed some light on the times, contexts, and modes of transmission of the *Hemerologia* during the middle ages, it becomes clear that the surviving evidence does not allow to determine with any certainty whether the *Hemerologia* originated in the Roman period, in late antiquity, or in the early middle ages. The question of the date of creation of the *Hemerologia* – in the form in which they are currently preserved – remains open.

5 The *Hemerologia*: Functions and Purposes

The question of whether the *Hemerologia* tables originated in the Roman or early medieval period is directly linked to another central issue: for what

65 Kubitschek, *Kalenderbücher*, pp. 79–81; Stern, *Calendars in Antiquity*, pp. 260–61.

66 Tihon, *Les 'Tables Faciles' de Ptolémée*, p. 52.

67 Marchetti, 'La trasmissione', p. 122.

68 Ibid., pp. 123–24.

purpose or purposes were the *Hemerologia* produced? The answer to this question is relatively straightforward if we assume that early medieval scholars composed the *Hemerologia*. Comparative calendar tables which permit the transposition of dates from local calendars of the eastern Roman provinces to the Julian calendar and vice versa would have been of service to medieval intellectuals as an additional aid to astrological and astronomical reckonings based on examples of dates and astronomical events from antiquity, as well as, presumably, for a variety of chronological calculations (including Computus and other liturgical calculations). On the other hand, if we assume that the *Hemerologia* originated in the Roman period, a univocal and definite answer to the question of their purposes and functions can hardly be provided. In his fundamental edition of the four preserved sets of calendar conversion tables, Wilhelm Kubitschek did not concern himself with what might have been the point of composing them in the first place. The first scholar to make assumptions on their possible functions and purposes seems to have been Sacha Stern. Assuming, as he does, that large comparative calendar tables of the type of the *Hemerologia* did exist in the Roman period, one can only agree on his plausible idea that these 'were probably intended for scribes and imperial administrators, to help them with the conversion of multiple calendar dates into one another'.[69] In this perspective, one can imagine that such conversion tables, perhaps written on scrolls or wax tablets or some other perishable material that has not been preserved over time, were at the disposal of provincial officials and governors in the Roman East.

Additional functions and purposes of these texts are just as plausible: as suggested again by Stern, the *Hemerologia* may have also served the needs of commercial and other private uses, or carried ideological and political significance, representing local adherence to the Roman imperial power. In this respect, one may think of the calendar of Asia that was adopted very early in celebration of the emperor Augustus; or the Metropolis inscription, whose large dimensions and central position within the public space of the city suggest that the monument was not merely aimed at facilitating conversions of dates of the calendar of the province of Asia into Julian dates and vice versa; presumably, it also functioned as a means to popularise the Julian calendar in the eastern provinces and as such it did carry ideological and political significance, as a tangible sign of the Roman presence in the Greek East.

In this respect, one might wonder whether the *Hemerologia* may be seen as an instrument for the fixation of provincial calendars and the imposition of conformity to the Julian calendar, or, on the contrary, as a 'celebration of

69 Stern, *Calendars in Antiquity*, p. 261.

diversity'. In this connection, it is deemed worth formulating some reflections on the social and political implications of the use of multiple dating systems in the Roman East. Let us consider the relatively numerous cases of double dates, that is, dates given according to the Julian calendar and a local calendar, in epigraphic and papyrological documents from the Eastern provinces of the Roman Empire, which are mostly documents of an official character. In a number of instances, dates are given in Latin as well as Greek, such as in *P.Flor.* 2.278 = *CPL* 145, col. II (a military text from Heliopolis dating to the late second or early third century CE, on which see further no. 25 in the list of new external sources below): *X Kal(endas) Aug(ustas)* Ἐπεὶφ κθ. This bilingual date gives the day of the month in both the Julian calendar (in Latin) and the Alexandrian calendar (in Greek), and translates as 'the 10th day before the calends of August, 29 Epeiph', both dates corresponding to 23 August. In cases like this, as well as when double dates are given in one language, either Latin or Greek, the Roman date is required to confer official status on the document, whereas the date given according to the local calendar is presumably for the benefit of those who were not particularly familiar with the Roman system and, at the same time, adds precision to the document itself. In general terms, it could be said that double dates, particularly bilingual ones, testify to the celebration of diversity and the multiculturalism that in many respects did characterise the Roman Empire. Quite different assumptions can be drawn from the case of official documents emanating from the Roman administration in the Eastern provinces dated exclusively according to the Julian calendar and the Roman consular system – which seem to have been the norm, judging from the available evidence. Whether translated into Greek or put into Latin, a Roman date in a Greek document from the East might indeed be seen as symbolic of Roman domination, and even more so if that date is also expressed in Latin.[70] That said, it should not be forgotten that after the introduction of the Julian calendar, the Roman East was characterised by the unique feature of having multiple dating systems in place. On the contrary, in the western provinces the new calendar rapidly replaced completely all pre-existing local calendars. This phenomenon is in itself revelatory of the complexity of the social and political dynamics associated to the Roman presence in the East. It is clear that the Romans acknowledged the rich cultural tradition of their Greek subjects in the East, including as regards their calendrical traditions. And, as considered earlier, it appears that the eastern provinces (at least the Greeks of the Province of Asia) adopted the Julian calendar or adapted their local calendars to the Julian as part of a political game to demonstrate their

70 Adams, *Bilingualism*, pp. 391–92.

loyalty to the Roman emperor. In this perspective, one can hardly conceive of the *Hemerologia* or more generally of the coexistence of the Julian calendar and a multiplicity of local calendars in the Roman East as a signal of Roman cultural and political hegemony on the eastern provinces.[71]

6 New External Evidence since Kubitschek's 1915 Study

The *Hemerologia* were edited by Wilhelm Kubitschek, who not only published the tables with transcriptions and critical commentary, but also compared the data of the *Hemerologia* to that of relevant epigraphic, literary, and documentary sources that were available to him in the early twentieth century. My research on the *Hemerologia* within the framework of the project *Calendars in Antiquity and the Middle Ages* included a re-assessment of the information provided by these manuscripts by taking into account newly discovered epigraphic and documentary evidence since Kubitschek's work over a century ago.[72] To this end, I carried out a systematic search for literary, epigraphic, and documentary evidence comprising a date given according to two or multiple calendars, as well as for sources providing equivalence among month names of different calendars, and related evidence. The current section, along with the paper itself, concludes with a full list of these new external sources accompanied by an indication of whether or not in each case the calendrical information is in agreement with that provided by the *Hemerologia*. One of the purposes of collecting and analysing these new data has been to establish to what extent the *Hemerologia* are consistent with external textual evidence from the Roman Empire, and how inconsistencies should be accounted for. While it appears that broad conclusions can hardly be drawn from the new evidence, to some extent one can still evaluate the significance of inconsistencies on a case-by-case basis. Even so, it must be borne in mind that in several cases discrepancies could be due to a wide range of factors, including ignorance of the local calendar on the part of Roman imperial administrators or, vice versa, ignorance of the Julian calendar on the part of the locals, as well as computational or scribal error. At least in some cases, however, inconsistencies between the *Hemerologia* and external evidence may also reflect regional diversity of calendars and/or diachronic change. Indeed, the historical reality behind the *Hemerologia* was probably more complex and flexible than the

71 See generally Stern, *Calendars in Antiquity*, pp. 260–62.

72 In addition to his 1915 publication, a number of relevant sources also appear in Kubitschek, 'Kalenderstudien' and id., 'Pamphylische Kalender'.

fixed schematism of the information provided by the four sets of calendar tables that are the focus of this paper: provincial calendars may have evolved and changed in the course of the Roman period, and it is clear that widespread calendars such as the calendar of Asia were not uniformly reckoned in any given period or geographic area – as is evident from the different types of Asian calendars listed in the *Hemerologia* themselves.

The large majority of these new sources include a double date or a date given according to multiple calendars.[73] More than half of these are dates expressed according to the Roman (= Julian) and the Egyptian (= Alexandrian) calendars. As noted above (p. 82), the latter is represented in the *Hemerologia* as the 'calendar of the Alexandrians'. About twenty new sources show a date given according to the Julian calendar and the calendar of Provincia Arabia, i.e. the calendar that was mainly used in the Province of Palaestina Tertia, formerly known as Province of Arabia (cf. above n. 7). A papyrus from Oxyrhynchus (no. 64) is dated according to the Egyptian calendar and the calendar of Provincia Arabia. The date of a funerary inscription from Avdat in the Province of Arabia (no. 68) is given according to the calendars of Gaza and Elousa, the latter corresponding to the calendar of Provincia Arabia. In twelve sources the date is expressed according to the Roman calendar and the calendar of the Province of Asia. We then have four cases of dates given according to the Roman calendar and the calendar of Antioch (cf. above n. 5); a text dated by the Roman calendar and the calendar of Gaza (no. 88); and a text dated by the Roman calendar and a calendar used in the province of Judaea/Palaestina (no. 89).

Out of a total of twenty-three sources including double dates that are not in agreement with the *Hemerologia*,[74] it is to be noted that in nine cases the discrepancy in the two dates is of one day.[75] Four of these nine sources involve equivalences between the Roman and Egyptian calendars;[76] obviously, this proportion may be simply reflecting the higher amount of sources comprising the Roman and Egyptian calendars. Also, in six further cases the discrepancy in the two dates appears to be of one month.[77] The relatively high number of

73 These are 88, out of a total of 104 new sources. Only sources providing fully preserved double dates have been taken into account. Documents including fragmentary double dates in which only one of the two dates is complete were normally restored on the basis of the equivalent date, and are therefore of no use for our present purposes.

74 These are nos. 1, 23, 27–29, 32, 47, 49–52, 56, 60, 63–64, 67, 74–75, 78, 80, 83, 87, 89. The percentage of new sources with double dates that are not in agreement with the Hemerologia is 26%.

75 Nos. 27–28, 32, 47, 50–51, 74–75, 87.

76 Nos. 27–28, 32, 47.

77 Nos. 23, 29, 60, 63, 67, 83.

one-day or one-month discrepancies is potentially significant, though it does not necessarily imply developments in the calendars involved: the possibility remains that these incongruities are due to recurrent computational or scribal errors.

The sources including double dates which were collected and analysed by Kubitschek in the early twentieth century do not dramatically change the picture drawn so far. These are twenty-five and provide a total of twenty-nine date equivalences. Of these, six are in disagreement with the *Hemerologia*. In the aforementioned passage by Epiphanius of Salamis giving the date of the Epiphany as well as the date of Jesus's baptism according to a number of different calendars (cf. above, pp. 89–90), all dates agree with the *Hemerologia* except for those given according to the Cappadocian calendar. In this specific case it is difficult to assess the extent of the deviation, due to the fact that the limited number of ancient sources pertaining to the Cappadocian calendar give different names for the months of this poorly attested calendar (cf. above n. 48). The remaining four cases comprise a two-day discrepancy in a date equivalence between the Julian calendar and the calendar of Gaza;[78] a one-month discrepancy in a date equivalence between the Julian and Pamphylian calendars, the latter corresponding to the calendar of Asia;[79] and, finally, two inscriptions providing date equivalences between the Julian calendar and the calendar of the ancient city of Tyras in Lower Moesia (northern coast of the Black Sea). Earlier scholars assumed that the calendar used in Tyras was the calendar of Asia, in which case there would be a five-day and a seventeen-day discrepancy in the two date equivalences.[80] The assumption that the calendar used in Tyras was the calendar of Asia is based on the fact that Tyras was a Milesian colony. In actual fact, however, we have no evidence of this, as neither of the Tyras inscriptions is anywhere near compatible with the calendar of Asia. Moreover, by the Roman period, the identity of Tyras as a Milesian colony may have been long forgotten; thereofore, there is no reason why, in the Roman period, Tyras

78 In a passage from the *Life of Porphyry of Gaza*, on which see Kubitschek, 'Kalenderstudien', p. 103 n. 18; ibid., p. 99.

79 In a papyrus (*BGU* 3.887), on which see Kubitschek, *Kalenderbücher*, p. 108.

80 The inscriptions are *IOSPE* I^2 2 and *IOSPE* I^2 3, on which see Kubitschek, *Kalenderbücher*, p. 111. The date equivalence in *IOSPE* I^2 3 has been read by V. Latyshev as πρὸ · ιγ' καλανδῶν Μαρτίων, Ληνεῶνος η' ('on the 13th day before the calends of March, on the 8th of Lenaion'). Kubitschek assumes that the local date should be read instead as Ληνεῶνος H$^{\alpha}$, that is, H ἀπιόντος, which indeed corresponds to the 13th day before the calends of March. Sacha Stern adds that the initial alpha of the word that follows the date equivalence (ἀνεστάθη) might have been confused with the alpha of the date (abbreviation of ἀπιόντος) and hence erroneously omitted.

should have adopted the calendar of Asia. It thus appears safer to conclude that the two dates in question do not correspond to any known calendar.

The list of new sources is arranged according to the provincial calendar which the texts attest; within each subdivision, the entries are ordered chronologically from the earliest to the most recent.[81] Unless otherwise stated, equations are in agreement with the *Hemerologia*.

Appendix: List of New Sources

1 *Double/Multiple Dates*

1.1 Egyptian (= Alexandrian) Calendar

1. Papyrus. Provenance unknown (Egypt), late first century BCE (*CPL* 247 = *CEL* 8 = *ChLA* 43.1241c). Private letter, dated according to the Julian and Egyptian calendars: *XIIII· K(alendas)· August(as)·* Ἐπεὶφ κζ. 'On the 14th day before the calends of August, the 27th of Epeiph.' The equation is not in agreement with the *Hemerologia*. There is a two-day discrepancy: the 14th day before the calends of August = 19 July; 27 Epeiph = 21 July. De Romanis points out that the equivalence is correct if we assume that the Egyptian date is in fact not an Alexandrian but an Old Egyptian one.[82] According to the reconstruction of the late first-century BCE Julian and Alexandrian calendars by Bennett, the equation 27 Epeiph = 19 July would have applied in the years 11, 8, and 7 BCE.[83] If the old Egyptian calendar was used in this papyrus, as De Romanis suggests, then the equation would have applied in the years 20–18 BCE.[84] Given that these solutions only apply to a limited number of years in the late first century BCE, it is plausible to suggest that this equation might be an error.
2. Papyrus. Provenance unknown, 62 CE (*CPL* 148). Birth certificate, dated according to the Julian and Egyptian calendars: *X K(alendas) Augustas, mense Epiph die XXIX*. 'On the 10th day before the calends of August, the 29th of the month Epeiph.'
3. Papyrus. Provenance unknown, 81 CE (the text is dated to 'the third year of the *theos* Titus') (*P.Lond.* 1.130). Horoscope dated according to the Alexandrian, Julian, and Old Egyptian calendars: Φαρμοῦθι ἕκτῃι, ὡς δὲ Ῥωμαῖοι ἄγουσι

81 Both no. 64 and no. 68 appear under 'Calendar of Provincia Arabia', though they are also relevant to the Egyptian calendar and the calendar of Gaza, respectively. No. 88, whose date is expressed according to the Egyptian, the Julian, and the Antiochene calendar, is listed among the sources providing evidence for the calendar of Antioch.

82 De Romanis, 'Lysas e il tempo', pp. 18–22.

83 Bennett, 'Early Augustan calendars' and 'Two Egyptian birthdays'.

84 Cf. Stern, *Calendars in Antiquity*, pp. 266–67, table 5.4.

καλάνδαις Ἀπριλείαις, κατ' ἀρχαίους δὲ Παχὼν νεομηνίᾳ εἰς τὴν δευτέραν. 'On the sixth of Pharmouthi; according to the Roman reckoning, on the calends of April, while according to the old (calendar), on Pachon the first to the second.'[85]

4. Wood tablet. Philadelphia, Egypt, 94 CE (*CPL* 104 = *ILS* 3.2 9059). Edict of Domitian on the military, dated according to the Julian and Egyptian calendars: *VI Non(as) Iulias, mense Epip die VIII*. 'On the 6th day before the nones of July, the 8th of the month Epeiph.'

5. Wood tablet. Alexandria, Egypt, 109 CE (*CPL* 150 = *BGU* VII 1691). Birth certificate, dated according to the Julian and Egyptian calendars: [*pr(idie)*] *K(alendas) Iu*[*l(ias)*], [*mense*] *Ephip die VI*. 'On the day before the calends of July, the 6th of the month Epeiph.'

6. Papyrus. Alexandria, Egypt, 119 CE (*BGU* 1.140). Imperial letter, dated according to the Julian and Egyptian calendars: πρίδιε νό[ν]ας Ἀουγο[ύσ]τας, ὅ ἐ̣σ̣τ̣ι̣ν̣ Μεσορὴ ια. 'On the day before the nones of August, which is the 11th of Mesore.'

7. Wood tablet. Arsinoites, Egypt, 128 CE (*CPL* 151). Birth certificate, dated according to the Julian and Egyptian calendars: *Idib(us) April(ibus), mense Pharmuthi die XVIII*. 'On the ides of April, the 18th day of the month Pharmouthi.'

8. Wood tablet. Philadelphia, Egypt, 131 CE (*CPL* 160). Birth certificate dated according to the Julian and Egyptian calendars: *VII Kal(endas) Ia*[*nuar(ias)*], *mense Choeac die XXX*. 'On the 7th day before the calends of January, the 30th day of the month Choiak.' As De Romanis explains, the date should be 27 December (instead of 26 December), as 131 CE was a leap year in the Alexandrian calendar.[86] This source, when considered along with nos. 16 and 20 below, raises the following question: is this just an error, or does it reveal, in fact, a certain scribal practice, namely that when a one-day discrepancy occurred between the Alexandrian and the Julian calendars (between the extra epagomenal day in August, and the bissextile day in the following February) – which only lasted for a period of six months, every four years – scribes sometimes did not bother to take account of it?[87]

9. Papyrus. Oxyrhynchus, Egypt, 134 CE (*P.Oxy.* 38.2857 = *ChLA* 47.1413). Roman will, dated according to the Julian and Egyptian calendars: πρὸ ις Καλανδ(ῶν) Ἰουνίων, Παχὼν κβ. 'On the 16th day before the calends of June, the 22nd of Pachon.'

10. Papyrus. Provenance unknown, 143 CE (*BGU* 1.113). Document connected with the army, dated according to the Julian and Egyptian calendars: ἀπὸ τῆς πρ[ὸ] ιε καλανδῶν Μ̣[αρτίω]ν̣, ἥτι[ς ἐστ]ὶ [το]ῦ [Μ]εχεὶρ κα, ἄ[χρι τ]ῆς π̣[ρὸ ιζ καλανδῶν Ἰου]νίων̣, [ἥτι]ς ἐστὶν μ[ηνὸς Παχὼν] κα. 'From the 15th day before the calends of

85 See Neugebauer and Van Hoesen, *Greek Horoscopes*, p. 24; Hagedorn and Worp, 'Wandeljahr', p. 247.

86 De Romanis, 'Lysas e il tempo', p. 17.

87 Sacha Stern (personal communication).

March, that is, the 21st of Mecheir, until the 17th day before the calends of June, that is the 21st of the month of Pachon.'

11. Wood tablet. Alexandria, Egypt, 144 CE (*CPL* 152). Birth certificate, dated according to the Julian and Egyptian calendars: *Idib(us) Octobr(ibus), mense Phaophi die XVIII*. 'On the ides of October, the 18th of the month Phaophi.'

12. Wood tablet. Provenance unknown, 145 CE (*CPL* 153). Birth certificate, dated according to the Julian and Egyptian calendars: *XI kạ[l] Iụḷias, [me]nse Paụṇ[i die] XX[VI]I*. 'On the 11th day before the calends of July, the 27th of the month Payni.'

13. Wood tablet. Alexandria, Egypt, 145 CE (*CPL* 154). Birth certificate, dated according to the Julian and Egyptian calendars: *XVI K(alendas) Iun(ias), mense Pachon die XXII*. 'On the 16th day before the calends of June, the 22nd of the month Pachon.'

14. Wood tablet. Karanis, Egypt, 145 CE (*CPL* 162) Birth certificate, dated according to the Julian and Egyptian calendars: *III· K(alendas)· Maias, mense Pachon die· IIII*. 'On the 3rd day before the calends of May, the 4th of the month Pachon.'

15. Inscription on stone. Rome, 146 CE (*IGUR* I 77 = *IG* XIV 1084). Dedication of the *Paianistai* of Serapis, dated according to the Julian and Egyptian calendars: τῇ πρ(ὸ) · α · νωνῶν Μαίων, ἥτις ἐστὶν κατὰ Ἀλεξανδρεῖς Παχὼν ια'. 'On the day before the nones of May, which is, according to the Alexandrians, the 11th of Pachon.'

16. Wood tablet. Alexandria, Egypt, 147 CE (*CPL* 155). Birth certificate, dated according to the Julian and Egyptian calendars: *IIII Kal(endas) Septembres, mense Mesore die intercalari VI*. 'On the 4th day before the calends of September, the 6th intercalary day of the month Mesore.'[88]

88 148 CE was a leap year, with a bissextile in February. This means that the preceding 1 Thoth, 147 CE, was postponed to 30 August through the intercalation of a 6th epagomenal (on 29 August). This appears to be the only attestation of the use of *intercalaris* to designate an epagomenal day in Latin non-literary sources. Bede (*De ratione temporum* 11) refers that the Egyptians 'call the remaining five days epagomenals – "intercalated" or "added". Every fourth year, they add a sixth day to these five, made up from the quarter-days' (*residuos quinque die ἐπαγομένας, vel intercalares, sive additos vocant, quibus etiam quarto anno diem sextum, qui ex quadrantibus confici solet, adnectunt*). Bede appears to be quoting Isidore of Seville (*De natura rerum* 4.7; cf. also 1.5), who in turn presumably draws on Macrobius (*Sat.* 1.15.1). The latter tells that the Egyptians 'return' the remaining five days to their year, and refers to the 6th epagomenal as *intercalaris* (*reliquos quinque dies anno suo reddunt, annectentes quarto quoque anno exacto intercalarem, qui ex quadrantibus confit*). In Latin inscriptions, *intercalaris* occurs mostly as part of date formulae according to the pre-Julian calendar. Greek documentary and epigraphic sources from Egypt show formulae such as Μεσορὴ ἐπαγομένων δ' to express epagomenal dates (*SB* 1.411). It thus seems that there was no standard Latin term for the epagomenals, and that *mense Mesore die intercalari VI* in our source reproduces the standard Greek formula.

17. Wood tablet. Alexandria, Egypt, 148 CE (*CPL* 156). Birth certificate, dated according to the Julian and Egyptian calendars: *III non(as) Novembr(es), mense Athyr VII*. 'On the 3rd day before the nones of November, the 7th of the month Hathyr.'
18. Wood tablet. Philadelphia, Egypt, 157 CE (*BGU* 7.1695 = *CPL* 223 = *AE* 1927, 180). Will of a soldier, dated according to the Julian and Egyptian calendars: *Nonis Octobr(ibus), mense Phaophi die X*. 'On the nones of October, the 10th day of the month Phaophi.'
19. Papyrus. Arsinoites, Egypt, 160 CE (*P.Ross.Georg.* 2.26). Roman will, dated according to the Julian and Egyptian calendars: πρὸ τεσσάρων καλανδ̣ῶν Φεβ[ρουα]ρ̣ί̣ω(ν), ὅς ἐστιν Μεχὶρ τετράς. 'On the fourth day before the calends of February, which is the fourth of Mecheir.'
20. Wood tablet. Alexandria, Egypt, 163 CE (*CPL* 157). Birth certificate, dated according to the Julian and Egyptian calendars: *X K(alendas) Dec(embres), mense Ath[y]r die XXV*. 'On the 10th day before the calends of December, the 25th of the month Hathyr.'[89]
21. Papyrus. Philadelphia, Egypt, 169 CE (*BGU* 7.1655). Opening of a Roman will, dated according to the Julian and Egyptian calendars: πρὸ ις Καλανδῶν Μ[α]ρτίων, μηνὸς ['Ελλ]ήνων Μεχεὶρ κ. 'On the 16th day before the calends of March, the 20th day of the Greek month Mecheir.' πρὸ γ Νονῶν 'Ιουνίων, Πα̣ο̣[ῦ]νι … [ἡμ]έρᾳ̣ θ. 'On the 3rd day before the nones of June, the 9th day of Payni.'
22. Wood tablet. Krokodilopolis/Ptolemais Euergetis, Egypt, 170 CE (*AE* 1906, 172 = *CPL* 215). Document regarding inheritance, dated according to the Julian and Egyptian calendars: *III Kal(endas) Octobr(es), mense Phaophi die II*. 'On the 3rd day before the calends of October, the 2nd day of the month Phaophi.'
23. Papyrus. Karanis, Egypt, 189 and 194 CE (*BGU* 1.326). Greek version of a Roman will, dated according to the Julian and Egyptian calendars. There are two different date equivalences. The first runs as follows: πρὸ ιε καλανδῶν Νοεμβρ[ι]ῶν, Ἁθὺρ κα. 'On the 15th day before the calends of November, the 21st of Hathyr.' There is a discrepancy of one month: the 15th day before the calends of November = 18 October; 21 Hathyr = 17 November. The equivalence is correct if we assume that the Egyptian date is correct and that the scribe accidentally wrote November instead of December in the Roman date.[90] The second equation is in agreement with the *Hemerologia*: πρὸ θ καλανδῶν Μαρτιῶν, Μεχεὶρ κζ. 'On the 9th day before the calends of March, the 27th of Mecheir.'
24. Wood tablet. Alexandria, Egypt, 198 CE (*SB* 3.6223 = *CPL* 202 = *AE* 1919, 23). Assignment of a guardian to a woman, dated according to the Julian and

89 There is no discrepancy in these dates, as 163 CE was an Egyptian leap year.
90 Cf. Rowlandson, *Women and Society*, p. 190; nos. 29, 60, 63, 67, 83.

Egyptian calendars: *VIIII Kal(endas) Octobr(es), mense Thoth die XXVI*. 'On the 9th day before the calends of October, the 26th day of the month Thoth.'

25. Papyrus. Heliopolites, Egypt, late second/early third century CE (*P.Flor.* 2.278 = *CPL* 145, col. II). Military letter, dated according to the Julian and Egyptian calendars: *X Kal(endas) Aug(ustas)*, Ἐπεὶφ κθ. 'On the 10th day before the calends of August, 29 Epeiph.'
26. Papyrus. Provenance unknown, late second or early third century CE (*P.Ryl.* 2.92). List of work-people, dated according to the Julian and Egyptian calendars: πρὸ ιδ καλανδῶν Ἰουνῶν, [Πα]χὼν κδ. 'On the 14th day before the calends of June, 24th of Pachon.' πρὸ ια καλανδῶ[ν] Ἰ[ουνίων], Παχὼν κζ. 'On the 11th day before the calends of June, 27th of Pachon.'
27. Wood tablet. Hermopolis, Egypt, 211 or 221 CE (*CPL* 172). Manumission dated according to the Julian and Egyptian calendars: *VII Kal(endas) Augustas, mense Mesore die I*. 'On the 7th day before the calends of August, the 1st of the month Mesore.' The equation is not in agreement with the *Hemerologia*. There is a discrepancy of one day: the 7th day before the calends of August = 26 July; 1 Mesore = 25 July.
28. Papyrus. Apollonopolites Heptakomias, Egypt, 212 CE (*P.Giss.* 40).[91] Edicts of Caracalla, dated according to the Julian and Egyptian calendars: πρὸ ε̣ εἰδῶν Ἰο̣υλίων, Ἐπεὶφ ις. 'On the 5th day before the ides of July, 16th Epeiph.' The equation is not in agreement with the *Hemerologia*. There is a discrepancy of one day: the 5th day before the ides of July = 11 July; 16 Epeiph = 10 July.[92]
29. Papyrus. Oxyrhynchus, Egypt, 224 or 225 CE (*P.Oxy.* 22.2348). Greek version of a Roman will, dated according to the Julian and Egyptian calendars. There are two different date equivalences; in the first one the Egyptian month day is fragmentary and has been restored on the basis of the Julian date: πρὸ ιβ καλανδῶν Α[ὐγ] ο̣ύστων, Ἐπεὶφ κ̣[ζ]. 'On the 12th day before the calends of August, the 27th of Epeiph.' The second equation is not in agreement with the *Hemerologia*: τῇ πρὸ μιᾶς εἰδῶν Ὀκτωβρίαν, Θὼθ ιε. 'On the day before the ides of October, the 15th of Thoth.' The day before the ides of October = 14 October; 15 Thoth = 12 September.

91 = Van Minnen, 'Three edicts of Caracalla?'.

92 Taking into consideration that the Roman date of Caracalla's edicts is confirmed by the Justinian Code as 11 July, Sijpensteijn ('Some remarks', p. 235) and Van Minnen ('Three edicts of Caracalla?', p. 217 n. 9) suggest two different (though equally plausible) explanations for the reasons that may have led to this error. The first attributes it to 'the different ways the Greek scribes used to express the day immediately before one of the three fixed points in a month (…) if the scribe took πρὸ α εἰδῶν Ἰουλίων as an equivalent of *pridie idus Iulias* he would arrive with πρὸ ε εἰδῶν Ἰουλίων at Epeiph 16.' According to Van Minnen, the scribe was 'probably distracted by the date in the next line (16 Mecheir)' and wrote 16 Epeiph instead of 17 Epeiph.

As already pointed out by M. Amelotti (reported in *P.Oxy*) and more recently restated by De Romanis,[93] there appears to be a discrepancy of one month: the equivalence between the two dates is correct if we assume that the scribe wrote October instead of September, whereupon the day before the Ides becomes the 12th.[94]

30. Wood tablet. Oxyrhynchus, Egypt, 242 CE (*CPL* 163) Birth certificate, dated according to the Julian and Egyptian calendars: *mense Phamenoth die XXI, XVI Kal(endas) April(es)*. 'On the 21st day of the month Phamenoth, the 16th day before the calends of April.'

31. Papyrus. Oxyrhynchites, Egypt, 245 CE (*P.Oxy.* 12.1466 = *ChLA* 46.1361 = *CPL* 204). Request for a guard by a woman, dated according to the Julian and Egyptian calendars: Παχὼν κϛ, πρὸ ιβ καλạνδῶν Ἰουνίων. 'On the 26th of Pachon, the 12th day before the calends of June.'

32. Papyrus. Antinoopolis, Egypt, 249 CE (*CPL* 216 and 216 appendix = *SB* 6.9298). Inheritance matter, dated according to the Julian and Egyptian calendars. πρὸ δέκα ὀκτὼ [καλ]ανδῷν Ὀκτωβρίων, Θὼθ ιη. 'On the 18th day before the calends of October, 18th of Thoth.' The equation is not in agreement with the *Hemerologia*. There is a discrepancy of one day: the 18th day before the calends of October = 14 September; 18 Thoth = 15 September.[95]

33. Papyrus. Oxyrhynchus, Egypt, 258 CE (*P.Oxy.* 9.1201). Succession to an inheritance, dated according to the Julian and Egyptian calendars: π[ρ]ὸ η καλ(ανδῶν) Ὀκτωβρίω[ν], Θὼθ κζ. 'On the 8th day before the calends of October, 27 Thoth.'

34. Papyrus. Karanis, Egypt, 302 CE (*P.Cair.Isid.* 41). Receipts for various taxes, dated according to the Julian and Egyptian calendars: α π̣[ρὸ] Καλενδῶν Ἰουνίων τοῦτ' ἔστιν Παῦνι ϛ. 'On the day before the calends of June, that is, the 6th of Payni.'

35. Papyrus. Karanis, Egypt, 309 CE (*P.Cair.Isid.* 9). Report of the *sitologoi*, dated according to the Julian and Egyptian calendars: εἰς τὴν πριδ(ιε) Καλ(ενδας) Δεκεμ(βρες), Χοι[ἀ]κ δ̣. 'Until the day before the calends of December, the 4th of Choiak.'

36. Papyrus. Theadelphia, Egypt, 310 CE (*P.Sakaon* 1 = *ChLA* 19.685). Declaration of persons, dated according to the Julian and Egyptian calendars: πρὸ γ καλανδῶν

93 De Romanis, 'Lysas e il tempo', p. 17.

94 Cf. sources nos. 23, 60, 63, 67, 83.

95 Sijpensteijn ('Some remarks', pp. 236–37) attributes this incongruity to 'the total equation of Roman and Egyptian months', while De Romanis ('Lysas e il tempo', p. 17) assumes that the one-day discrepancy between the Roman and the Egyptian date is due to the attribution of 31 days to a Roman month that had 30 days. According to Sacha Stern (personal communication), the error comes from assimilation of the two dates (18 for both). Cf. no. 49 below.

Μαρτίων, ἥτις ἐστὶν Φαμενὼθ γ. 'On the 3rd day before the calends of March, that is the 3rd of Phamenoth.'

37. Papyrus. Kellis, Egypt, 310 CE (*P.Kellis* 1.40). Loan of money dated according to the Julian and Egyptian calendars: πρὸ δ Εἰδῶν Ἰουλίων ὅ ἐστιν Ἐφὶφ ιη κατ' Ἕλ̣λ̣ην̣(ας). 'On the 4th day before the ides of July, which is 18 Epeiph according to the Greeks.'[96]

38. Papyrus. Theadelphia, Egypt, 312 CE (*P.Flor.* 1.36 = *ChLA* 25.778). Petition dated according to the Julian and Egyptian calendars: *XVI Kal(endas) Septembres*, Μεσορὴ κδ. 'On the 16th day before the calends of September, the 24th of Mesore.'

39. Papyrus. Karanis, Egypt, 315 CE (*P.Cair.Isid.* 74 = *P.Merton* 2.91). Petition dated according to the Julian and Egyptian calendars: *VI Kal(endas) Ianuari(as)*, Χοιὰκ λ. 'On the 6th day before the calends of January, the 30th of Choiak.' The equation is in agreement with the *Hemerologia*, as 315 CE was a leap year.

40. Papyrus. Krokodilopolis/Ptolemais Euergetis, Egypt, 318/320 CE (*P. Ryl.* 653). Judicial proceedings, dated according to the Julian and Egyptian calendars: *die III Nonas Iunias*, Παῦνι θ. 'On the 3rd day before the nones of June, 9 Payni.'

41. Papyrus. Arsinoites, Egypt, 321 CE (*P.Thead.* 13 = *ChLA* 41.1204). Judicial proceedings, dated according to the Julian and Egyptian calendars: *die pridie Idus Dec[em]bres*, Χοιὰκ ις. 'On the day before the ides of December, the 16th of Choiak.'

42. Papyrus. Hermoupolis Magna, Egypt, probably 323 CE (*P.Herm.* 18). Record of official proceedings, dated according to the Julian and Egyptian calendars: πρὸ η ε̣ἰ̣δ̣ῶν Δεκεμβρίων, Χοιὰκ θ. 'On the 8th day before the ides of December, the 9th of Choiak.'

43. Papyrus. Alexandria, Egypt, 350 CE (*P.Abinn.* 63 = *ChLA* 18.661). Legal text, dated according to the Julian and Egyptian calendars: *die Idus Novembr(es)* Ἀθὺρ ιζ. 'On the ides of November, the 17th of Hathyr.'

44. Church Council, Ephesus, Province of Asia, 431 CE (*ACO* 1.1.3, p. 53, l. 9; 1.1.3, p. 53, l. 11; 1.1.3, p. 60, l. 7; 1.1.7, p. 84, l. 33). Four dates given according to the Julian and Egyptian calendars: τῆι πρὸ ἓξ Εἰδῶν Ἰουλίων, ἥτις ἐστὶ κατ' Αἰγυπτίους Ἐπὶφ ⟨ις⟩. 'On the sixth day before the ides of July, which is the 16th of Epeiph according to the Egyptians.' τῆι πρὸ πέντε Εἰδῶν Ἰουλίων, ἥτις ἐστὶ κατ' Αἰγυπτίους Ἐπὶφ ⟨ιζ,⟩. 'On the fifth day before the ides of July, which is the 17th of Epeiph according to the Egyptians.' τῆι πρὸ δέκα Καλανδῶν Ἰουλίων, Παυνὶ εἰκάδι ὀγδόηι κατ' Αἰγυπτίους. 'On the tenth day before the calends of July, which is the 28th of Payni according to the Egyptians.' τῆι πρὸ δεκαμιᾶς Καλανδῶν Αὐγούστων, ἥτις ἐστὶ κατ' Αἰγυπτίους Ἐπιφὶ ⟨κη⟩. 'On the eleventh day before the calends of August, which is the 28th

96 The formula κατ' Ἕλληνας refers to the Alexandrian calendar. Cf. Hagedorn and Worp, 'Wandeljahr', esp. p. 245; Jones, *Astronomical Papyri*, vol. 1, p. 12.

of Epeiph according to the Egyptians.' The equations are in agreement with the *Hemerologia*.

45. Church Council, Chalcedon, Bithynia, 451 CE (*ACO* 2.1.1, p. 77, l. 13; 2.1.1, p. 189, l. 33). Two dates given according to the Julian and Egyptian calendars: τῆι πρὸ ἓξ Εἰδῶν Αὐγούστων, ἥτις ἐστὶν κατ' Αἰγυπτίους Μεσορὶ ⟨ιε⟩. 'On the sixth day before the ides of August, which is the 15th of Mesore according to the Egyptians.' τῆι πρὸ ἕνδεκα Καλανδῶν Αὐγούστων, ἥτις ἐστὶ κατ' Αἰγυπτίους Ἐπιφὶ ⟨κη⟩. 'On the eleventh day before the calends of August, which is the 28th of Epeiph according to the Egyptians.' The equations are in agreement with the *Hemerologia*.

46. Papyrus. Herakleopolis, Egypt, 461 CE (*P.Oxy.* 16.1878 = *ChLA* 47.1408). Report of proceedings, dated according to the Julian and Egyptian calendars: *die Kal(endas) Septembr(es)*, Θὼθ δ. 'On the calends of September, the 4th of Thoth.'

47. Inscription on a marble slab. Atripalda, Italy, 465 CE (*AE* 2008, 338 = *SEG* 58, 1079). Epitaph dated according to the Julian and Egyptian calendars. The inscription is Latin, except for the final part of the dating formula (comprising of indiction and Egyptian month date), which is written in Greek. The equation is not in agreement with the *Hemerologia*. The date of death is noted solely as *IIII idus Februa⟨ria⟩s*, that is, 10 February. The date of deposition is given as *III idus Februarias*, μεχὶρ ιη'. 'On the 3rd day before the ides of February, 18th of Mecheir.' There is a discrepancy of one day, as the Julian date equals 11 February, while 18 Mecheir ought to correspond to 12 February.[97]

48. Papyrus. Oxyrhynchus, Egypt, c. 480 CE (*P.Oxy.* 16.1876 = *ChLA* 47.1406). Report of proceedings for debt, dated according to the Julian and Egyptian calendars: *ḍịe pridie Idus Novembr(es)* Ἁθῦρ ις. 'On the day before the ides of November, the 16th of Hathyr.'

49. Papyrus. Oxyrhynchus, Egypt, 497 CE (*P.Oxy.* 16.1982). Receipt dated according to the Julian and Egyptian calendars: [Φα]ῶφι δ, *octombrio* δ. 'On the 4th of Phaophi, the 4th of October.' The equation is not in agreement with the *Hemerologia*. 4 Phaophi corresponds to 1 October.[98]

---

97 Perrin wonders whether the Egyptian date refers to the date on which the inscription was engraved; Perrin presumably assumes that the engraver only intended to record the date of deposition, in the Julian and Egyptian calendars, but when writing the Egyptian date, he mistakenly entered the date of engraving, which is a plausible explanation for the one-day discrepancy; M.Y. Perrin, in *AE* 2008, 338. Solin notes that the whole text, including the Egyptian date, was inscribed by the same letter-cutter. He also points out that there are no signs that the deceased had eastern origins; Solin, *San Modestino*, p. 225. The Egyptian date remains elusive. Whatever the reason for the presence of the Egyptian date, the one-day discrepancy between the Julian and the Egyptian date is likely to be a simple error in the conversion of the dates.

98 According to Sijpesteijn, this text represents 'the best documentary evidence for the total equation of Roman months with Egyptian months'; Sijpesteijn, 'Some remarks', p. 236.

1.2 Calendar of Provincia Arabia

50. Papyrus. Maoza, Province of Arabia, 125 CE (*P.Yadin* 1.14). Summons dated according to the Julian calendar and the calendar of Provincia Arabia: πρὸ τεσσάρων εἰδῶν Ὀκτωβρίω[ν, κατὰ] δὲ τὸν ἀριθμὸν τῆς [ἐπαρχείας Ἀραβίας ἔτους εἰκοστοῦ] μηνὸς Ὑπερβερεταίου λεγ[ομένου Θεσρεὶ τετάρτῃ καὶ εἰ]κας. 'On the fourth day before the ides of October, and according to the compute of the province of Arabia year twentieth, on the twenty-fourth of the month Hyperberetaios called Thesrei'. The equation is not in agreement with the *Hemerologia*. There is a discrepancy of one day: the 4th day before the ides of October = 12 October; 24 Hyperberetaios = 11 October.[99]

51. Papyrus. Maoza, Province of Arabia, 125 CE (*P.Yadin* 1.15). Deposition dated according to the Julian calendar and the calendar of Provincia Arabia: πρὸ τεσσ[άρων εἰ]δῶν Ὀκτωβρίων, κατὰ δὲ τὸν ἀριθμὸν τῆς ἐπαρχείας Ἀραβίας ἔτους εἰκοστοῦ μηνὸς Ὑπερβερεταίου λεγομένου Θεσρεὶ τετάρτῃ καὶ εἰκάς. 'On the fourth day before the ides of October, and according to the compute of the province of Arabia year twentieth, on the twenty-fourth of the month Hyperberetaios called Thesrei'. This document was written by the same scribe immediately after *P.Yadin* 1.14 (= no. 50), and presents the same date equation (4th day before the ides of October and 24 Hyperberetaios), which is not in agreement with the *Hemerologia*.

52. Papyrus. Rabbath Moaba, Province of Arabia, 127 CE (*PXHev/Se* 62). Land declaration, dated according to the Julian calendar and the calendar of Provincia Arabia: πρὸ τριῶν ε[ἰ]δῶν Δεκεμβρίων, κατὰ δὲ τὸν τῆς ν[έας] ἐπαρχείας Ἀραβίας ἀριθμὸν ἔτους δευτέρου εἰκοστοῦ μηνὸς Ἀπελλαίου ὀκτοκαιδεκά[τῃ]. 'On the third day before the ides of December, and according to the computation of the new province of Arabia year twenty-second, on the eighteenth day of the month Apellaios'. The equation is not in agreement with the *Hemerologia*. There is a discrepancy of seven days: the 3rd day before the ides of December = 11 December; 18 Appellaios = 4 December.[100]

Contra Sacha Stern (personal communication) who observes that there is far too little evidence to claim that there ever was a total equation between the two calendars. In his opinion, it is far more likely that this is an error due to assimilation. Cf. no. 32 above.

99 Sijpesteijn and Lewis agree that the discrepancy in these dates may be ascribed to the scribes in the new province Arabia having difficulties with the relatively new computational system of the Romans; Sijpesteijn, 'Some remarks', p. 240; Lewis, *P.Yadin* 1.14.

100 According to Lewis, the only conclusion that can be drawn from the date inconsistency here and in a few further documents from the same dossier is that 'more than twenty years after Rome's annexation of the province of Arabia local scribes were still having trouble equating the months and days of the old (Macedonian) and the new (Roman) calendar'; Lewis, 'A Jewish landowner'. Sacha Stern (personal communication) suggests that 18 Appellaios could very well be a lunar date. The pre-106 CE Nabatean calendar was

53. Papyrus. Rabbath Moaba, Province of Arabia, 127 CE (*P.Yadin* 1.16). Registration of land, dated according to the Julian calendar and the calendar of Provincia Arabia: πρὸ τεσσάρων νωνῶν Δεκεμβρίων, κατὰ δὲ τὸν τῆς νέας ἐπαρχείας Ἀραβίας ἀριθμὸν ἔτους δευτέρου εἰκοστοῦ μηνὸς Ἀπελλαίου ἑκκαιδεκάτῃ. 'On the fourth day before the nones of December, and according to the computation of the province of Arabia year twenty-second, on the sixteenth of the month Apellaios.'

54. Papyrus. Maoza, Province of Arabia, 128 CE (*P.Yadin* 1.17). Deposit dated according to the Julian calendar and the calendar of Provincia Arabia: πρὸ δέκα καλανδῶν [illegible] 'On the tenth day before the calends of March, and by the compute of the new province of Arabia year twenty-second, on the sixth of Dystros.'

55. Papyrus. Maoza, Province of Arabia, 128 CE (*P.Yadin* 1.19). Deposit dated according to the Julian calendar and the calendar of Provincia Arabia: π̣ρ̣[ὸ ἑκ]κ̣α̣ί̣δ̣ε̣κ̣α̣ κ̣α̣λ̣α̣ν̣δ̣ῶ̣ν̣ Μαίων, κ̣α̣τ̣ὰ̣ [τὸν] ἀ̣ρ̣[ιθ]μ̣ὸ̣ν̣ [τῆς] ν̣έ̣α̣ς̣ ἐ̣π̣α̣ρ̣[χείας ἔτους τρίτου εἰ]κ̣[οστο]ῦ̣ Ξ̣α̣νδικοῦ ἕκ[τ]η καὶ εἰκάδ[ι]. 'On the sixteenth day before the calends of May, and by the compute of the new province of Arabia year twenty-third, on the twenty-sixth of Xandikos.'

56. Papyrus. Maoza, Province of Arabia, 129 CE (*PXHev/Se* 64). Deed of gift, dated according to the Julian calendar and the calendar of Provincia Arabia: πρὸ π̣έ̣ν̣τ̣ε̣ εἰδῶν [Νοουεμβρίων κατὰ τὸν ἀριθμὸν τῆς νέας ἐπαρχείας Ἀραβία]ς ἔτους τετάρτου καὶ εἰκοστοῦ Δ̣ε̣ί̣ου κ. 'On the fifth day before the ides of November, and according to the computation of the new province of Arabia year twenty-fourth, on the twentieth of Dios.'. The equation is not in agreement with the *Hemerologia*. There is a discrepancy of three days: the 5th day before the ides of November = 9 November; 20 Dios = 6 November.[101]

57. Papyrus. Maoza, Province of Arabia, 130 CE (*P.Yadin* 1.20). Concession of rights, dated according to the Julian calendar and the calendar of Provincia Arabia: π̣ρ̣ὸ̣ ιγ καλανδῶν Ἰουλίων, κατὰ τὸ⟨ν⟩ ἀριθμὸν τῆς νέας ἐπαρχίας Ἀραβίας ἔτους πέμτου καὶ εἰκοστοῦ Δαισίου λ. 'On the 13th day before the calends of July, and by the compute of the new province of Arabia year twenty-fifth, on the 30th of Daisios.'

58. Papyrus. Maoza, Province of Arabia, 130 CE (*P.Yadin* 1.21). Purchase of a date crop, dated according to the Julian calendar and the calendar of Provincia Arabia: πρὸ τριῶν εἰδῶν Σεπτεμβρίων, κατὰ τὸ⟨ν⟩ ἀριθμὸν τῆς νέας ἐπαρχίας Ἀραβίας ἔτους πέμτου καὶ εἰκοσ̣τ̣οῦ Γορπιαίου τετάρτῃ καὶ εἰκάς. 'On the third day before

lunar; if this lunar calendar was still in use in 127 CE, the scribe might have confused it with the new calendar of Provincia Arabia.

101 Regarding the reading Δ̣ε̣ί̣ου κ, Cotton (ad *PXHev/Se* 64) points out 'there may not be enough room after the kappa for another letter', which confirms that we have to assume that there is a discrepancy of three days between the Julian and the local calendars.

the ides of September, and by the compute of the new province of Arabia year twenty-fifth, on twenty-fourth of Gorpiaios.'

59. Papyrus. Maoza, Province of Arabia, 130 CE (*P.Yadin* 1.22). Sale of a date crop, dated according to the Julian calendar and the calendar of Provincia Arabia: πρὸ τριῶν εἰ̣δ̣[ῶν Σεπτεμβρί]ων, κατὰ τὸ⟨ν⟩ ἀριθμὸν τῆς νέας ἐπαρχίας Ἀ[ρ]α̣β̣[ί]α̣ς [ἔτους πέμ⟨π⟩του] κ̣α̣ὶ εἰκοστοῦ Γορπιαίου τετάρτῃ καὶ κας. 'On the third day before the ides of September, and by the compute of the new province of Arabia year twenty-fifth, on twenty-fourth of Gorpiaios.'

60. Papyrus. Maoza, Province of Arabia, 130 CE (*P.Yadin* 1.23). Summons dated according to the Julian calendar and the calendar of Provincia Arabia: π]ρ̣ὸ̣ δ̣ε̣κ̣α̣πέντε καλανδῶν Δεκενβρίω̣ν̣ Δ̣[ίο]υ πρώτῃ. 'On the fifteenth day before the calends of December, on the first of Dios.' The equation is not in agreement with the *Hemerologia*: the 15th day before the calends of December (17 November) does not correspond to 1 Dios but to the 1st of the next month, Apellaios.[102]

61. Papyrus. Maoza, Province of Arabia, 131 CE (*PXHev/Se* 65 = *P.Yadin* 1.37). Marriage contract, dated according to the Julian calendar and the calendar of Provincia Arabia: πρὸ ἑπτὰ εἰ̣δ̣[ῶν Αὐγούστων, κατὰ δὲ] τὸν τῆς [νέ]ας ἐ̣παρ̣χ̣είας Ἀραβίας ἀριθμ̣ὸν ἔτο[υς ἕκτου καὶ εἰκοστοῦ μην]ὸς Λῴ[ο]υ̣ ἐ̣ν̣ν̣ε̣ακαιδεκ̣ά̣[τῃ. 'On the seventh day before the ides of August, and according to the computation of the new province of Arabia year twenty-sixth, on the nineteenth of the month Loos'.

62. Papyrus. Maoza, Province of Arabia, 132 CE (*P.Yadin* 1.27). Receipt dated according to the Julian calendar and the calendar of Provincia Arabia: [π]ρ̣ὸ̣ δεκατεσσάρων καλανδῶν Σεπτεμ̣[β]ρίων, κατὰ τὸν τῆς νέας ἐ̣[πα]ρ̣χ̣ί̣ας Ἀραβίας ἀριθμὸν ἔτους ἑβδόμου εἰ̣κ̣οστο̣[ῦ] μ̣ηνὸς Γορπιαίου πρώτ̣[ῃ. 'On the fourteenth day before the calends of September, and by the compute of the new province of Arabia year twenty-seventh, on the first of the month Gorpiaios.'

63. Papyrus. Maoza, Province of Arabia (written), 132 CE (*P.Yadin* 35). Summons dated according to the Julian calendar and the calendar of Provincia Arabia. The equation is not in agreement with the *Hemerologia*. The text is fragmentary but there are legible traces of the Roman month September (which can also indicate a date in August) and of the month Panemos of the calendar of Provincia Arabia. However, neither August nor September coincide with Panemos, which ran from 20 June to 19 July.[103]

64. Papyrus. Oxyrhynchus, Egypt, c. 265 CE (*P.Oxy.* 42.3054). Registration of sale of a slave, dated according to the Egyptian calendar and the calendar of Provincia

102 Lewis is certain that this is a *lapsus calami* or *memoriae*; Lewis, *P.Yadin* 1.23. Cf. nos. 23, 29, 63, 67, 83.

103 Cf. nos. 23, 29, 60, 67, 83: an explanation similar to the one suggested in those cases might apply to this one as well.

Arabia. While the month-days are missing, the concordance between the 'Arabian' (Loos) and the Egyptian (Payni) months is not in agreement with the *Hemerologia*.

65. Papyrus. Petra, Palaestina Tertia, 538 CE (*P.Petra* 1.3). Request for transfer of taxation, dated according to the Julian calendar and the calendar of Provincia Arabia: πρὸ ἐννέα Καλανδῶν Σεπ̣τεμβρίων, μηνὸς Γορπιαίου̣ ἕκτῃ. 'On the ninth day before the calends of September, the sixth of Gorpiaios.'

66. Papyrus. Petra, Palaestina Tertia, 544 CE (*P.Petra* 3.23). Request for transfer of [illegible] Arabia: π̣ρ̣ὸ δεκαὲξ καλανδῶν Σεπτεμβρίων, μηνὸς Λῴου εἰκ[ο]στῇ ἐ̣ν̣άτῃ. 'On the sixteenth day before the calends of September, the twenty-ninth of the month Loos.'

67. Papyrus. Petra, Palaestina Tertia, 559 CE (*P.Petra* 3.25). Request for transfer of taxation, dated according to the Julian calendar and the calendar of Provincia Arabia: πρὸ τρειῶ̣ν̣ καλανδῶν Ἰανουαρίων, [μ]ηνὸς Περ̣ι̣τίου πε̣[ν]τ̣εκ̣αιδ̣εκάτῃ. 'On the third day before the calends of January, the fifteenth of the month Peritios.' The equation is not in agreement with the *Hemerologia*: the 3rd day before the calends of January = 30 December; 15 Peritios = 30 January. As in a few other cases,[104] there is a one-month discrepancy. It is plausible to assume that the scribe confused the Roman months January and February: in this case, the correct Julian date would be the 3rd day before the calends of February.

68. Inscription on a marble slab. Avdat, Province of Palaestina Tertia, 576 CE (*SEG* 28, 1395 = *SEG* 31, 1400). Epitaph dated according to the calendars of Gaza and Elousa (= calendar of Provincia Arabia): τῇ κατὰ Γάζ(αν) μη(νὶ) Ἀπελλαίῳ κδ', κατὰ δὲ Ἐλούσην Αὐδοναίῳ δ'. 'According to (the calendar of) Gaza, on the 24th of the month Appellaios, while according to (the calendar of) Elousa, on the 4th of Audynaios'.[105]

69. Papyrus. Elousa, Palaestina/Judaea, 590 CE (*P.Ness.* 3.29) Summons dated according to the Julian calendar and the calendar of Bostra (= the calendar of Provincia Arabia): πρὸ δέκα καλανδῶν Ἰανουαρίων, μηνὸς Αὐδναίου ἑβδό]μ̣ῃ. 'On the tenth day before the calends of January, the seventh of Audynaois.'

70. Papyrus. Nessana, Palaestina/Judaea, 596 CE (*P.Ness.* 3.30). Inheritance dated according to the Julian calendar and the calendar of Provincia Arabia: Εἴδαις Σεπτεμβρίαις μηνὸς Γορπιαίου εἰκάδι ἕκτῃ. 'On the ides of September, the twenty-sixth of Gorpiaios.'

104 See previous note.

105 On the dates see Meimaris, *Chronological Systems*, p. 254 n. 352.

1.3 Calendar of the Province of Asia

71. Inscription on a marble cinerary chest. Sardis, Lydia, early first century CE?[106] (*AE* 1995, 1455 = *SEG* 45, 1652). Epitaph dated according to the calendar of Asia and the Julian calendar: μη(νὸς) Ξανδικοῦ η' Καλ(άνδαις) Μαρτί(αις). 'On the 8th of the month Xandikos, on the calends of March.'

72. Inscription on a marble stele. Metropolis, Lydia, 14–37 CE (*AE* 1999, 1538 = *SEG* 49, 1523). Epigraphic '*hemerologion*' equating dates of the Julian calendar and of the calendar of Asia (cf. full discussion of this document above, pp. 91–3). The surviving equivalences are as follows: 7 October = 14 (Dios), 5 November = 13 (Apellaios), 5 December = 12 (Audynaios), 5 January = 12 (Peritios).[107]

73. Inscription on a marble block. Acmonia, Phrygia, 64 CE (*AE* 2006, 1427 = *SEG* 56, 1489). Honorific decree dated according to the Julian calendar and the calendar of Asia: πρὸ δεκάπεντε Καλάνδων Ὀκτοβρείων, μηνὸς Ὑπερβερεταίου ϛ' ἀπίοντος. 'On the fifteenth day before the calends of October, on day 6 ἀπίοντος of the month Hyperberetaios.'[108]

74. Inscription on a marble block. Acmonia, Phrygia, 68 CE (*AE* 2006, 1426 = *SEG* 56, 1490). Honorific decree dated according to the Julian calendar and the calendar of Asia: πρὸ ἓξ εἰδῶν Ἀπρειλίων, μηνὸς Ἀρτεμεισίου ἑπτακαιδεκάτῃ. 'On the sixth day before the ides of April, the seventeenth of the month Artemisios.' The equation is not in agreement with the *Hemerologia*. There is a discrepancy of one day: the 6th day before the ides of April = 8 April; 17 Artemisios = 9 April.[109]

75. Inscription on a marble block. Acmonia, Phrygia, 85 CE (*IGR* 4.661; *SEG* 13, 542 = *MAMA* VI List 148, 159). Decree concerning a testament, dated according

---

106 In the *editio princeps*, P. Herrmann dates the inscription to 9 BCE or slightly later on palaeographical grounds and on the basis of the alleged date of introduction of the calendar of Asia in the province. This dating, however, is problematic, not only because, as established above, the calendar of Asia was most likely instituted in 8 BCE, but also because the Julian calendar at this point in time was still in a state of disruption. The Julian calendar as we know it only begins from 1 CE, following Augustus' reforms starting in 8 BCE (cf. Stern, *Calendars in Antiquity*, table 5.4). On the basis of these circumstances, and given that palaeography is in itself an uncertain criterion, it appears more likely that the inscription was engraved in the early first century CE. This supposed dating makes the epitaph from Sardis the earliest extant case of a double date equating the Julian and the newly established Asian calendar.

107 The names of the months in the local calendar do not appear on the stone; they are presumably implicit.

108 It must be observed that the *ed.pr.* gives an incorrect reading of a part of the dating formula. This has been rectified by G. Petzl in *SEG*. Cf. Thonemann, 'Calendar', p. 127 n. 22.

109 Even if this was simply an error by the letter-cutter or by whoever drafted the text to be inscribed on the marble block, which might well be the case, there is no reason to emend the text of the inscription, as G. Petzl suggests in *SEG* 56, 1490. Cf. Thonemann, 'Calendar', p. 127 n. 23.

to the Julian calendar and the calendar of Asia: πρὸ τριῶν Νωνῶν Μαρτίων, μηνὸς Ξανδικοῦ τρισκαι[δεκάτου]. 'On the third day before the nones of March, the thirteenth of the month Xandikos.' The equation is not in agreement with the *Hemerologia*. There is a discrepancy of one day: the third day before the nones of March = 5 March; 13 Xandikos = 6 March.[110]

76. Inscription on a limestone stele. Kaunos, Caria, 111 CE (*IKaunos* 34). Approval of a private foundation, dated according to the Julian and Kaunian calendars (= calendar of Asia): [πρὸ ... |κα]λ(ανδῶν) Σεβάστου ἢ δ' ἀπίοντος Κορύμ[βου].[111] 'On [illegible]' As observed by Thonemann,[112] Sebastos here must be a literal translation into Greek of the Roman month-name Augustus; on this basis, if we assume that Korymbos corresponded to Loos in the calendar of Asia, the Kaunian date (the 4th day of waning Korymbos) ought to correspond to the 13th day before the calends of August, i.e., according to the *Hemerologia*, 20 July.[113]

77. Inscription on a marble slab. Pergamon, Mysia, 129/138 CE (*IGR* 4.353 = *IPergamon* 2.374). Calendar of rituals to be performed by the *hymnodoi* of Augustus and Roma. The following correspondences between the calendar of Asia and the Julian calendar/Roman festivities are given: 'first day of the month Kaisar = birthday of Augustus (= 23 September); month Peritios = calends of January; penultimate day of the month Hyperberetaios = birthday of the *Augusta* (i.e. Livia, 21 September).'[114]

78. Inscription on a marble base. Amastris, Province of Bithynia-Pontos, 155 CE (*SEG* 35, 1327 = *SEG* 40, 1163).[115] Funerary epigram, dated according to the Julian and the Amastrian calendar (= calendar of Asia):[116] πρὸ α' καλ(ανδῶν) Σεπτεμβρίων, Λώου ζι'. 'On the day before the calends of September, on 17th Loos.' The equation is not in agreement with the *Hemerologia*. There seems to be a discrepancy

---

110 Laffi, 'Le iscrizioni', p. 80, and Thonemann, 'Calendar', p. 127 (T8) suggest that this discrepancy can be explained by assuming that at Acmonia 31-day months were counted 'Sebaste, day 2, day 3, etc.', instead of 'Sebaste, day 1, day 2, etc.', as is the case in other places of the province of Asia.

111 Though the date equation is fragmentary, it is included in this list in view of the significance of this source, given the scarcity of evidence relating to the Kaunian calendar.

112 Thonemann, 'Calendar', p. 127 (T10).

113 It is unclear why Thonemann (ibid.) gives 19 July as the corresponding date.

114 Cf. Laffi, 'Le iscrizioni', p. 76; Thonemann, 'Calendar', p. 127 (T11), and 130–31. According to the *Hemerologia*, the calends of January fell on 8 Peritios in the calendar of Asia, and not on the 9th of that month, as Thonemann states.

115 = Marek, *Stadt, Ära und Territorium*, pp. 168–69, Kat. Amastris no. 44.

116 One wonders why the calendar used here is that of the province of Asia rather than the calendar of Bithynia. On the calendar in use in the Province of Bithynia-Pontos, see Kubitschek, *Kalenderbücher*, pp. 97–99.

of seven weeks: the day before the calends of September = 31 August; 17 Loos = 10 July.[117]

79. Inscription on a marble sarcophagus. Metropolis, Lydia, 168 CE (*AE* 1995, 1469 = *SEG* 45, 1598). Instructions for a burial place, dated according to the Julian calendar and the calendar of Smyrna (= calendar of Asia):[118] πρὸ ἓξ εἰδῶν Ἰουνίων, μη(νὸς) Στρατον[ικ]εῶνος ἑκκαιδεκάτῃ. 'On the sixth day before the ides of June, on the sixteenth of the month Stratonikeon.'

80. Inscription on a marble slab. Laodikeia on the Lykos, Phrygia, third century CE (*MAMA* 6.18 = *ILaodikeia* 85). Epitaph, in which the annual crowning of the tomb is set to occur on a date given according to the Julian and Laodikean (= Asiatic) calendars: τῇ πρὸ α' καλανδῶν Νοεμβρίων, μη(νὸς) γ' ιη'. 'On the day before the calends of November, the 18th of the 3rd month.' The equation is not in agreement with the *Hemerologia*, where 31 October corresponds to 8 Apellaios or the 8th of month 2 in Asian calendars.[119]

81. Anonymous, *In sanctum pascha*, 387 CE.[120] The date of the Epiphany (6 January) is given according to the calendar of Asia: ἡμέρᾳ ὡρισμένῃ τρισκαιδεκάτῃ τετάρτου μηνὸς κατὰ Ἀσιανούς. 'On the thirteenth day of the fourth month according to the Asians.' Easter falls: μηνὸς ἑβδόμου κατὰ Ἀσιανούς. 'In the seventh month [= Artemision] according to the Asians'.[121]

82. Alexander Monachus, *Laudatio Barnabae*, second half of the sixth century CE.[122] The alleged date on which Barnabas was martyred is given according to the Julian, eastern Cypriot (of Salamis), and Asian/Paphian calendars: κατὰ μὲν Ῥωμαίους τῇ πρὸ τριῶν εἰδῶν Ἰουνίων, κατὰ δὲ Κυπρίους Κωνσταντιεῖς μηνὶ Μεσωρὶ τοῦ καὶ δεκάτου ἑνδεκάτῃ, κατὰ δὲ Ἀσιανοὺς ἤτοι κατὰ Παφίους μηνὶ Πληθυπάτῳ τοῦ καὶ ἐννάτου ἐννεακαιδεκάτῃ. 'According to the Romans on the third day before the ides of June, according to the Cyprians from Constantia [= Salamis], on the

117 Thonemann assumes that this incongruity is due to the fact that in the mid-second century CE the Amastrians employed a lunisolar calendar; 'Calendar', p. 130. According to Sacha Stern (personal communication), 17 is really excessive for a lunar date. The latest possible lunar date on 31 August 155 CE would be 16, and earlier dates (15 or 14) would be more likely.

118 This is a copy of a document that was deposited in the Museion of Smyrna.

119 Perhaps ιη (18) is a scribal error for η (8). Thonemann's idea that the calendar used at Laodikeia in the third century CE was lunar or lunisolar does not fit with the fact that the correspondence of 31 October with 18 Month 3 was expected to recur on an annual basis, which implies that the local calendar was solar and in synchronism with the Julian calendar; Thonemann, 'Calendar', p. 130. Cf. also Laffi, 'Le iscrizioni', pp. 79–80.

120 Floëri and Nautin, *Homélies pascales*, p. 113 (ch. 3); p. 119 (ch. 9).

121 Cf. Laffi, 'Le iscrizioni', p. 78; Thonemann, 'Calendar', p. 128 (T16).

122 van Deun, *Hagiographica Cypria*, p. 120 lines 846–50.

eleventh day of the tenth month Mesore, and according to the Asians or the Paphians, on the nineteenth of the ninth month Plethypatos.'[123]

83. *Martyrium prius Andreae*, eighth century CE.[124] The date on which Andrew was martyred is given according to the calendar of Asia and the Julian calendar: κατὰ Ἀσιανοὺς μηνὸς Περιτ(ίου) ἕκτῃ, κατὰ δὲ Ῥωμαίους μηνὶ Νοεμβρίῳ λʹ. 'According to the Asians, on the sixth of the month Peritios, according to the Romans, on the 30th of November.' This apparent equation is not in agreement with the *Hemerologia*, which equal 6 Peritios to 29 December, rather than 30 November.[125]

1.4 Calendar of Antioch

84. Papyrus. Dura-Europos, Syria, 232 CE (*P.Dura* 30). Marriage contract, dated according to the Julian calendar and the calendar of Antioch: Καλάνδαι Ὀκτώβριαι, μηνὸς [Ὑ]π̣ε̣ρ̣[β]ε̣ρεται̣ί̣ο̣υ ν̣ο̣υ̣μηνίᾳ̣. 'On the calends of October, on the first day of the month Hyperberetaios.'

85. Leather strip. Appadana, Syria Mesopotamia (found), Markopolis, Osrhoene (written), 249 CE (*SB* 24 16167 = *P.Euphrates* 6). Sale of a slave, dated according to the Julian calendar and the calendar of Antioch: πρὸ ὀκτὼ Εἰδῶν Νοεμ[βρ]ί̣ων ἥτ̣ις ἐστὶν μηνὸς Δ̣[ί]ου ἕκτῃ. 'On the eighth day before the ides of November, that is, on the sixth of the month Dios.'

86. Papyrus. Dura-Europos, Syria, 251 CE (*P.Dura* 29). Contract dated according to the Julian calendar and the calendar of Antioch: πρὸ ϛʹ Νωνῶν Ὀκτωβρ̣[ί]ῳ̣ν̣, μηνὸς Ὑπερβερεταίου δ̣ευτ̣έ̣ρᾳ̣. 'On the 6th day before the nones of October, on the second day of the month Hyperberetaios.'

87. Papyrus. Dura-Europos, Syria, 254 CE (*P.Dura* 32).[126] Divorce dated according to the Julian calendar and the calendar of Antioch: μηνὸς Ξα̣[ν]θικοῦ ε̣ἰκασενά̣[τῃ], [πρὸ] δύο Καλανδῶν [Μ]αείω. 'On the 29th of the month Xandikos, the day before the calends of May.' The equation is not in agreement with the *Hemerologia*. There is a discrepancy of one day: 29 Xandikos = 29 April; the day before the calends of May = 30 April.[127]

---

123 Cf. Laffi, 'Le iscrizioni', p. 78; S. Stern, 'A "Jewish" birth record'; Thonemann, 'Calendar', p. 129 (T22).

124 Prieur, *Acta Andreae*, vol. 2, p. 703 (ch. 19).

125 Thonemann, 'Calendar', p. 129 (T24) observes that there is some confusion in the manuscript tradition of the *Martyrium prius Andreae*, with one manuscript giving 'month of Audynaios' instead of '6 Peritios' as the equivalent of 30 November. The Asian month Audynaios did partly overlap with the Julian November. Cf. sources nos. 23, 29, 60, 63, 67.

126 Cf. Feissel, Gascou, and Teixidor, 'Documents', p. 54.

127 The formula πρὸ δύο is odd (its only other occurrence is in *P.Cair.Isid.* 87, also dated πρὸ δύο Καλανδῶν Μαίων); the Greek formula normally used to translate the Roman *pridie*, 'the day before', is πρὸ μιᾶς. In the *Hemerologia*, the last day of the month as well as the day before a fixed date in the Julian calendar are consistently expressed as α; β

1.5 Calendar of Gaza

88. Papyrus. Petra, Palaestina Tertia (found), Gaza, Palaestina Prima (written), 538 CE (*P.Petra* 1.2). Agreement concerning inherited property, dated according to the Julian and Gaza calendars:[128] πρὸ ἓξ εἰδῶν Μαειων, μηνὸς [Ἀ]ρτεμισίου πεν[τ]εκαιδεκάτ[η]. 'On the sixth day before the ides of May, the fifteenth of the month Artemisios.'

1.6 Calendar Used in Provincia Judaea

89. Papyrus. Murabba'at, province of Judaea, 124 CE (*SB* 12.10305 = *P.Mur.* 2.115). Marriage agreement, dated according to the Julian calendar and the calendar in use in the province of Judaea/Palaestina: πρὸ ιδ κ(αλανδῶν) Νοεμβρίων, Δύστρου ιε. 'On the 14th day before the calends of November, the 15th of Dystros.' The equation is not in agreement with the *Hemerologia*: the Julian date corresponds to 19 October; the Macedonian month of Dystros normally corresponded to February/March. Even if we take into consideration the different placement of months in different calendars, whereby Dystros could fall on December/January (in Egypt), May (Sidon), February or March, it does not seem possible to make it coincide with October.[129]

2 *Month Correspondences and Month-Name Lists*

90. *Scholia vetera in Aratum*, second/third century CE.[130] Concordances among Egyptian and Julian months.
91. Inscription on a stone tablet. Nineveh? Seleukia? Third century CE (*IK Estremo oriente* 68). It lists the Macedonian months in the following order: Audynaois, Peritios, Dystros, Xandikos, Artemisios, Daisios, Panemos, Loos, Gorpiaios, Apellaios, Dios, Hyperberetaios. The list seems to follow the order of months in the calendar of Antioch (starting with Audynaios = January); however, the order of the last three months does not match that sequence: Gorpiaios should

is in fact omitted throughout, the penultimate day of the month being expressed as γ. Sijpesteijn suggests that πρὸ δύο in *P.Dura* 32 could in fact refer to 29 April; Sijpesteijn, 'Some remarks', p. 235. It is indeed plausible to assume some disagreement or confusion on the part of the scribes on how to translate into Greek this particular Julian date. If we follow this argument, the double date appearing on this document is in agreement with the *Hemerologia*.

128 On the calendar of Gaza, see Meimaris, *Chronological Systems*, p. 119.

129 Even assuming that Δύστρου is an error for Δίου, which in the calendar of Asia overlaps partly with the Julian month of October, the date equation would be incorrect (19 October = 26 Dios).

130 Martin, *Scholia in Aratum vetera, Schol.* nos. 265, 287, 300, 315, 332, 409, 443, 462, 513.

be followed by Hyperberetaios, Dios, and Apellaios.[131] Interestingly, the same, 'incorrect' order of months appears in another source (see below, no. 93).

92. Inscription on marble wall of public building (*Bouleuterion*). Stratonikeia, Caria, Imperial period (*IStratonikeia* 1044).[132] Verse inscription meant to act as an aid to memorise the lengths of the Julian months.[133] The sequence begins with Dios (= approx. October) and thus appears to correspond to the months order in the calendar of the province of Asia.[134]

93. Inscription on a bronze tessera. Provenance unknown, now in Hungary, third/ [illegible] ical formulae, which are followed by a list of Macedonian month-names; the order is the same as in no. 91, i.e. it seems to correspond to the months sequence in the calendar of Antioch, except for Hyperberetaios and Apellaios, which are swapped. Should we conclude that these two sources attest to the existence of a calendar that presented this particular order of the Macedonian month-names, instead of assuming an error in the sequence?

94. Papyrus. Fayum, Egypt, fourth century CE (*Fay.* 135 V = *C. Gloss. Biling.* 1 11) Concordance of Julian and Egyptian month-names.

95. Athanasius of Alexandria, *De synodis Arimini in Italia et Seleuciae in Isauria*, 359 CE.[135] Concordance of Julian (September), Egyptian (Thoth), and Macedonian (Gorpaios) months.

96. Inscription on a copper sphere. Chevroches, France, second half of the fourth century CE (*AE* 2006, 822 = *SEG* 56, 1168). Astrological inscription including Egyptian months, zodiac signs, and Julian months. The order followed is that of the Egyptian year, from Thoth (September) to Mesore (August).

97. Ostracon. Provenance unknown, fourth/fifth century CE (*SB* 26 16521). Concordance of Julian and Egyptian month-names.

98. *Fasti* or *Laterculus Polemii Silvii* (written in Gaul), 448/9 CE (*IIt* 13.2.43).[136] Preserved in a single, twelfth-century manuscript, the calendar offers useful

---

131 In *IK Estremo oriente*, Canali De Rossi assumes that the letter-cutter may have got confused by the omission of Hyperberetaios, which he added on the left hand margin. Canali De Rossi refers to no. 13 in *IK Estremo oriente*, a rock-cut inscription from Armaouira (Armenia) dating to the early second century BCE. In this case the Macedonian months are listed in the order of the original Macedonian calendar (beginning the year with Dios = October).

132 = Merkelbach and Stauber, *Steinepigramme*, no. 02/06/20.

133 Each line has as many letters as the number of days of each of the twelve months of the Julian year.

134 Cf. Thonemann, 'Calendar', p. 126 (T1).

135 Opitz, *Athanasius Werke*, vol. 2.1, pp. 231–78 (ch. 12.1).

136 Cf. Salzman, *On Roman Time*, pp. 242–46; Lehoux, *Astronomy*, pp. 161–62, 311, 324; Paniagua, 'New perspectives'; id., *Polemii Silvii Latercvlvs*; Weidemann and Weidemann, *Römische Staatskalender*.

information on every month, including festivity days, the due ephemerids, and the weather forecast. Each month is preceded by an introduction that provides the etymology of that month's name and the denominations of the month in different calendars: Julian, Hebrew, Egyptian, Athenian, and Greek (= Macedonian month-names).[137]

99. Papyrus. Lycopolis, Lower Egypt, fifth century CE (*P.Acad. inv.* 1 r°). Concordance of Egyptian, Julian, and Macedonian month-names. As argued above (p. 95), the calendar labelled as 'according to the Asians' (κατὰ Ἀσιανούς) probably corresponds to the calendar marked as 'of the Hellenes' (Ἑλλήνων) on the *Hemerologia*, which identifies with the calendar of Antioch. If that were the case, the month concordances would be in agreement with the *Hemerologia*.

100. Papyrus. Provenance unknown, sixth/seventh century CE (*P. Iand.* inv. 654). Concordance of Julian and Egyptian month-names (cf. above n. 46).

101. Parchment. Provenance unknown, sixth/seventh century CE.[138] Concordance of Egyptian and Bithynian month-names. The order of the Bithynian months corresponds roughly to the order of the Roman months (January to December). The missing months after Aphrodisios are, according to the *Hemerologia*: Demetrios, Heraios, Hermaios, Metroos. The order of the Egyptian months, however, does not match that of the Bithynian months, and begins with Pharmouthi.

102. Papyrus. Provenance unknown, late seventh/early eighth century CE (*P.Rain.Cent.* 31) Concordance of Julian, Cappadocian, and Egyptian month-names (cf. above n. 48).

103. Mosaic inscription. Khirbat Dariya (church of SS. Kosmas and Damianos), Province of Arabia or Palaestina Secunda, date unknown (*SEG* 57, 1845). It lists the Macedonian month-names. The order of the months is in agreement with that shown by the *Hemerologia* for the calendar of the *Hellenes*, corresponding to the so-called calendar of Antioch (cf. above n. 5).[139]

104. Astrological manuscript. Date unknown (*CCAG* 9.1, pp. 128–37). Equation of Julian, Macedonian (= calendar of Antioch), and Egyptian months.

137 The order of the Macedonian months corresponds to that of the calendar of Antioch.

138 Montserrat, 'Fragment'.

139 Other (fragmentary) mosaic inscriptions listing the Macedonian month-names in the order of the calendar of Antioch are *IGLS* III.1.808 (Antioch – first half of the second century CE), *IGerasa* 295 (Gerasa – sixth century CE) and 307 (Gerasa – 531 CE).

References

J.N. Adams, *Bilingualism and the Latin Language* (Cambridge: Cambridge University Press, 2003).

G. Almagno and G.L. Gregori, 'L'istituzione e la ricorrenza del *dies natalis* augusteo nella documentazione epigrafica', *Maia* 68 (2016), pp. 446–59.

G. Almagno and G.L. Gregori, *Roman Calendars: Imperial Birthdays, Victories and Triumphs* (Beau-Bassin: LAP-Lambert Academic Publishing, 2019).

[illegible]

C. Bennett, 'The early Augustan calendars in Rome and Egypt', *Zeitschrift für Papyrologie und Epigraphik* 142 (2003), pp. 221–40.

C. Bennett, 'The two Egyptian birthdays of Augustus', *Zeitschrift für Papyrologie und Epigraphik* 161 (2007), pp. 195–98.

D. Bianconi, 'Il *Laur. Plut.* 28.26 ovvero la storia di Bisanzio nelle storia di un codice', in M. D'Agostino (ed.), *Alethes Philia. Studi in onore di Giancarlo Prato* (Collectanea, 23; Spoleto: Fondazione Centro italiano di studi sull'alto Medioevo, 2010), pp. 39–63.

E.J. Bickerman, *Chronology of the Ancient World* (London: Thames & Hudson, 1980).

B. Buxton and R. Hannah, '*OGIS* 458, the Augustan calendar, and the succession', in C. Deroux (ed.), *Studies in Latin Literature and Roman History, XII* (Brussels: Éditions Latomus, 2005), pp. 290–306.

A. Cutler and J.-M. Spieser, *Byzance médiévale: 700–1204* (Paris: Gallimard, 1996).

F. de Romanis, 'Lysas e il tempo: ulteriori considerazioni su *AEp*, 1954, 121a', *Epigraphica* 63 (2001), pp. 9–36.

B. Dreyer and H. Engelmann, 'Augustus und Germanicus im ionischen Metropolis', *Zeitschrift für Papyrologie und Epigraphik* 158 (2006), pp. 172–82.

V. Ehrenberg and A.H.M. Jones, *Documents Illustrating the Reigns of Augustus and Tiberius* (Oxford: Oxford University Press, 1955).

H. Engelmann, 'Inschriften aus Metropolis', *Zeitschrift für Papyrologie und Epigraphik* 125 (1999), pp. 137–46.

D. Feeney, *Caesar's Calendar: Ancient Time and the Beginnings of History* (Berkeley: University of California Press, 2007).

D. Feissel, J. Gascou, and J. Teixidor, 'Documents d'archives romains inédits du Moyen Euphrate (IIIe s. après J.-C.)', *Journal des savants* (1997), pp. 3–57.

F. Floëri and P. Nautin, *Homélies pascales* (Sources Chretiennes 27, 36, 48; Paris: Editions du Cerf, 1950–1957).

J.-L. Fournet and J. Gascou, 'Un lot d'archives inédit de Lycopolis (Égypte) à l'Académie des inscriptions et belles-lettres', *Comptes rendus de l'academie des inscriptions et belles lettres* 3–4 (2008), pp. 1066–69.

A. Fraschetti, *Roma e il Principe* (Rome: Laterza, 2nd ed., 2005).

C. Grassien, 'Deux hymnes et une litanie chrétiennes byzantines conservés par le P.Rainer Cent. 31 et cinque autres témoins', *Tyche* 12 (1997), pp. 51–84.

V. Grumel, *Traité d'études byzantines. I. La chronologie* (Paris: Presses Universitaires de France, 1958).

D. Hagedorn and K. Worp, 'Das Wandeljahr im Römischen Ägypten', *Zeitschrift für Papyrologie und Epigraphik* 104 (1994), pp. 243–55.

R. Hannah, *Greek and Roman Calendars: Constructions of Time in the Classical World* (London: Duckworth, 2005).

A. Heller, 'Domination subie, domination choisie: les cités d'Asie Mineure face au pouvoir romain, de la République à l'Empire', *Pallas. Revue d'Études Antiques* 96 (2014), pp. 217–32.

G.F. Hill, *The Life of Porphyry of Gaza* (Oxford: Clarendon Press, 1913).

F. Hurlet and A. Suspène, 'Le proconsul et le prince. À propos des portraits monétaires des proconsuls d'Afrique et d'Asie sous le Principat d'Auguste', in R. Baudry and S. Destephen (eds), *La société romaine et ses élites: hommages à Élizabeth Deniaux* (Paris: Picard, 2012), pp. 73–90.

T. Janz, 'The scribes and the date of the Vat. gr. 1291', *Miscellanea Bibliothecae Apostolicae Vaticanae*, x (Vatican City: Biblioteca Apostolica Vaticana, 2003), pp. 159–80.

A. Jones, *Astronomical Papyri from Oxyrhynchus (P. Oxy. 4133–4300a)* (Philadelphia: American Philosophical Society, 1999).

F. Kirbihler, 'César, Auguste et l'Asie: continuités et évolutions de deux politiques', in O. Devillers and K. Sion-Jenkis (eds), *César sous Auguste* (Bordeaux: Ausonius Éditions, 2012), pp. 125–44.

F. Kirbihler, 'Ruler cults and imperial cults at Ephesos: first century BCE to third century CE', in D. Schowalter et al. (eds), *Religion in Ephesos Reconsidered. Archaeology of Spaces, Structures, and Objects* (Leiden: Brill, 2019), pp. 195–210.

F. Kirbihler and L. Zabrana, 'Archäologische, epigraphische und numismatische Zeugnisse für den Kaiserkult im Artemision von Ephesos. Der Kult der Dea Roma und des Divus Iulius unter dem Triumvirat', *Jahreshefte des Österreichen Archäologischen Institutes in Wien* 83 (2014), pp. 101–31.

W. Kubitschek, 'Kalenderstudien', *Jahreshefte des Österreichischen Archäologischen Institutes in Wien* 8 (1905), pp. 87–118.

W. Kubitschek, 'Der pamphylische Kalender', *Wiener Studien. Zeitschrift für klassische Philologie* 34 (1912), pp. 347–51.

W. Kubitschek, *Die Kalenderbücher von Florenz, Rom und Leyden* (Vienna: Hölder, 1915).

U. Laffi, 'Le iscrizioni relative all'introduzione nel 9 a.C. del nuovo calendario della provincia d'Asia', *Studi classici e orientali* 16 (1967), pp. 5–98.

D. Lehoux, *Astronomy, Weather, and Calendars in the Ancient World: Parapegmata and Related Texts in Classical and Near Eastern Societies* (Cambridge: Cambridge University Press, 2007).

J. Lempire, *Le commentaire astronomique aux* Tables Faciles *de Ptolémée attribué à Stéphanos d'Alexandrie* (Publications de l'Institut Orientaliste de Louvain 68; Louvain-La-Neuve: Peeters Publishers, 2016).

N. Lewis, 'A Jewish landowner from the Province of Arabia', *Scripta Classica Israelica* 8–9 (1985–8), pp. 132–37.

N. Lewis, *The Documents from the Bar Kokhba Period in the Cave of Letters: Greek Papyri* (Jerusalem: Israel Exploration Society, 1989).

H. Lietzmann, *Zeitrechnung der römischen Kaiserzeit: des Mittelalters und der Neuzeit* [illegible]

F. Marchetti, 'La trasmissione della cultura scientifica greca a Bisanzio: codici di medicina e astronomia della Biblioteca Medicea Laurenziana', in M. Bernabò (ed.), *Voci dell'Oriente. Miniature e testi classici da Bisanzio alla Biblioteca Medicea Laurenziana* (Firenze: Polistampa, 2011), pp. 121–32.

C. Marek, *Stadt, Ära und Territorium in Pontus-Bithynia und Nord-Galatia* (Tübingen: Wasmuth, 1993).

J. Martin, *Scholia in Aratum vetera* (Stuttgart: Teubner, 1974).

Y.E. Meimaris, in collaboration with K. Kritikakou and P. Bougia, *Chronological Systems in Roman-Byzantine Palestine and Arabia: The Evidence of Dated Greek Inscriptions* (Athens: National Hellenic Research Foundation, 1992).

Y.E. Meimaris and K.I. Kritikakou-Nikolaropoulou, *Inscriptions from Palaestina Tertia, Ia: The Greek Inscriptions from Ghor Es-Safi (Byzantine Zoora)* (Meletemata 41; Athens: National Hellenic Research Foundation, 2005).

R. Mercier, *Πτολεμαίου Πρόχειροι Κανόνες. Ptolemy's Handy Tables volume 1b. Tables A1–A2. Transcription and Commentary* (Publications de l'Institut Orientaliste de Louvain 59b; Louvain-La-Neuve: Peeters, 2011).

R. Merkelbach and J. Stauber, *Steinepigramme aus dem griechischen Osten. Band I: Die Westküste Kleinasiens von Knidos bis Ilion* (Stuttgart: Teubner, 1998).

F. Millar, *Rome, the Greek World, and the East, Vol. 1: The Roman Republic and the Augustan Revolution* (Chapel Hill: University of North Carolina Press, 2002).

D. Montserrat, 'A Fragment of a Monastic Duty Roster?', *The Bulletin of the American Society of Papyrologists* 27 (1990), pp. 163–68.

A. Mosshammer, *The Prologues on Easter of Theophilus of Alexandria and [Cyril]* (Oxford: Oxford University Press, 2017).

O. Neugebauer and H.B. van Hoesen, *Greek Horoscopes* (Philadelphia: American Philosophical Society, 1959).

G.H. Opitz, *Athanasius Werke* (Berlin: Walter de Gruyter, 1940), vol. 2.1, pp. 231–78.

P. Orsini, 'Genesi e articolazioni della "maiuscola liturgica"', in A. Bravo Garcia (ed.), *The Legacy of Bernard de Montfaucon: Three Hundred Years of Studies on Greek Handwriting. Proceedings of the Seventh International Colloquium of Greek Palaeography (Madrid – Salamanca, 15–20 September 2008)* (Turnhout: Brepols, 2010), pp. 17–35, 669–82.

D. Paniagua, 'New perspectives for the Laterculus of Polemius Silvius', in D. Hernández de la Fuente (ed.), *New Perspectives on Late Antiquity* (Newcastle: Cambridge Scholars Publishing, 2011), pp. 393–406.

D. Paniagua, *Polemii Silvii Latercvlvs* (Rome: Sede dell'Istituto Palazzo Borromini, 2018).

S.R.F. Price, *Rituals and Power: The Roman Imperial Cult in Asia Minor* (Cambridge: Cambridge University Press, 1984).

R.M. Price and M. Gaddis, *The Acts of the Council of Chalcedon: Volume 2* (Liverpool: Liverpool University Press, 2007).

J.-M. Prieur, *Acta Andreae*, 2 vols (Turnhout: Brepols, 1989).

M. Ross, 'An introduction to the horoscopic ostraca of Medinet Madi', *Egitto e Vicino Oriente* 29 (2006), pp. 147–80.

J. Rowlandson, *Women and Society in Greek and Roman Egypt: A Sourcebook* (Cambridge: Cambridge University Press, 1998).

J. Rüpke, *Kalender und Öffentlichkeit. Die Geschichte der Repräsentation und religiösen Qualifikation von Zeit in Rom* (Berlin: Walter de Gruyter, 1995).

J. Rüpke, *The Roman Calendar from Numa to Constantine. Time, History, and the Fasti* (Chichester: Wiley-Blackwell, 2011).

M. Salzman, *On Roman Time. The Codex-Calendar of 354* (Berkeley: University of California Press, 1990).

A.E. Samuel, *Greek and Roman Chronology: Calendars and Years in Classical Antiquity* (Munich: C.H. Beck, 1972).

A.E. Samuel, 'Calendars and time-telling', in M. Grant and R. Kitzinger (eds), *Civilization of the Ancient Mediterranean: Greece and Rome Vol. 1* (New York: Scribner's , 1988), pp. 389–95.

R.K. Sherk, *Roman Documents from the Greek East. Senatus Consulta and Epistulae to the Age of Augustus* (Baltimore: Johns Hopkins Press, 1969).

R.K. Sherk, *Rome and the Greek East to the Death of Augustus* (Cambridge: Cambridge University Press, 1984).

P.J. Sijpesteijn, 'Some remarks on Roman dates in Greek papyri', *Zeitschrift für Papyrologie und Epigraphik* 33 (1979), pp. 229–40.

H. Solin, *San Modestino e l'Abellinum Cristiana* (Avellino: ISSR G. Moscati, 2013).

I. Spatharakis, 'Some observations on the Ptolemy MS. Vat. gr. 1291: its date and the two initial miniatures', *Byzantinische Zeitschrift* 71 (1978), pp. 41–49.

S. Stern, 'A "Jewish" birth record, *sambat-*, and the calendar of Salamis', *Zeitschrift für Papyrologie und Epigraphik* 172 (2010), pp. 105–14.

S. Stern, *Calendars in Antiquity. Empires, States, and Societies* (Oxford: Oxford University Press, 2012).

S. Stern, 'Calendars, politics, and power relations in the Roman Empire', in J. Ben-Dov and L. Doering (eds), *The Construction of Time in Antiquity. Ritual, Art, and Identity* (Cambridge: Cambridge University Press, 2017), pp. 31–49.

P. Thonemann, 'The calendar of the Roman province of Asia', *Zeitschrift für Papyrologie und Epigraphik* 196 (2015), pp. 123–41.

A. Tihon, 'Les *Tables Faciles* de Ptolémée dans les manuscrits en onciale (IXe–Xe siècles)', *Revue d'histoire des textes* 22 (1992), pp. 47–87.

A. Tihon, 'Theon of Alexandria and Ptolemy's Handy Tables', in N.M. Swerdlow (ed.), *Ancient Astronomy and Celestial Divination* (Cambridge, Mass.: MIT Press, 1999), pp. 357–69.

A. Tihon, 'Les Tables faciles de Ptolémée: une édition critique', in C. Burnett et al. (eds), [illegible] 2004), pp. 223–46.

A. Tihon, *Πτολεμαίου Πρόχειροι Κανόνες. Les 'Tables Faciles' de Ptolémée volume 1a. Tables A1–A2. Introduction, édition critique* (Publications de l'Institut Orientaliste de Louvain 59a; Louvain-La-Neuve: Peeters, 2011).

G.J. Toomer, *Ptolemy's Almagest* (*Translated and Annotated*) (London: Duckworth, 1984).

P. van Deun, *Hagiographica Cypria* (Turnhout: Brepols, 1993).

P. van Minnen, 'Three edicts of Caracalla? A new reading of *P.Giss.* 40', *Chiron* 46 (2016), pp. 205–21.

K. Weidemann and M. Weidemann, *Römische Staatskalender aus der Spätantike: die von Furius Dionisius Filocalus und Polemius Silvius überlieferten römischen Staatskalender und deren historische Einordnung* (Mainz: Verlag des Römisch-Germanischen Zentralmuseums, 2016).

S.J. Weinstock, *Divus Iulius* (Oxford: Clarendon Press, 1971).

M. Whitby, *The Ecclesiastical History of Evagrius Scholasticus* (Liverpool: Liverpool University Press, 2000).

T. Witulski, *Kaiserkult in Kleinasien. Die Entwicklung der kultisch-religiösen Kaiserverehrung in der römischen Provinz Asia von Augustus bis Antoninus Pius* (Göttingen: Vandenhoeck & Ruprecht, 2007).

P. Zanker, *The Power of Images in the Age of Augustus* (Ann Arbor: University of Michigan Press, 1988).

CHAPTER 3

The Origin(s) of the Medieval Calendar Tradition in the Latin West

Immo Warntjes

1 Introduction

Ever since the invention of the derogatory term 'Middle Ages' by fifteenth-century Humanists, the period between the end of the Western Roman Empire and the emergence of a new Empire under Charlemagne has been variously interpreted. The Mediterranean as the centre of the ancient world is the focal point of any of these views. Gibbon considered the old world order to have collapsed with the fall of Rome in the fifth century,[1] while Pirenne stressed that the truly seismic shift in the Mediterranean was the rise of Islam in the seventh.[2] Since the 1960s, scholars have tended to downplay the importance of watersheds and opt for more generic models of transition.[3] From the perspective of the history of science, certainly a substantial shift occurred in the sixth century in what was to become the Latin West. Greek lost its currency as the language of learning, and texts written in that language were known only by name in the following five centuries until their recovery through Arabic channels in the so-called Renaissance of the twelfth century.[4] The centres of learning shifted from secular institutions to the rapidly growing monasteries, which monopolized access to knowledge until the rise of the universities in

1 Gibbon, *History*. I gratefully acknowledge that this research was funded by the Irish Research Council as part of the Laureate Consolidator Award Scheme.

2 Pirenne, *Mahomet et Charlemagne*.

3 One of the earliest examples is the volume White (ed.), *The Transformation of the Roman World*; more recently, see, e.g., Wickham, *The Inheritance of Rome*; Wood, *The Transformation of the Roman West*; and the review of the concept by Pohl, 'The transformation of the Roman world revisited'.

4 Haskins, *The Renaissance of the Twelfth Century*, an influential and sweeping overview that should be read in connection with his detailed analyses collected in *Studies in the History of Mediaeval Science*. For an excellent recent overview, see the two chapters 'Translation and transmission of Greek and Islamic science to Latin Christendom' and 'The twelfth-century renaissance' by Burnett in Lindberg and Shank (eds), *The Cambridge History of Science, vol. 2: Medieval Science*.

 | DOI:10.1163/9789004459694_005

the eleventh century.[5] For six centuries, from the sixth to the eleventh, learning was therefore decidedly Christian in character in the Latin West.[6]

The calendar tradition fits this pattern well. In this article, the term calendar will be used for a tabulation of the 365 days of the Julian calendar year in grid format, with one line for each calendar date, organized according to the 12 months of the Julian calendar year.[7] The format is well-known to a twenty-first-century audience, since it has not changed since the days of Julius Caesar in 46 BC: We still have the same months with the same number of days (January 31, February 28, March 31, April 30, May 31, June 30, July 31, August 31, September 30, October 31, November 30, December 31). What has changed is the place of the bissextile or leap-day, traditionally on 24 February, but today added to the end of February as its 29th day.[8] And, more importantly, the configuration of dates in each month; while we today are used to a consecutive count of days from 1 to 28, 30, or 31, the Julian calendar months were designed around three marker day, the Calends (invariably the first day of the month), the Nones (fifth or seventh day of the month), and the Ides (thirteenth or fifteenth day of the month). From these marker days, the days were counted backwards (e.g.: Ides of March = 15 March, second day before the Ides of March = 14 March, etc.).

Roman calendar representations in grid format are older than Caesar's reform, as evidenced by the famous *Fasti antiates maiores*. The evidence becomes more numerous from Augustus' reign. However, none of these inscriptions, paintings, or mosaics survive in full.[9] The oldest calendar of the Julian type that is transmitted complete, in later manuscript copies, formed part of

---

5 See especially the classic by Riché, *Les écoles et l'enseignement dans l'Occident chrétien.*

6 See especially the studies by Contreni, e.g. 'The Carolingian Renaissance', pp. 725–47; 'Learning for God', pp. 94–112.

7 It is important to stress here the difference between a martyrology and a calendar: a martyrology lists saints's feastdays by calendar date; typically, one line is not enough for each date to include all the saints of that date, so that the visual format of a martyrology resembles a prose text more than the tabular format of a calendar; also, if no saint was listed for a given Julian calendar date, that date was not recorded. Early medieval calendars cut the list of saints short, concentrating only on the most important ones in the view of the author; but a calendar always provided a full grid of all 365 days. In this definition, a list of exclusively Christian feastdays by calendar dates is a martyrology; if additional elements are included (typically the computistical, astronomical, and prognostic data discussed below) and the format is a recognizable grid, it is a calendar.

8 The minimal change (at least in application) from the Julian to the Gregorian calendar by omitting three bissextile days in a given 400-year period (in years of full hundreds not divisible by 4) is obviously of no relevance here.

9 Degrassi, *Inscriptiones Italiae 13*, for earlier calendar fragments; see also the summary in Salzmann, *On Roman Time*, pp. 7–8.

TABLE 3.1 Grid of the Julian calendar

	Jan	Feb	Mar	Apr	May	Jun	Jul	Aug	Sep	Oct	Nov	Dec
1	Kal	Kal	Kal	Kal	Kal	Kal	Kal	Kal	Kal	Kal	Kal	Kal
2	IV No	IV No	VI No	IV No	VI No	IV No	VI No	IV No	IV No	VI No	IV No	IV No
3	III No	III No	V No	III No	V No	III No	V No	III No	III No	V No	III No	III No
4	II No	II No	IV No	II No	IV No	II No	IV No	II No	II No	IV No	II No	II No
5	Nonae	Nonae	III No	Nonae	III No	Nonae	III No	Nonae	Nonae	III No	Nonae	Nonae
6	VIII Id	VIII Id	II No	VIII Id	II No	VIII Id	II No	VIII Id	VIII Id	II No	VIII Id	VIII Id
7	VII Id	VII Id	Nonae	VII Id	Nonae	VII Id	Nonae	VII Id	VII Id	Nonae	VII Id	VII Id
8	VI Id	VI Id	VIII Id	VI Id	VIII Id	VI Id	VIII Id	VI Id	VI Id	VIII Id	VI Id	VI Id
9	V Id	V Id	VII Id	V Id	VII Id	V Id	VII Id	V Id	V Id	VII Id	V Id	V Id
10	IV Id	IV Id	VI Id	IV Id	VI Id	IV Id	VI Id	IV Id	IV Id	VI Id	IV Id	IV Id
11	III Id	III Id	V Id	III Id	V Id	III Id	V Id	III Id	III Id	V Id	III Id	III Id
12	II Id	II Id	IV Id	II Id	IV Id	II Id	IV Id	II Id	II Id	IV Id	II Id	II Id
13	Idus	Idus	III Id	Idus	III Id	Idus	III Id	Idus	Idus	III Id	Idus	Idus
14	XIX Kl	XVI Kl	II Id	XVIII Kl	II Id	XVIII Kl	II Id	XIX Kl	XVIII Kl	II Id	XVIII Kl	XIX Kl
15	XVIII Kl	XV Kl	Idus	XVII Kl	Idus	XVII Kl	Idus	XVIII Kl	XVII Kl	Idus	XVII Kl	XVIII Kl
16	XVII Kl	XIV Kl	XVII Kl	XVI Kl	XVII Kl	XVI Kl	XVII Kl	XVII Kl	XVI Kl	XVII Kl	XVI Kl	XVII Kl
17	XVI Kl	XIII Kl	XVI Kl	XV Kl	XVI Kl	XV Kl	XVI Kl	XVI Kl	XV Kl	XVI Kl	XV Kl	XVI Kl
18	XV Kl	XII Kl	XV Kl	XIV Kl	XV Kl	XIV Kl	XV Kl	XV Kl	XIV Kl	XV Kl	XIV Kl	XV Kl
19	XIV Kl	XI Kl	XIV Kl	XIII Kl	XIV Kl	XIII Kl	XIV Kl	XIV Kl	XIII Kl	XIV Kl	XIII Kl	XIV Kl
20	XIII Kl	X Kl	XIII Kl	XII Kl	XIII Kl	XII Kl	XIII Kl	XIII Kl	XII Kl	XIII Kl	XII Kl	XIII Kl
21	XII Kl	IX Kl	XII Kl	XI Kl	XII Kl	XI Kl	XII Kl	XII Kl	XI Kl	XII Kl	XI Kl	XII Kl
22	XI Kl	VIII Kl	XI Kl	X Kl	XI Kl	X Kl	XI Kl	XI Kl	X Kl	XI Kl	X Kl	XI Kl
23	X Kl	VII Kl	X Kl	IX Kl	X Kl	IX Kl	X Kl	X Kl	IX Kl	X Kl	IX Kl	X Kl
24	IX Kl	VI Kl	IX Kl	VIII Kl	IX Kl	VIII Kl	IX Kl	IX Kl	VIII Kl	IX Kl	VIII Kl	IX Kl
25	VIII Kl	V Kl	VIII Kl	VII Kl	VIII Kl	VII Kl	VIII Kl	VIII Kl	VII Kl	VIII Kl	VII Kl	VIII Kl
26	VII Kl	IV Kl	VII Kl	VI Kl	VII Kl	VI Kl	VII Kl	VII Kl	VI Kl	VII Kl	VI Kl	VII Kl
27	VI Kl	III Kl	VI Kl	V Kl	VI Kl	V Kl	VI Kl	VI Kl	V Kl	VI Kl	V Kl	VI Kl
28	V Kl	II Kl	V Kl	IV Kl	V Kl	IV Kl	V Kl	V Kl	IV Kl	V Kl	IV Kl	V Kl
29	IV Kl		IV Kl	III Kl	IV Kl	III Kl	IV Kl	IV Kl	III Kl	IV Kl	III Kl	IV Kl
30	III Kl		III Kl	II Kl	III Kl	II Kl	III Kl	III Kl	II Kl	III Kl	II Kl	III Kl
31	II Kl		II Kl		II Kl		II Kl	II Kl		II Kl		II Kl

the so-called Chronograph of 354 (i.e. four centuries after Caesar's reform), which also contained, among other items, a depiction of the four most important cities of the late Roman Empire (Rome, Alexandria, Constantinople, and Trier), a list of birthdays of Roman emperors, depictions of the planetary weekdays and zodiac signs with commentaries, a list of consuls from the foundation of the Republic (509 BC) to AD 354, a list of Easter dates AD 312–411, a list of Roman prefects for AD 254–354, a list of death dates and burial places of popes organized by calendar dates starting with Christmas, a list of Christian mar-[illegible] a list of popes from Peter to Liberius with commentaries.[10] The calendar itself had the dates centred (see Figure 3.1).[11] To the left were three alphabetic columns, representing the synodic lunar month, the seven- and the eight-day week respectively. To the right one finds important information for the respective dates, ranging from heathen and imperial feast or memorial days through senate meetings to astronomical and astrological data. Thus, the calendar itself shows no Christian features, but the context of its production with papal and martyr lists as well as Easter dates certainly does.

A second calendar from Late Antiquity survives, composed by one Polemius Silvius in AD 448/9 for Bishop Eucherius of Lyon. Polemius also added texts to his calendar, most notably a list of emperors and Roman provinces. But in his work, the calendar took centre stage.[12] Polemius had limited interest in astronomical or technical calendrical phenomena. There are no alphabetic columns representing the synodic lunar month or the weekday, his calendar immediately displays the Julian calendar dates as the first column. This left considerable space to the right, which Polemius populated not only with important Roman events, but also with major Christian ones, like the crucifixion and resurrection of Christ on 25 and 27 March respectively. Rather than recording the entry of the sun in the respective zodiac signs, Polemius had a strong interest in information related to agriculture: weather prognostics, solstices and equinoxes, beginnings of the seasons.

---

10 No manuscript transmitting the entire Chronograph of 354 survives, its original extent needs to be reconstructed from the just over 10 codices that evidently contain part of the work. See now Burgess, 'The Chronograph of 354'.

11 The standard edition of the calendar of 354 (in parallel with the one by Polemius Silvius to be discussed shortly) is Mommsen, *Corpus Inscriptionum Latinarum*, vol. 1, pp. 334–57. A new edition with extensive commentary has been produced by Divjak and Wischmeyer, *Das Kalenderbuch von 354*, vol. 1, pp. 157–344; for criticism of this edition see Burgess, 'The new edition of the Chronograph of 354', pp. 391–401, with response by the authors in the same journal issue (*Zeitschrift für antikes Christentum* 21), pp. 416–18.

12 For the ed. of Polemius' calendar, see previous note. It survives in only one manuscript, Brussels KBR 10615–729, 93r–95r (Trier, saec. XII).

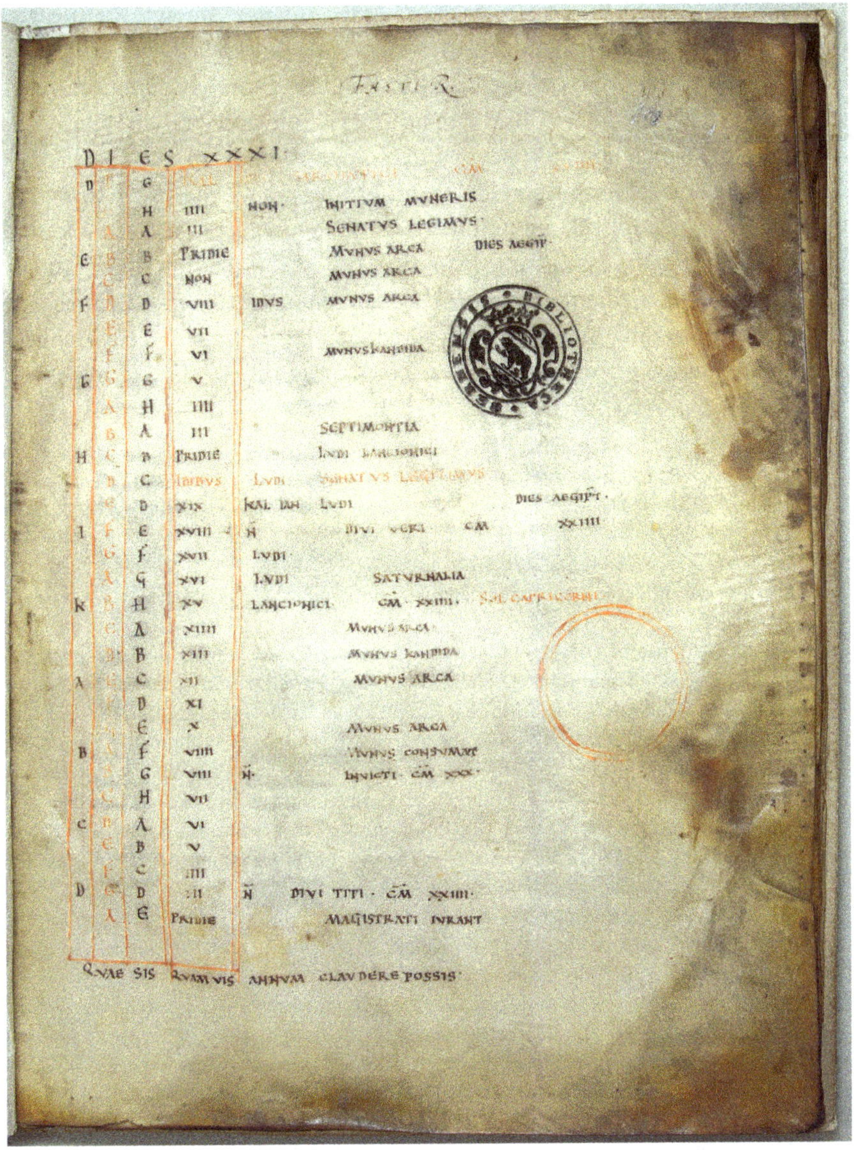

FIGURE 3.1 Burgerbibliothek Bern, Cod. 108, fol. 1r: December page of the Calendar of 354
WITH PERMISSION OF BURGERBIBLIOTHEK BERN

After Polemius Silvius's calendar of AD 448/9, the next surviving witness to this genre is Willibrord's calendar preserved in Paris BnF Lat. 10837.[13] Willibrord grew up in Northumbria, received his education in the strongly Roman-minded monastery of Ripon under Wilfrid (later of York)'s tutelage.[14] When Wilfrid was expelled from Northumbria in AD 678, his student Willibrord left for Ireland,[15] whose monasteries boasted an older tradition and more expertise in Christian learning.[16] From Ireland, more precisely the monastery of Rath Melsigi, Willibrord initiated his Frisian mission in AD 690.[17] A mission needed texts. [illegible] the basics for keeping track of liturgical time: an Easter table,[19] a basic collection of calendrical algorithms (for, e.g., calculating the AD date of any given

13 Paris BnF Lat. 10837, 34v–40r (Echternach?, early eighth century; available online at: https://gallica.bnf.fr/ark:/12148/btv1b6001113z.r=10837?rk=150215;2, accessed 13 August 2020). A facsimile edition of the calendar was produced by Wilson, *The Calendar of St. Willibrord*.

14 For the relationship between Wilfrid and Willibrord, see especially Stephen of Ripon, *Vita Wilfridi* ch. 26 (ed. and trans. Colgrave, *The Life of Bishop Wilfrid by Eddius Stephanus*, pp. 52–53). Willibrord's 'Romanism' has recently been under severe scrutiny; cf. especially van Berkum, 'Willibrord en Wilfrid'; the overview by Honée, 'St Willibrord in recent historiography'; and Ziegler, 'The Ripon connection?'. Note that even if Willibrord is, rightly, pulled out of a rather one-dimensional Anglo-Saxon/Roman perspective, this does not affect Willibrod's views on Easter, which were strongly and exclusively Roman/Dionysiac.

15 Cf. especially Hauck, *Kirchengeschichte Deutschlands*, vol. 1, pp. 434–35; Wampach, *Geschichte der Grundherrschaft Echternach im Frühmittelalter*, vol. 1, pp. 20–24; Levison, 'St. Willibrord and his place in history'; repr. in id., *Aus rheinischer und fränkischer Frühzeit*, pp. 314–29, esp. 315–16. Critical about this reconstruction of events, van Berkum, 'Willibrord en Wilfrid', pp. 385–91; Ziegler, 'The Ripon connection?'.

16 It is enough here to refer to Aldhelm's and Bede's famous quotes in *Epistulae ad Wihtfridum* and *ad Ehfridum* (ed. Ehwald in MGH Auct. ant. vol. 15, pp. 479–80, 486–94) and *Historia ecclesiastica*, ch. III 27 (Plummer, *Baedae opera historica*, vol. 1, p. 192) respectively.

17 Bede, *Historia ecclesiastica*, ch. V 9–10 (Plummer, *Baedae opera historica*, vol. 1, pp. 296–301). The year is suggested by Bede, *Historia ecclesiastica*, ch. V 8 (Plummer, *Baedae opera historica*, vol. 1, p. 294), and is explicit in the marginal note to the November page of Willibrord's calendar (Paris BnF Lat. 10837, 40v).

18 Cf. the well-established *Handbibliothek* of Willibrord's contemporary Boniface as reconstructed by Schüling, 'Die Handbibliothek des Bonifatius'.

19 The same manuscript that contains Willibrord's calendar (Paris BnF Lat. 10837) also includes an Easter table, which falls into four successive stages of composition: 1) AD 684–702: fol. 44r; this single sheet Willibrord brought from Ireland to the Continent; cf. Ó Cróinín, 'Rath Melsigi, Willibrord, and the earliest Echternach manuscripts'; repr. in id., *Early Irish History*, pp. 145–72 (with an appendix by Fanning), pp. 155–56; 2) AD 703–721: fol. 40v; 3) AD 722–759: fol. 41r–v; 4) AD 760–797: fol. 43r–v.

year, the weekday and lunar age of any given calendar date),[20] and the aforementioned calendar.

The focus of the calendar itself, in the first and subsequent hands, is strongly martyrological, not computistical. It reflects Willibrord's Northumbrian and Irish background, commemorating, e.g., the Deiran Kings Edwin (†AD 633) and Oswine (†AD 651), the Bernician Kings Oswald (†AD 642) and Ecgfrith (†AD 685), Bishop Cuthberht of Lindisfarne (†AD 687), the Irish Saints Patrick (†fifth cent.?), Brigit (†sixth cent.?), and Columba (†AD 593), and also the Irish missionary Bishop Áedán of Lindisfarne (†AD 651). His debt to the Roman Christianisation of Anglo-Saxon England is highlighted by recording Bishops Paulinus of York (†AD 644) and Theodore of Canterbury (†AD 690). His own missionary interest is reflected in the listing of the two Hewalds who were slaughtered when trying to convert the Saxons shortly before Willibrord came to the Continent. The computistical information is very limited: the original composition included only synodic lunar letters and a numerical weekday sequence in the left-hand columns. The Isidorian beginnings of the seasons (with their calculated lengths) and a note for the bissextile day on 24 February were added by a slightly later hand. In Willibrord's design, the martyrological entries had centre stage, the calendar tradition had turned fully Christian.

From Willibrord's calendar onwards, there is no break in the medieval calendar tradition, they become more frequent in the eighth century, and then mushroomed in the ninth. Though Willibrord's calendar is the first witness of this genre in the Middle Ages, it is not considered the 'prototype' for the medieval calendar tradition. According to Arno Borst, who collated 252 calendars from the eighth to the twelfth centuries in a monumental 3-volume study published in 2001,[21] three aspects define an early medieval calendar:

a. Liturgical / martyrological entries
b. (Bedan) computistical information
c. Astronomical data (especially Pliny)

In Borst's opinion, the first calendar to show all three features fully developed was a calendar of AD 789, produced in the royal abbey of Lorsch, preserved in a Prüm manuscript of the AD 840s.[22] Borst argues that this type of calendar radiated out from the Frankish heartland (with the Aachen palace school at

20 London BL Cotton Caligula A XV, 73r–80r (north-eastern France?, AD 743?); see Warntjes, 'Computus Cottonianus'.

21 Borst, *Reichskalender*, pp. XIII–XX, LIII, 54–333; Borst added 13 more calendars in *Streit*, pp. 129–58.

22 Id., *Kalenderreform*, esp. pp. 245–99.

its centre) to all areas of the Empire as part of Charlemagne's grand design of centralizing power and reforming the education system.

Borst's theory was soon challenged, principally by pushing the construction of the 'prototype' calendar back in time and space. Brigitte Englisch places its design in Cologne in AD 740, while Paul Meyvaert credited the Anglo-Saxon priest-monk Bede with its invention as a prefix to his computistical textbook *De temporum ratione* of AD 725.[23] All three theories (Borst's, Englisch's, Meyvaert's) have their problems, but here is not the space to discuss this in detail.[24] What is important for the present article is the agreement of all three most recent scholars in the existence of a 'prototype' calendar and their assessment of the general genesis of such a prototype, wherever it was first produced.

---

23 Englisch, *Zeiterfassung*; Meyvaert, 'Discovering the calendar'. Borst responded to this criticism in his beautiful little book on *Der Streit um den karolingischen Kalender*.

24 Borst's model of calendrical thought radiating out from the Carolingian centre has been criticised for the related umbrella subject, computus; see Dobcheva, 'The umbrella of Carolingian computus'; Warntjes, 'Köln'; Palmer, 'Computus after the paschal controversy of AD 740'; id., 'Calculating time and the end of time'. Englisch's theory of a crisis of time-reckoning in Francia in the AD 740 providing the stimulus for the early medieval calendar tradition is based on three observations, all of them wrong: 1) The decrees of the Council of Soissons of 744 (MGH Conc. vol. 2,1, p. 33) were signed off on 2 March, *luna* 14 (full moon), which was the factual moon contradicting the calculated one (pp. 84–85); the factual moon was full, however, on 3 March (http://www.astropixels.com/ephemeris/phasescat/phases0701.html, accessed 13 August 2020), and thus coincided with the calculated one according to both Victorius and Dionysius; 3 March is also the date of the decrees of the council of Soissons of AD 744, as the best manuscript (V) explicitly recorded this date (already pointed out by Krusch, 'Das Datum des Concils von Soissons 744 März 3.'); 2) A total solar eclipse was visible in Francia on 1 April 744 (pp. 87–88); however, this eclipse was not visible in western Europe, its path of totality was from central Africa through Pakistan to Mongolia and Korea (https://eclipse.gsfc.nasa.gov/SEsearch/SEsearchmap.php?Ecl=07400401, accessed 13 August 2020; this is already evident from Oppolzer's 1887 *Canon der Finsternisse* that Englisch cites); 3) Englisch claims that her key witness, the calendar in Cologne DB 83-II, 72v–76r (Cologne, AD 805), contains a substantial Victorian computistical element, if it should not be classified as Victorian (pp. 33–40); this is not the case (see n. 51). Meyvaert's theory is problematic on at least three levels: 1) Bede mentions the calendar prefixed to his *De temporum ratione* explicitly only in three chapters (19, 23, 41), indicating that it incorporated sidereal and synodic lunar letters, and a reference to the bissextile day on 24 February; there is no conclusive direct evidence that anything else was part of Bede's calendar; 2) some of the computistical features discussed by Meyvaert do not represent Bedan thought (e.g., Bede does not mention any of the moveable feasts dependent on Easter, like Pentecost; Bede favours the *saltus lunae* on 25 November, not 21 March); 3) the study misses an analytical corrective: the computistical calendar entries are compared to Bede's *De temporum ratione* only, not considering the earlier and contemporary Irish and Frankish texts.

The starting point of this narrative is the Gregorian mission under Augustine, sent from Rome to Anglo-Saxon England (more precisely Kent) in AD 596. This initiated a gradual south to north spread of 'Roman' Christianity in Britain, which eventually clashed in Northumbria with a 'Celtic' version of Christianity, introduced there from the Irish monastery of Iona (off the cost of modern-day south-western Scotland) in AD 635.[25] The problem was resolved at the famous Synod or Council of Whitby of AD 664, which brought 'Roman' unity to all of Anglo-Saxon England, or so Bede and Stephen of Ripon want to make us believe.[26] Instrumental in this process were Benedict Biscop, founder of Bede's monastery of Wearmouth, and the aforementioned Wilfrid. Both of them frequently travelled to Rome, and returned from the Holy See with the latest publications and the classics to stock their newly created libraries at home. Bede, when preparing his *opus magnum*, the *Historia ecclesiastica gentis Anglorum*, of AD 731 or shortly thereafter, drew on similar channels of book transfer from Rome to Northumbria. This import of Roman literature also included a calendar similar to that of Polemius Silvius. Bede, in his History of the Abbots of Wearmouth and Jarrow, relates that Benedict Biscop had brought a *Codex cosmographiorum* from Rome, which Abbot Ceolfrith then presented to King Aldfrith of Northumbria in exchange for a considerable portion of land, presumably in the AD 690s.[27] This codex contained a calendar, of which the details are not known; the inference is that it must have resembled the

25 The terms 'Celtic' and 'Roman' are obviously problematic in this context, as the early Middle Ages neither knew a monolithic 'Celtic Church', nor was 'Roman dogma' considered binding by the Western churches. As this article is principally concerned with calendrical/computistical theory, the use of Easter calculations may illustrate this point: The 'Celtic' Easter reckoning, called *latercus* and invented by Sulpicius Severus in Gaul, was abandoned by the Celtic *gentes* of Britain and Ireland not at one synod binding for all, but rather in a slow process from southern Ireland in AD 633 to northern Ireland towards the end of the seventh century, Pictland in AD 710, Iona in AD 716, and the Britons of Wales and elsewhere in AD 768; cf. p. 94 n. 176. Roman 'unity' remained a myth as long as the Victorian and the Dionysiac reckoning, both claiming papal sanction, co-existed until the end of the eighth century.

26 Our knowledge of the Synod of Whitby derives exclusively from Stephen's and Bede's highly biased narratives: Stephen of Ripon, *Vita Wilfridi*, ch. 10 (Colgrave, *Life of Bishop Wilfrid*, pp. 20–23); Bede, *Historia ecclesiastica*, ch. III 25 (Plummer, *Baedae opera historica*, vol. 1, pp. 181–89). For the political background, see Abels, 'The council of Whitby'. For the theological/scientific debate, see now especially Holford-Strevens, 'Marital discord in Northumbria'; Dailey, 'To choose one Easter from three'. A good overview is provided by Corning, *The Celtic and Roman Traditions*, pp. 112–29.

27 Bede, *Historia abbatum*, ch. 15 (ed. Plummer, *Baedae opera historica*, vol. 1, pp. 379–80; see now the new edition and translation by Grocock and Wood, *Abbots of Wearmouth and Jarrow*, pp. 56–59). For the *Codex cosmographiorum*, see Meyvaert, 'Discovering', pp. 16–25; and also Henderson, *Vision and Image in Early Christian England*, pp. 94–96.

one composed by Polemius Silvius. This model informed the Northumbrian engagement with calendars in the first half of the eighth century, especially Bede. Directly from Northumbria, this calendar tradition entered the Frankish kingdoms, possibly as part of Alcuin's travel library in the AD 770s, probably earlier. The exact shape and form of this tradition remains debatable. But the principal early medieval route of transmission of the calendar genre therefore was from Rome through Northumbria to the Frankish heartland.

This theory leaves little room for alternative readings of the origin and [illegible] tradition the second well-established route of transmission for late antique and early medieval texts: from Visigothic Spain to Ireland, and from Ireland to Northumbria on the one hand, and through north-eastern France or Brittany to the Loire valley and the Frankish heartland and then further south to modern-day Switzerland and northern Italy on the other.[28] I will deal with this question in a separate publication. Second, for the creation of an early medieval calendar, no Roman model was needed. It is this second aspect that the present article is concerned with.

2 Calendrical Knowledge in Seventh-Century Ireland

It is a still very popular, wrong assumption that medieval calendrical science, called by its Latin term computus or computistics in English, began with Bede (†AD 735). Bede was no isolated genius. Quite the contrary, he drew on a decades-old tradition establish in Ireland.[29] Ireland benefitted from a substantial import of texts from Visigothic Spain, and by the AD 630s, as Cummian's letter attests, had a more or less full set of late antique texts dealing with the question of how to best calculate the date of Easter and related issues.[30] The

28 For this route of transmission (or parts of it), especially for computistical texts, see Jones, *Bedae opera de temporibus*, pp. 105–13; Ó Cróinín, 'The Irish provenance of Bede's computus', repr. in id., *Early Irish History*, pp. 173–90; id., 'Bede's Irish computus', ibid., pp. 201–12; Warntjes, *Munich Computus*, pp. XCVII–CVI; id., 'The continuation of the Alexandrian Easter table'.

29 For pre-Bedan Irish computistics, see especially the overviews in Warntjes, 'Irische Komputistik'; id., 'Seventh-century Ireland'; id., 'Computus as scientific thought'. For Bede's debt to Irish thought, see now Mac Carron, 'Bede, Irish computistica and Annus Mundi'; ead., *Bede and Time*, pp. 20–47. Cf. also my review of Kendall and Wallis, *Bede*, in *The Medieval Review*: https://scholarworks.iu.edu/journals/index.php/tmr/article/view/17613, accessed 13 August 2020.

30 Cummian, *Epistola de controversia paschali*, ll. 208–220 (ed. and trans. Walsh and Ó Cróinín, *Cummian's Letter*, pp. 84–87).

principal challenge of Easter calculation was to press synodic lunar months (the period from one new or full moon to the next) of *c.*29.530589 day-length into a mathematical model that would align with the Julian calendar of 365 1/4 days. The solution was a 19-year lunar cycle of 6939 3/4 days (19 Julian calendar years of 365 1/4 days: 19×365 1/4 = 6939 3/4 days; 235 lunar months averaging at *c.*29.530851, distributed in the following way: 19 lunar years of 12 lunar months alternating between 30 and 29 days + 4 3/4 bissextile days + 7 intercalated lunar months of 30 days called embolisms – 1 day called *saltus lunae*: 19×6×30 + 19×6×29 + 4 3/4 + 7×30–1 = 6939 3/4 days). That is to say that after 19 years, the lunar age of any given Julian calendar date returns to its initial value (e.g., the lunar age on 1 January was *luna* 5 in AD 700 and returned to this value only in AD 719, 738, etc.). Thus, for a full understanding of the underlying theory, the (solar) Julian calendar and the 19-year lunar cycle needed to be understood. The 19-year lunar cycle posed the biggest challenge here, as the only texts available detailing its constructions were Dionysius' faulty and Paschasinus' incomplete descriptions.[31]

For the Julian calendar, the authoritative text was Macrobius' *Saturnalia*, probably composed in the AD 430s or shortly thereafter (i.e. roughly contemporary with Polemius Silvius' calendar). The passages dealing with the Julian calendar (*Sat.* 1.12.2–1.15.20) circulated independently of the main text in a recension called *Disputatio Chori et Praetextati*.[32] It was an integral part of the (loosely defined) corpus of computistical texts that reached Ireland before Cummian's days in the AD 630s.[33] The *Disputatio* outlines the genesis of the Julian calendar, as perceived by Macrobius: Romulus, the founder of Rome, had designed a calendar of 10 months and 304 days. This calendar is important because it foreshadowed the Julian calendar of seven centuries

31 The computistica of Dionysius Exiguus are edited by Krusch, 'Studien', pp. 59–86; for Dionysius' problematic description of the 19-year cycle, see his *Epistola ad Bonifatium et Bonum* (Krusch, 'Studien', p. 83). Paschasinus' *Epistola ad Papam Leonem* is edited by Krusch, *Studien*, pp. 245–50: here chapter 1, p. 248. Cf. Warntjes, 'Cradle', pp. 50–51, 65–66.

32 The *editio princeps* of these excerpts (with translation and commentary) by Holford-Strevens has just appeared in print: *The Disputatio Chori et Praetextati*. The most comprehensive studies before this edition are the transcript of these excerpts from the famous codex Padua BA I 27, 66r–71v by Carton, *Three Unstudied Manuscripts*, pp. 161–203; and Arweiler, 'Zu Text und Überlieferung', who, unfortunately, shows limited awareness of key literature on this text and therefore misses crucial manuscript witnesses. The literature on the *Disputatio* is conveniently listed in Warntjes, *Munich Computus*, p. LXVII. I thank Leofranc Holford-Strevens for granting me access to his edition of the *Disputatio* before publication, and for providing me with a copy of the relevant pages in Carton's study.

33 See n. 30 above. Whether these texts arrived in Ireland as a corpus, or were there assembles as such, remains a matter of debate.

later. The 10 months were labeled March, April, May, June, Quintilis, Sextilis, September, October, November, December. Of these, April, June, Sextilis (= August), September, November, and December consisted of 30, March, May, Quintilis (= July), and October of 31 days.[34] Romulus' calendar was reformed by his successor, Numa Pompilius, who added 50 days, resulting in a 354-day year in line with Greek lunar theory. In this process, he introduced two new months (January and February), and reshuffled the number of days per months. March, May, Quintilis (= July), and October remained unaltered [illegible] (= August), September, November, and December of 29 days. In preference for odd numbers, Numa finally added one more day to this year (for a total of 355 days), to the month of January (now of 29 days).[35] The final reform then happened in the first century BC under Gaius Caesar and his *scriba* Marcus Flavius: Following Egyptian custom of computing the length of the year on the basis of the sun's course through the zodiac rather than on the lunar phases, Caesar added 10 1/4 days to Numa's 355. Again, March, May, Quintilis (= July), and October remained unaltered at 31 days, but January, Sextilis (= August), and December were upgraded to the same number, while one day was added to April, June, September, and November for a total of 30 each. The quarter day accumulated to a full day in four years, to be placed before the fifth last day of February (*sextus dies ante Kalendas Martias*). Finally, Quintilis was renamed to July in honour of Ceasar, Sextilis was turned into August in honour of the Emperor Augustus.[36]

Macrobius was also explicit about the structure of each month. Right at the beginning of the *Disputatio*, when discussing Romulus' calendar, it is argued that *hodie* ('today'):

a) March, May, Quintilis (= July), October have the Nones on the seventh day of the month.
b) all the others (a statement that includes January and February, which did not exist in Romulus' framework, but did at the time of the author – *hodie*) have the Nones on the fifth day of the month.[37]

At the end of the text, Macrobius lets Praetextatus argue that the Ides fall on the ninth day after the Nones in every month.[38]

---

34 *Disputatio Chori et Praetextati*, chs. 1, 10 (Holford-Strevens, *Roman Calendar for Beginners*, pp. 52–55, 64–67).
35 Ibid. chs. 11–12 (Holford-Strevens, *Roman Calendar for Beginners*, pp. 66–69).
36 Ibid. chs. 17–18 (Holford-Strevens, *Roman Calendar for Beginners*, pp. 78–83).
37 Ibid. ch. 1 (Holford-Strevens, *Roman Calendar for Beginners*, pp. 88–91).
38 Ibid. ch. 23 (Holford-Strevens, *Roman Calendar for Beginners*, pp. 54–55).

TABLE 3.2 The development of the number of days of the Roman calendar months in three stages, according to Macrobius. The four months that remained unaltered are in bold.

	Romulus	Numa	Caesar
January	–	29	31
February	–	28	28
March	**31**	**31**	**31**
April	30	29	30
May	**31**	**31**	**31**
June	30	29	30
Quintilis (= July)	**31**	**31**	**31**
Sextilis (= August)	30	29	31
September	30	29	30
October	**31**	**31**	**31**
November	30	29	30
December	30	29	31

This leaves the number of calends after the Ides for each month. Macrobius did not really have to be explicit about these, since the place of the Nones (either on the fifth or the seventh day) in the respective month, followed by the Ides nine days later (i.e. either on the 13th or 15th day of the month), leaves a clearly defined number of days, depending on the total number of days of the respective month, which Macrobius had explicitly mentioned earlier in his text. But these numbers could also more directly be reconstructed from the *Disputatio*: At the beginning of the text, Macrobius outlines that the months with the Nones on the seventh day (March, May, Quintilis (= July), October) have 17 calends after the Ides, and these months remained unaffected by the later reforms.[39] Later on, it is argued that all months to which Caesar added days originally had 17 calends after the Ides, but this changed to 18 calends for months to which Caesar added one day (April, June, September, November), to 19 calends for months to which Caesar added two day (January, August, December).[40]

The *Disputatio Chori et Praetextati* is a good example of how patchy the information available to seventh-century Irish monks was, both for solar and especially lunar theory. The data needed to get extracted, reconstructed,

39 Ibid. chs. 1 with 18 (Holford-Strevens, *Roman Calendar for Beginners*, pp. 54–55, 80–83).

40 Ibid. ch. 18 (Holford-Strevens, *Roman Calendar for Beginners*, pp. 82–83).

systematize, and added to. This endeavour led to the invention of a new genre of text, the computistical textbook. Three such textbooks survive in full from pre-Bedan Ireland.[41] Of these, only one is datable, the so-called Munich Computus of AD 718/9. Internal evidence suggests that the *Computus Einsidlensis* was composed slightly earlier, and *De ratione conputandi*, the most sophisticated of the three outshining Bede's rather wordy work in precision and clarity, slightly later.[42] This order is confirmed by their discussions of the Julian calendar. The principle structure is the same in all three texts:[43]

1) definition of the terms *Kalendae*, *Nonae*, *Idus*, principally based on the *Disputatio Chori*;
2) classification of the months in 2 groups according to the number of days before the Nones (4 or 6 when referring to the nones proper, 5 and 7 when referring to the total number of days before the Nones, including the Calends), in 4 groups according to the number of calends after the Ides (19, 18, 17, 16);
3) concordances of the months, according to the total number of days, the number of days before the Nones, and the number of calends after the Ides:
   a. January, August, December: 31, 4 (or 5), 19
   b. March, May, July, October: 31, 6 (or 7), 17
   c. April, June, September, November: 30, 4 (or 5), 18
   d. February: 28, 4 (or 5), 16;
4) illustration of the structure of individual months.

This is followed by weekday calculations within this Julian calendar structure.

---

41 There is a good chance to find more pre-Bedan Irish computistical texbooks in the thousands of medieval computistical manuscripts, in full or in parts. Fragments of what appears to have been a full textbook have survived in Harvard HL Typ 613, 7r–v; Regensburg SB Frag. 1ar–v, 1dr–1ev. The Irish computistica incorporated in the scientific encyclopedia preserved in Vatican BAV Reg. Lat. 123 (Ripoll, saec. XI) may also stem from a full textbook; see Bisagni, 'The newly-discovered Irish and Breton computistica'.

42 The Munich Computus is edited and translated in Warntjes, *Munich Computus*; *De ratione conputandi* is edited by Dáibhí Ó Cróinín in Walsh and Ó Cróinín, *Cummian's Letter*, pp. 99–213; an *editio princeps* of the *Computus Einsidlensis* by Tobit Loevenich (Trinity College Dublin) is in preparation; the sole manuscript of this text is Einsiedeln SB 321 (647), pp. 82–125 (Lake Constance region?, c.AD 874–892), accessible online at: https://www.e-codices.unifr.ch/de/list/one/sbe/0321, accessed 13 August 2020; for this text, see especially Bisagni and Warntjes, 'Early Old Irish material'. For the chronological order of these three texts, see especially Warntjes, *Munich Computus*, pp. CXXXIII–CLII, CXCI–CCI.

43 *Computus Einsidlensis* (Einsiedeln SB 321 [647], pp. 94–96); Munich Computus chs. 24–28 (Warntjes, *Munich Computus*, pp. 70–81; in slightly different order); *De ratione conputandi* chs. 30–32 (Walsh and Ó Cróinín, *Cummian's Letter*, pp. 142–44).

The earliest two of these three textbooks, the *Computus Einsidlensis* and the Munich Computus, include a limited vernacular element in the otherwise Latin text. Evidently, these texts were written by Irish authors for an Irish audience. The vernacular was often introduced when explaining particularly complex concepts, which had never before been expressed in Latin. At least these authors had no exemplar to work from when they formulated these ideas. Interestingly, Old Irish was used in two of the four passages outlined above, which clearly indicates that both the available sources and the structure of the Julian calendar with its counter-intuitive backward count from the marker days Calends, Nones, and Ides, was considered a challenge by the receiving society. The author of the Munich Computus reverts to his native tongue when explaining the concordances of the months (item 3), probably because he was used to express this in Irish in the classroom.[44] More impressive is the Einsiedeln computist's paralleling in two columns of Old Irish ordinal numbers with the first 13 Julian calendar dates of January (item 4; see Figure 3.2). The vernacular here serves two purposes: 1) clearly separating the two columns; 2) making sure that the reader / student fully understands how the Julian calendar's backward count translates into the continuous forward count that Irish monks would have been used to.[45]

At the same time as the Irish computistical textbooks, around AD 700, the information for the construction of a calendar grid was also systematized on the Continent. One of the earliest stages may be represented in MS Paris BnF Lat. 14086.[46] The text reads in full:

> Ianuarius dies XXXI: [Kalendas] Ianuarias, IIII Nonas, VIII Idus, XVIIII Kalendas Februarias
> Februarius dies XXVIII: [Kalendas] Februarias, IIII Nonas, VIII Idus,, XVI Kalendas Martias
> Martius dies XXXI: Kalendas Martias, VI Nonas, VIII Idus, XVIII Kalendas Aprilis

44 Munich Computus ch. 28 (Warntjes, *Munich Computus*, pp. 80–81). See Bisagni and Warntjes, 'Latin and Old Irish in the Munich Computus', pp. 17–28.

45 Einsiedeln SB 321 (647), p. 97. See Bisagni and Warntjes, 'Early Old Irish material', pp. 96–102.

46 Paris BnF Lat. 14086, 1v–2r (Langres from a Luxeuil exemplar?, c.AD 700; accessible online at: https://gallica.bnf.fr/ark:/12148/btv1b105110031.r=14086?rk=21459;2, accessed 13 August 2020). Cf. especially Piper, *Karls des Grossen Kalendarium*, pp. 60–67; Salmon, 'Le martyrologe-calendrier'; Krusch, 'Chronologisches aus Handschriften', pp. 91–93; Borst, *Kalenderreform*, pp. 213–14.

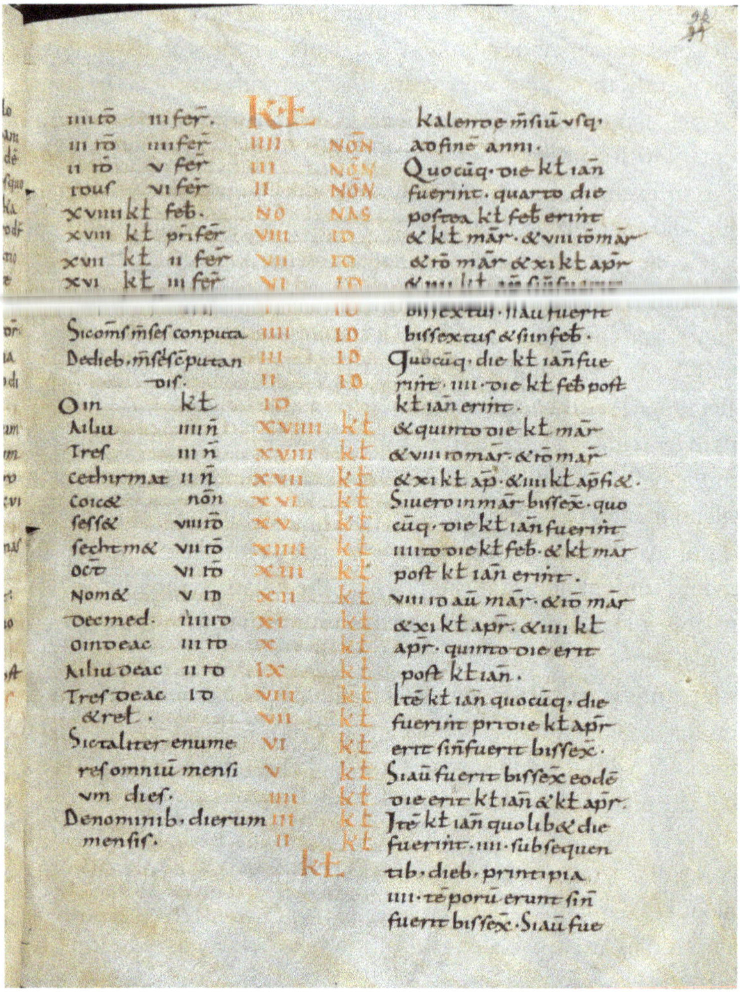

FIGURE 3.2 Einsiedeln, Stiftsbibliothek, Cod. 321 (647), p. 97, with the paralleling of Old Irish ordinals and Julian calendar dates in the left-hand column
WWW.E-CODICES.CH

Old Irish	translation	Julian calendar dates
oin	the first	Kalendae
ailiu	the second	IIII Nonas
tres	the third	III Nonas
cethirmat	the fourth	II Nonas
coicet	the fifth	Nonae
sesset	the sixth	VIII Idus
sechtmet	the seventh	VII Idus
octmet	the eighth	VI Idus
nomet	the ninth	V Idus
decmed	the tenth	IIII Idus
oindeac	the eleventh	III Idus
ailiu deac	the twelfth	II Idus
tres deac	the thirteenth	Idus

> Aprilis dies XXX: Kalendas Aprilis, IIII Nonas, VIII Idus, XVIII Kalendas Madias (sic)
> Maius dies XXXI: Kalendas Maias, VI Nonas, VIII Idus, XVII Kalendas Iunias
> Iunius dies XXX: Kalendas Iunias, IIII Nonas, VIII Idus, XVIII Kalendas Iulias
> Iulius dies XXXI: Kalendas Iulias, VI Nonas, VIII Idus, XVII Kalendas Agustas
> Agustus dies XXXI: Kalendas Agustas, IIII Nonas, VIII Idus, XVIIII Kalendas Septembris
> September dies XXX: [Kalendas] Septembris, IIII Nonas, VIII Idus, XVIII Kalendas Octobris
> October dies XXXI: [Kalendas] Octobris, VI Nonas, VIII Idus, XVII Kalendas Novembris
> November dies XXX: [Kalendas] Novembris, IIII Nonas, VIII Idus, XVIII Kalendas Decembris
> December dies XXXI: Kalendas Decembris, III Nonas, VIII Idus, XVIIII Kalendas Ianuarias

A more compressed and systematized version of this passage can be found later in the eighth century (see Figure 3.3):[47]

> Ianuarius, Agustus et December: dies XXXI, IIII Nonas, VIII Idus, XVIIII Kalendas.
> Martius, Maius, Iulius e[t] October: dies XXXI, VI Nonas, VIII Idus, XVII Kalendas.
> Aprilis, Iunius, September et November: dies XXX, IIII Nonas, VIII Idus, XVIII Kalendas.
> Ferbuarius (sic): dies XXVIII, IIII Nonas, VIII Idus, XVI Kalendas.

By AD 760 at the latest, this information was turned into a standardized wording that followed the Irish precursors.[48] It remained extremely popular throughout the entire early Middle Ages, as it features in virtually every computistical manuscript that survives.

47 St Gall SB 225, p. 120 (St Gall, AD 773?); accessible online at: https://www.e-codices.unifr.ch/de/list/one/csg/0225 (accessed 22 September 2020). For the computistical section of this codex, see especially Springsfeld, 'Beschreibung'.

48 The first datable text that contains this systematization is *Lect. comp.* ch. I 9 of AD 760, edited by Borst, *Schriften*, p. 552.

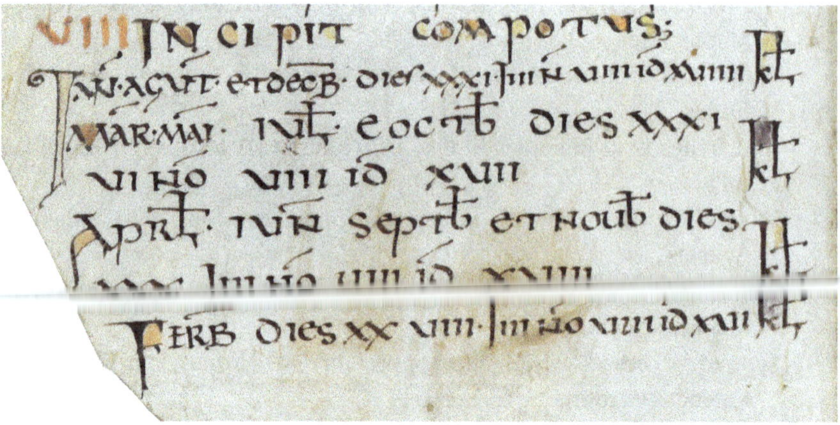

FIGURE 3.3 St Gall, Stiftsbibliothek, Cod. 225, p. 120: classification of the Julian calendar months
WWW.E-CODICES.CH

With this information at hand, it certainly was no problem, in the seventh century, to compose the 365-entry grid divided into 12 months needed for producing a calendar. How this grid was then populated with additional data was obviously up to the composer. Of the three categories outlined by Borst to define an early medieval calendar, the first, liturgical and martyrological entries, appeared, according to the Christian purpose the calendar had to serve, the most essential to include. The inclusion of computistical material was at least very intuitive, as the calendar grid provided excellent visualization for the placements of, e.g., the bissextile day and especially the embolisms, but obviously also, and more importantly, for the Julian calendar limits for Easter Sunday and for the related beginning of Lent.

A word needs to be said here about the computistical material found in early medieval calendars. First, all medieval calendars are Dionysiac. In Late Antiquity and the early Middle Ages, various competing methods of calculating Easter existed.[49] By AD 800, the Alexandrian method as translated into Latin by Dionysius Exiguus was unanimously followed throughout Christendom (and remained so until the Gregorian calendar reform of AD 1582). Rome accepted it by the AD 640s or 650s, Anglo-Saxon England in the second half of the seventh century, Ireland from the AD 680s in a slow process ending in AD 716, Pictland in AD 710, Wales in AD 768.[50] In Francia, it was introduced through Willibrord's

---

49 The best study of these reckonings, though outdated, is Schwartz, 'Christliche und jüdische Ostertafeln'. See now Mosshammer, *Easter Computus*, pp. 3–316; Stern, *Calendars in Antiquity*, pp. 380–430; and the excellent survey by Holford-Strevens, 'Paschal lunar calendars'.

50 A guide to further literature on these processes is provided by Warntjes, *Munich Computus*, pp. XXXV–XLI.

mission of AD 689 and by AD 800 all parts of Charlemagne's Empire seem to have accepted it, replacing Dionysius' main rival, Victorius of Aquitaine's Easter table composed in AD 457 and officially sanctioned by the Frankish Church at the Council of Orleans in AD 541. Some calendars display Victorian data. In all of these instances, however, the favoured Easter reckoning is, without exception, the Dionysiac one, the Victorian material is only introduced as comparative information.[51] As far as can be assessed from the surviving witnesses, there is no such thing as a Victorian calendar. The medieval calendar tradition is intrinsically connected to the Dionysiac reckoning.

But Dionysius, in his own writings, remained silent about crucial information. In fact, he was a rather limited computist himself. Crucially, he gave no indication of the placements of the embolisms and the *saltus lunae*.[52] With no authoritative account on these questions, early medieval computists designed their own regional customs.[53] In this sense, the Bedan placements of the embolisms was just one of many options. Bede's writings became standard by the early ninth century. Non-Bedan placements of the embolisms in particular, as they are sometimes found in early medieval calendars, therefore appear to indicate an older computistical stratum.

The last of Borst's three criteria, astronomical entries, depended heavily on Pliny in the early medieval calendar tradition. At what date the Pliny reception set in in Ireland is debatable. Certainly, the three Irish computistical

51 Englisch, *Zeiterfassung*, pp. 28–40, argues that the calendar in Cologne DB 83-II, 72v–76r (Cologne, AD 805) belongs to a stage when the Dionysiac and the Victorian reckonings were compared closely in Francia, reflected in the fact that, in this calendar, details of both methods of calculating Easter can be found side-by-side without commitment for either. This is not the case, the Cologne calendar is indisputably Dionysiac. Englisch's argument hinges on the embolismic lunar months recorded in this calendar, which she believes are Victorian. They are not (note that Englisch had to posit 'modification' of the Victorian system, a shift of the *saltus*; all of this hypothetical construction is not necessary, the embolisms are perfectly Dionysiac, just not Bedan). Dionysius had not stipulated dates for the intercalation of the embolismic lunar months. This led to varying customs in the Latin West in the seventh and eighth centuries (all within the Dionyisac framework), of which Bede was just one of many. Because of her lack of engagement with the computistical literature of the early eighth century other than Bede, Englisch does not entertain the possibility that the embolismic lunar months in the Cologne calendar reflect non-Bedan Dionysiac practice, which is exactly what they do. The only Victorian information in the Cologne calendar is the reference to the *saltus* on 17 November, which contradicts Englisch's supposition that the author of the Cologne calendar worked from a 'modification' of the Victorian reckoning that placed the *saltus* elsewhere.

52 Cf. the previous note and Warntjes, *Munich Computus*, pp. XLV–XLVI.

53 The two regions that developed the most distinctive features are Ireland and Visigothic Spain. For Ireland, see Warntjes, *Munich Computus*, pp. LXIX–LXIII, CXXXIII–CLII, CLIX–CLXVIII; id., 'Irische Komputistik', pp. 16–21. A full-scale study of Visigothic computistica is currently prepared by Alden Mosshammer and myself.

textbooks introduced above did not know Pliny. First traces may be found in the Irish eclipse prediction of AD 754.[54] More generally, the Irish computistical textbooks are almost completely silent about astronomical phenomena. If, therefore, a calendar was produced in Ireland in the second half of the seventh century or the first half of the eighth, one would expect the first two of Borst's characteristics (liturgical / martyrological and computistical entries), but not necessarily astronomical phenomena.

What is important here, though, is the realisation that by the mid-seventh century [illegible] for the creation of substantial calendars which only lacked the third of Borst's characteristics.

3 The Zürich (Nivelles) Calendar

That a calendar could be created from the information available in the seventh century without the need for a Roman model, can best be illustrated by means of an indicative example. In order to understand the early history of medieval calendars, it is essential to closely study the earliest exponents of this genre. In eighth-century Latin manuscripts, the following calendars can be found:[55]

---

54 The Irish author of the eclipse prediction of AD 754, when explaining the reasons why a solar eclipse may not be visible to the observer, uses the exact same arguments, in the exact same order, as Pliny in *Naturalis Historia* chs. II 10, II 71. For this text, see Warntjes, 'An Irish eclipse prediction of AD 754'; and Nothaft, *Walcher of Malvern*, p. 39, on the relevant passage. More generally on the reception of Pliny in early medieval Ireland, see Borst, *Das Buch der Naturgeschichte*, pp. 95–97; the only earlier trace of Pliny in Ireland appears to be the use of the unusual Plinian term *mentagra* by Cummian in AD 634 and in Laidcen's *Lorica* of shortly thereafter.

55 This is not to say that eighth-century calendars do not survive in later manuscripts. Good examples are: Borst's key witness, the *Reichskalender* of AD 789, which survives in a Prüm manuscript of *c.*AD 840 (Berlin SBPK Phillipps 1869, 1r–11v); the calendar in Trento MP 1590, 204r–217v (northern Italy, *c.*AD 835), which can be dated internally to AD 772 (though Borst, in my opinion unjustified, doubts such an early date; *Reichskalender*, p. 201). Note also that AD 800 is obviously a rather artificial cut-off point, though one supported by palaeographical practice (Lowe's *Codices Latini Antiquiores* operate with AD 800 as the end date of the catalogue); the first two decades of the ninth-century, i.e. the 20 years immediately after the cut-off point chosen here, saw the composition (or copying) of numerous calendars (in the order in which they appear in Borst, *Kalenderreform*, XVIII–XXVI): Vatican BAV Pal. lat. 1449, 3r–8v (Lorsch, *c.*AD 810) – A2; Vatican BAV Pal. lat. 1448, 6r–11v (Trier, AD 810) – A3; Cologne DB 83–II, 72v–76r (Cologne, AD 805) – A4; Cologne DB 103, 3r–8v (Cologne, AD 798) – A5; Vatican BAV Pal. lat. 1447, 11v–17v (Mainz, AD 808–13) – a1; Munich BSB Clm 14456, 48r–53v (Regensburg, AD 810–23) – a2; Berlin SBPK Phillipps 1831, 1r–6r (Verona, AD 800–10) – B1; Florence BML Plut.36.39, 1r–6v (Verona, AD 816/7) – B3; Rome BC 641, 76r–81v (Montecassino, AD 811–2) – B4; Milan BA H 150 inf., 55r–56v

Anglo-Saxon

Paris BnF Lat. 10837: this is the famous Willibrord calendar discussed above.

Ilmmuenster Kalenderfragment (Munich HAS Raritätenselekt 108, now lost): This is a one-folio binding fragment in Anglo-Saxon script, covering 3–24 May and 4–24 June only.[56] On these 43 lines, no less than 5 Anglo-Saxons are recorded by the original hand (the Northumbrian kings Ecgfrith and Osric, Ecgfrith's Queen Æthelthryth, Bishop Eadberht of Lindisfarne and Archbishop of York John of Beverley). All these entries point to Northumbria. Traditionally, Northumbria consisted of two parts, the northern Bernicia, and the southern Deira. 4 of the 5 Northumbrians recorded belong to a Bernician context (only John of Beverley belongs rather to a Deiran context), as does the mention of the Irish saint Columba of Iona,[57] from where Bernicia received Christianity. Thus, a Bernician origin of the calendar would seem most likely, but Mechthild Gretsch has recently argued for Continental composition (in the circle of Boniface) on linguistic grounds.[58] Of the entries that have survives, the latest is that of Osric of Northumbria (†729), while the earliest addition in a later hand is that of Boniface (†754), pointing to AD 729 × 754 as the time of writing. No particular interest in computistical and astronomical events is discernible, though it should be noted that May and June do not usually abound with computistical data.[59] The fragment is now lost, 'last seen in 1939'.[60]

Walderdorffsches Kalenderfragment: The library of the count of Walderdorff at Castle Kautzenstein near Regensburg houses the fragment of a calendar in

(Bobbio, c.AD 810) – b1; Berlin SBPK Phillipps 1831, 135r–137v (Verona, AD 818/9) – b2; Rome BV E 26, 3r–8v (Lyon, AD 814–6) – C1; Madrid BN 3307, 5r–6v, 79r–80v (Murbach, c.AD 820) – D1; Munich BSB Clm 210, 7v–15v (Salzburg, AD 810–8) – E1; Vienna ÖNB 387, 7v–15v (Salzburg, AD 810–30) – E2.

56 This calendar was transcribed by Bauerreiß, 'Ein angelsächsisches Kalendarfragment', pp. 178–79; this transcription was copied by Gamber, *Bonifatius-Sakramentar*, pp. 49–52; only the entries by the original hand were copied from Bauerreiß's transcription by Grosjean, 'Fragment', p. 322.

57 Columba's death is recorded two days too early; for discussion of this misdating, see Grosjean, 'Fragment', pp. 328–31, 340.

58 Gretsch, 'Æthelthryth of Ely', pp. 170–73.

59 Computistical elements are the column of synodic lunar letters and the Bedan beginning of summer on 9 May (the CX days given here for the summer period is clearly a scribal error for XC, not given by Bede, *De temporum ratione*, ch. 35, but correctly recorded in later calendars; cf. Borst, *Reichskalender*, pp. 871, 873–74; that this error was introduced by the original scribe and not by Bauerreiß in his transcription is evident from the facsimile in *CLA* vol. 9, facing p. 3).

60 *CLA* vol. 9, p. 3 (no. 1236).

Anglo-Saxon script.[61] Of the same manuscript, two more fragments survive, both preserving excerpts of the same sacramentary.[62] According to Bischoff, the script is Northumbrian.[63] Since the early twentieth century, a narrative has been constructed that it was Boniface who donated the sacramentary with calendar to the diocese of Regensburg at its foundation in AD 739.[64] The earliest datable addition to the calendar is that of Duke Theobald, who died in AD 719, presumably in Regensburg.[65] Place of rule, death, and burial may be the reason why Theobald's name was added to the calendar some 20 years after his death, after Boniface, as th[illegible], had given the codex to the newly founded bishopric. But I see no reason why this entry could not have been written earlier, closer to the duke's death. Certainly Boniface was active in Bavaria from at least AD 724, and there is no certainty that this calendar was originally owned by him. Of the original entries, the youngest layer points to Anglo-Saxon England, with the commemoration of Bishop Paulinus of York (†644) on 14 October and Archbishop Honorius of Canterbury (†653) on 30 September. Paulinus seems to be the more indicative of these two figures, as he was the most influential person in Christianising southern Northumbria, the kingdom of Deira. Unfortunately, only the months July to October survive of this calendar. But from what is preserved, nothing speaks against the possibility that this calendar was originally composed in Deira in the last few decades of the seventh century, earlier than or contemporary to Willibrord's calendar (and then later copied in the early eight century to be sent to the Continent). Maybe the sacramentary and calendar was given to Boniface by Willibrord when Boniface, after three years with his Northumbrian mentor, left in AD 721 to embark on his own missionary expedition. Like Willibrord's calendar, the Walderdorffer fragment contains the sequence of synodic lunar letters in the left-hand column. Additionally, it lists the Isidorian beginning of autumn (25 August), and the 'Latin' autumn equinox (24 September). Since the Ilmmünster calendar already contained Bedan information of which the Walderdorffer fragment is ignorant, I would be inclined to consider Walderdorff older than Ilmmünster.

---

61 This calendar fragment was first edited by Siffrin, 'Walderdorffer Kalenderfragment', pp. 204–9; a more accessible edition with full facsimile is provided by Gamber, *Bonifatius-Sakramentar*, pp. 53–59.

62 Berlin SBPK Lat. Fol. 877; Regensburg BZB Cim. 1. See especially Fingernagel, *Die illuminierten lat. Handschriften*, p. 119.

63 *CLA* vol. 8, p. 9 (no. 1052); Bischoff, *Die südostdeutschen Schreibschulen 1*, pp. 183–84, with additions in id., *Die südostdeutschen Schreibschulen 2*, p. 235.

64 This is the suggestion of Count Hugo of Walderdorff as related by Siffrin, *Walderdorffer Kalenderfragment*, pp. 212–13, and has often been repeated since. More sceptical is *CLA* vol. 8, p. 9 (no. 1052); Frank, 'Briefe des heiligen Bonifatius', pp. 76–79.

65 For the hand of this addition see Bischoff, *Die südostdeutschen Schreibschulen 1*, p. 184.

Irish

St Gall SB 51, 239–240 (Palimpsest): This manuscript is one of the oldest and most famous Irish Gospel codices that survive on the Continent. Pages 239–240 are a palimpsest, and the original text that was scraped off the vellum to reuse it for the Gospels was, in fact, a calendar. Unfortunately, the deletion was done so thoroughly that hardly anything remains decipherable. But since the Gospels were written in Ireland in a hand of the late eighth century, this palimpsest bears witness to the existence of a calendar in Ireland at around the middle of the eighth century.[66]

Zürich ZB Rhen. 30, 166v–169v: This calendar will be discussed in more detail below.

Frankish

Paris BnF nal 1203, 121v–124r: This famous evangelistary was written by a certain Godescalc in AD 781–83, probably at Worms, as the dedication poem to Charlemagne and his wife Hildegard at the end of the codex suggest.[67] An evangelistary is a collection of excerpts from the Gospels to be read in mass on sun- and feast days, starting with Christmas and systematically and linearly progressing through the year. Because of this format, it was only appropriate to accompany this text with a calendar and Easter table, here the Dionysiac one for AD 779–816. The pages of this *Prachtcodex* were purple, the script in gold and black. In the calendar, the martyrological entries and the grid of days were gold, the remaining data black. The latest martyr mentioned is Boniface (†AD 754), but for the seventh century also includes the Irish Kilian of Würzburg. Godescalc did not provide any information relevant for Easter calculation in the calendar proper, probably because of the following Easter table. The data provided was rather conservative, pre-Bedan, with the Isidorian beginnings of the seasons and the 'Latin' spring equinox and summer solstice. It also included the number of calendar and lunar days for each months, and the weekday sequence in the left-hand margin.

66 *CLA* vol. 7, p. 20 (no. 901); Kenney, *Sources*, p. 649 (no. 486), entertains the possibility that this codex was written on the Continent; see now the convenient entry with further literature in Ó Corráin, *Clavis litterarum Hibernensium*, pp. 483–88. For the calendar, see especially Borst, *Kalenderreform*, p. 194. The manuscript is available online at: https://www.e-codices.unifr.ch/en/csg/0051/239/0/Sequence-258, accessed 14 August 2020.

67 For the manuscript, calendar, and Easter table, see especially Piper, *Karls des Grossen Kalendarium*, pp. 8–19 (description of MS, including dating), 19–35 (edition of calendar and Easter table), 36–37 (edition of dedication poem), 38–50 (discussion of calendar). For the calendar in particular, see Borst, *Kalenderreform*, pp. 199–200 (with further literature). The manuscript is available online at: https://gallica.bnf.fr/ark:/12148/btv1b6000718s, accessed 14 August 2020.

Italian (Montecassino)

Paris BnF Lat. 7530, 277v–280r: The codex consists of principally grammatical content. Towards the end (fols 276r–290v), a computistical section is inserted, including a Dionysiac Easter table for AD 779–835 and a calendar. Because of this Easter table, which technically comprises three 19-year lunar cycles, the manuscript (and with it the calendar) is dated the first of these cycles, AD 779–797.[68] The early Beneventan hand points to Montecassino, as does the purview of the calendar with its six entries of abbots of Montecassino. [illegible], the *astronomica* and *computistica* listed in this calendar are Carolingian standard: number of solar and lunar days per months, hours of daylight and night-time at the end of each month, entrance of the sun into the zodiac signs, Bedan beginnings of the seasons with calculated length, 'Greek' spring equinox (21 March) but the 'Latin' summer solstice (24 June), Dionysiac start and end date for Easter new moon (8 March–5 April) and Easter Sunday (22 March–25 April), seats of the epacts (22 March) and concurrents (24 March).

Cava de' Tirreni BA 2, 70r–72v: The codex Paris BnF Lat. 7530 just discussed was produced principally as a collection of important grammatical texts. At the same time, the monastic community of Montecassino also composed its own, more comprehensive version of Isidore of Seville's *Etymologiae*.[69] In book 6 of his 20-volume work, Isidore included a discussion of Easter, which was very limited in scope and soon outdated. The authorities in Montecassino decided to replace Isidore's chapter by an impressive 12-folios long computistical fomulary. This included a calendar and, like BnF Lat. 7530, a Dionysiac Easter table starting in AD 779, but extending to AD 873 (as Isidore's orginal Easter table also covered 95 years). Again, the Easter table suggest compilation of the manuscript in the period AD 779–797, which can be further narrowed down to AD 779–792 on the basis of one of the calendrical algorithms. The calendar itself is very plain. Compared to the one in BnF Lat. 7530, it adds the alphabetical sequences in the left-hand column for the sidereal and synodic lunar months, but otherwise only records (besides martyrological entries, of course) the Bedan beginnings of the seasons (omitting spring), the range of the

---

68 *CLA* vol. 5, p. 15 (no. 569). For the manuscript, see especially Holtz, 'Le Parisinus Latinus 7530'; Brown, 'Where have all the grammars gone?'. For the calendar, see especially Loew (later Lowe)'s thesis *Kalendarien aus Monte Cassino*, pp. x, 4–6, 12–34 (edition), 39 (list of abbots), 42, 64, 69–83, Taf. II; Morin, 'Les quatre plus anciens calendriers'; Borst, *Kalenderreform*, pp. 196–98. The manuscript is available online at: https://gallica.bnf.fr/ark:/12148/btv1b84900617, accessed 14 August 2020.

69 *CLA* vol. 3, p. 3 (no. 284). For the caldendar, see especially Loew, *Kalendarien aus Monte Cassino*, pp. x, 1–4, 12–36 (edition), 39, 42, 64, 69–83, Taf. I; Morin, 'Les quatre plus anciens calendriers'.

Easter new moon (8 March – misplaced 4 rather than 5 April), only the lower limit of Easter Sunday (22 March; the upper limit, 25 April, was probably not recorded because the upper limit of the Easter new moon was misplaced), and the 'Greek' spring equinox (21 March) but 'Latin' summer solstice (24 June).

If these calendars are to be compared with Irish computistical knowledge towards the end of the seventh century, only one of the calendars listed qualifies, the calendar in Zürich ZB Rh. 30.[70] An interest in matters Irish (or even an Irish background) can fairly easily be detected in calendars with a martyrological / liturgical layer, as this would include Irish saints. The Zürich calendar contains three entries of four Irish saints:

> 16 January: Fursa (†c.649) and Foillan (†c.655)
> 1 February: Brigit (sixth century?)
> 17 March: Patrick (fifth century)

Of the Zürich calendar, only seven pages have survived. It starts with Jesus' birth on 25 December (the so-called nativity-style dating of the beginning of a year, a feature of pre-Carolingian calendars), breaks off after 10 April, resumes with 28 May, but then finally breaks off with 13 July. This does not allow for a full-scale interpretation of the evidence, but it provides enough insights for an indicative study. Concerning the Irish saints listed, it would have been of considerable interest if Columbanus was remembered on 23 November, or Gallus on 16 October.[71] What is remarkable is the exclusion of Columba on 9 June.[72] The reason for this may be found in the calendar's veneration for Fursa and Foillan. Bede, and the *Vita Fursei* that Bede worked from, only provide the most rudimentary outline of Fursa's life:[73] Fursa was of noble birth, left Ireland

70 The Zürich/Nivelles calendar is edited in Hänggi and Schönherr, *Sacramentarium Rhenaugiense*, pp. 294–99; for a short commentary with further literature, see the convenient entry in Borst, *Kalenderreform*, pp. 194–96.

71 The calendars recording Gallus' death on 16 October are numerous; see Borst, *Reichskalender*, pp. 1403–4. Columbanus is mentioned in one early calendar, Trento MP 1590, 216r (here, and only here, labeled abbot of Bobbio), and later principally in the Lake Constance area and northern Italy; see Borst, *Reichskalender*, pp. 1517, 1519–20.

72 Columba is commemorated by Willibrord and in the *Ilmmuenster Kalenderfragment* (see above near nn. 20 and 56), the early Trento calendar (fol. 210r), and a considerable number of later calendars; see Borst, *Reichskalender*, pp. 972–73.

73 Bede, *Historia ecclesiastica*, ch. III 19 (Plummer, *Baedae opera historica*, vol. 1, pp. 163–68). Bede worked from the *Transitus Fursei*; the chapters of this work relevant for Fursa's life have been extracted by Bruno Krusch in MGH SS rer. Merov. vol. 4, pp. 434–40; see now also the English translation of the entire *Transitus Fursei* by Rackham, *Transitus beati Fursei*.

(which part of Ireland is not specified) and travelled through the region of the Britons to East Anglia, where he preached the Gospel under King Sigbert. When pagan opposition grew in East Anglia, he left for Francia, where he was gracefully received by the under-aged Neustrian king Clovis II and, more importantly, his mayor of the palace Erchinoald. Erchinoald granted Fursa land to build a monastery in Lagny-sur-Marne, and the mayor of the palace took control of the body at Fursa's death so that its burial in Péronne (thenceforth known as *Peronna Scottorum*, no doubt also due to frequent visits by Irish pilgrims) would establish a [illegible]

The Irish entries in the Zürich calendar, therefore, tell a distinct story. The interest in the *regiones Scottorum* lies in Leinster, where the cult of Brigit flourished in Kildare and where the first record of Patrick, in Cummian's letter of AD 632, can be located.[74] Cummian wrote to abbot Ségéne of Iona to announce that the southern Irish churches would thenceforth follow 'Roman' Easter practices (at this time the system of Victorius of Aquitaine) against the traditional *latercus* which Iona clung to until AD 716.[75] This explains the exclusion of Columba, with the Columban community considered the principal adversaries in what became to be known as the Insular Easter controversy.[76] Cummian may well be connected to a circle of scholars in the eastern midlands, in the border territory between Munster and Leinster,[77] which saw the production of the cosmological work *De mirabilibus sacrae scripturae* in AD 654[78] and through which the loosely-defined Sirmond corpus of computistical texts was transmitted in AD 658.[79] Also, the lost Victorian computus of AD 689 can be located here,[80] and likewise the three computistical textbooks mentioned above appear to have been produced in such an intellectual

74 Cummian, *Epistola De controversia paschali*, ll. 208–9 (Walsh and Ó Cróinín, *Cummian's Letter*, pp. 84–85). Cf. Zimmer, 'Keltische Kirche', pp. 208, 210, 219; trans. by Meyer as *Celtic Church*, pp. 9–10, 18–19, 47–49.

75 For these Easter reckonings, see nn. 25 and 100.

76 For the Insular Easter controversy, see the pioneering (though outdated) studies by Krusch, 'Die Einführung des griechischen Paschalritus', pp. 141–69; Schmid, *Die Osterfestberechnung*; Gougoud, *Christianity in Celtic Lands*, pp. 185–201; Grosjean, 'Recherches'. More recently, see Corning, *The Celtic and Roman Traditions*; Charles-Edwards, *Early Christian Ireland*, pp. 391–415. The best study is Masako Ohashi's 1999 Nagoya Ph.D. thesis *Impact of the Paschal Controversy*. See now also Warntjes, 'Victorius vs Dionysius'.

77 See Grosjean, 'Sur quelques exégètes'; Ó Cróinín, 'Irish provenance'.

78 See Esposito, 'On the pseudo-Augustinian treatise', pp. 196–202; Kenney, *Sources*, pp. 275–77; MacGinty, 'The Irish Augustine', pp. 71–79; Ó Cróinín, 'Bischoff's Wendepunkte', pp. 212–20; the literature on this work is conveniently listed in Ó Corráin, *Clavis litterarum Hibernensium*, pp. 738–41.

79 Ó Cróinín, 'Irish provenance', pp. 177–85.

80 Warntjes, *Munich Computus*, pp. CXXIV–CXXVI.

environment in the 'Roman' southern half of Ireland.[81] Fursa and Foillan, in the AD 630s, moved to East Anglia presumably from this background.[82] When they eventually settled in northern-eastern France, connections were maintained between Ireland and their new foundations of Fosse and Langny-sur-Marne, but also *Peronna Scotorum* and Nivelles, as, e.g., evidenced by the Irish exile of Dagobert II of Neustria (orchestrated by the Austrasian mayor of the palace Grimoald, the brother of Gertrudis discussed below).[83]

The Frankish entries in the calendar point to the same region within the Frankish kingdoms, while also suggesting a foreign background of the older of its two layers. This older layer consists of the following popular Frankish saints: Genoveva of Paris (†c.502; 3 Jan), Hilary of Poitiers (†367/8; 13 Jan), Remigius of Reims (†c.533; 13 and 15 Jan), Vedastus of Arras (†c.540; 6 Feb), Germanus of Paris (†576; 28 May), Medardus of Nyon (†c.560; 8 Jun), and Martin of Tours (†c.397; 4 Jul). Of these, the Martin entry suggests original composition outside of what used to be Roman Gaul. It is phrased as if it commemorated Martin's death on this day, though 4 July rather referred to Martin's consecration as bishop and translation of his remains (he died on 8 November). Such a mistake is likely to have happened far removed from Tours, and, in fact, the entry's explicit geographic reference *in Gallis* may be an indication of composition outside of formerly Roman Gaul. This older layer of Frankish saints, up to c.AD 600, was probably copied directly from a late-sixth-century recension of the *Martyrologium Hieronymianum*.[84]

---

81 Ibid., pp. LIV–LV, LXXVIII–XCVI, CLII, CXCI–CXCII.

82 For the chronology of Fursa's departure from Ireland, see especially Plummer, *Baedae opera*, vol. 2, p. 173. For Fursa being potentially connected, intellectually, to the circle of Leinster scholars mentioned above, see Grosjean, 'Sur quelques exégètes', p. 75 n. 2; id., 'Chronologie de S. Feuillen', pp. 381–82 (p. 392 for Fursa's departure in the early AD 630s). Ó Riain, 'Les Vies de saint Fursy', pp. 405–10; id., *Dictionary*, pp. 357–59, argues for an eastern Ulster background of Fursa, which would firmly place him in Ireland's north. From the historical context of Fursa's later life in East Anglia and Francia, there can hardly be any doubt that Fursa followed 'Roman' practices (more specifically the Victorian Easter reckoning) introduced in southern Ireland in the early AD 630s. If Fursa was from the north, he may have belonged to a group of northern clerics favouring southern practice, and self-imposed exile abroad may have been a viable option for such clerics.

83 For the Irish background to Dagobert's exile, see especially Picard, 'Church and politics'; Ó Cróinín, 'Merovingian politics'. For Lagny, Fosse, Peronne, and Nivelles, see also Dierkens, *Abbayes*, pp. 293–96, 303–14; id., 'Prolégomènes', vol. 2, pp. 385–88.

84 The *Martyrologium Hieronymianum* survives in various recensions, for which see the neat overviews in Ó Riain, *Feastdays of the Saints*, pp. 2–8; and Borst, *Kalenderreform*, pp. 202–4. The oldest surviving recension precedes Willibrord's calendar in Paris BnF Lat. 10837, 2r–33v. All seven saints of the older Frankish layer in the Zürich/Nivelles calendar can be found in the *Martyrologium Hieronymianum* (edited by De Rossi and Duchesne

The five seventh-century Frankish entries in the Zürich calendar betray a more regional purview. Only one of these four saints, Sulpicius of Bourges (†646/7; 17 Jan), falls outside of the core region of the remaining entries, north-eastern Neustria. Sulpicius' inclusion can be explained by his active personal network facilitated by Chlothar II's palace school, which, besides Sulpicius, included later bishops in north-eastern France (Audoenus of Rouen, Paulus of Verdun, and others); this, no doubt, helped spreading Sulpicius' cult in this region, as most prominently manifested in his late seventh-century *Vita*.[85] The [illegible] compiler of the Zürich calendar is evidenced by the remaining four entries: Aldegundis of Maubeuge (†c.690; 30 Jan); Amandus of Maastricht (†c.680; 6 Feb); translation of Gertrudis of Nivelles (10 Feb); Gertrudis of Nivelles (†659; 17 Mar). The two entries of Gertrudis are the only ones in the calendar with an explicit place reference, which makes composition in Nivelles likely. St Amand and Maubeuge are less than 50 km from Nivelles, which explains the very early inclusion of Aldegundis and Amandus (Amandus was also instrumental in the foundation of Nivelles).[86] The latter two are the last datable saints in the calendar, which points to original composition at c.AD 700 or shortly thereafter (Borst suggests c.AD 750[87]). The script of the manuscript is dated to saec. VIII–IX by Lowe.[88]

Finally, it may be worth stressing that there is not a single Anglo-Saxon entry in this calendar. One may argue that Willibrord's death (7 Nov) was recorded on the lost November page, and that the omission of Boniface (5 Jun) was due to this calendar being composed before the latter's death in AD 754. Still, comparable Anglo-Saxon calendars of roughly the same time (Willibrord, Ilmmünster and Walderdorff) record numerous seventh-century, especially Northumbrian, kings and ecclesiasts.[89] The only entries of the Nivelles calendar that could potential be connected to an Anglo-Saxon milieu are those of Gregory the Great (†604), his death on 12 March and his ordination in Rome on 29 March. Gregory, famously, sent the first missionaries to Anglo-Saxon England, and his cult was

---

in *Acta Sanctorum Novembris 2.1*, pp. 1–195, with the relevant entries on pp. 5, 8, 17, 67, 76, 86), though Vedastus only in the Bern codex. The phrase *in Gallis* in the Martin of Tours entry on 4 July may be a remnant of this source; still, copying this phrase was a deliberate choice that reflects the context of the copyist.

85 The *Vita Sulpicii episcopi Biturigi* is edited by Krusch in MGH SS rer. Merov. vol. 4, pp. 364–80.

86 For Maubeuge, Nivelles, and Amandus, see especially the studies by Dierkens (with further literature): 'Saint Amand'; 'Notes biographiques'; and 'Maubeuge et Nivelles'.

87 Borst, *Kalenderreform*, p. 194.

88 *CLA* vol. 7, p. 50 (no. 1019).

89 See above, near nn. 20 and 56.

well established in Northumbria towards the end of the seventh century.[90] At the same time, Gregory may also have received some veneration in Ireland, especially among the southern Irish churches. The southern Irish sent a delegation to Rome when the Easter controversy escalated in the late AD 620s and early 630s.[91] This is a clear indication of the acceptance of Rome as the highest authority in spiritual matters. Gregory himself was heavily involved in the Easter question only some 30 years earlier, when he received a rather audacious letter by the Irish Abbot Columbanus arguing for the superiority of 'Irish' over 'Roman' / Frankish practices.[92] No response by Gregory is recorded, but it can easily be imagined that the southern Irish would cherish his memory as a guardian of Roman authority against northern Irish traditionalists in the Easter controversy of the remaining decades of the seventh century.

Thus, the Zürich calendar reveals a seventh-century (southern) Irish background and an early eighth-century composition in Nivelles with a particularly strong focus on its local monastic network. The Irish or Irish-influenced foundations of Lagny-sur-Marne, Fosse, and Péronne provide the link between Ireland and the Frankish heartland. Nivelles itself boasted an import of scholars from 'regions across the sea' according to the *Vita Geretrudis* of c.AD 670, which surely included Ireland.[93] The computistica in the Nivelles calendar, therefore, are quite likely to reflect Irish thought imported to Francia towards the end of the seventh century.

4 The Computistical Data of the Zürich (Nivelles) Calendar

In fact, a comparison of the computistical data of the Nivelles calendar to the pre-Bedan Irish computistical textbooks leads to an almost perfect match. The only exception is the list of lunar letters in the left hand column.

90 One of the earliest Saint's Lives composed in Anglo-Saxon England that has survived is the *Vita Gregorii* of AD 704×714 by a Whitby anonymous. It is edited and translated by Colgrave, *Earliest Life of Gregory the Great.*

91 Cummian, *Epistola de controversia paschali*, ll. 259–288 (Walsh and Ó Cróinín, *Cummian's Letter*, pp. 90–95).

92 Columbanus, *Epistula* I (ed. and trans. Walker, *Sancti Columbani Opera*, pp. 2–13).

93 *Vita Geretrudis* ch. 2 (edited by Krusch in MGH SS rer. Merov. 2, vol. 457). *Transmarinae regiones* can only refer to Britain and/or Ireland. Prinz interprets this phrase to refer to Ireland; Prinz, *Frühes Mönchtum im Frankenreich*, pp. 186, 661. Certainly, the calendar under discussion here points solely to Ireland, not Britain.

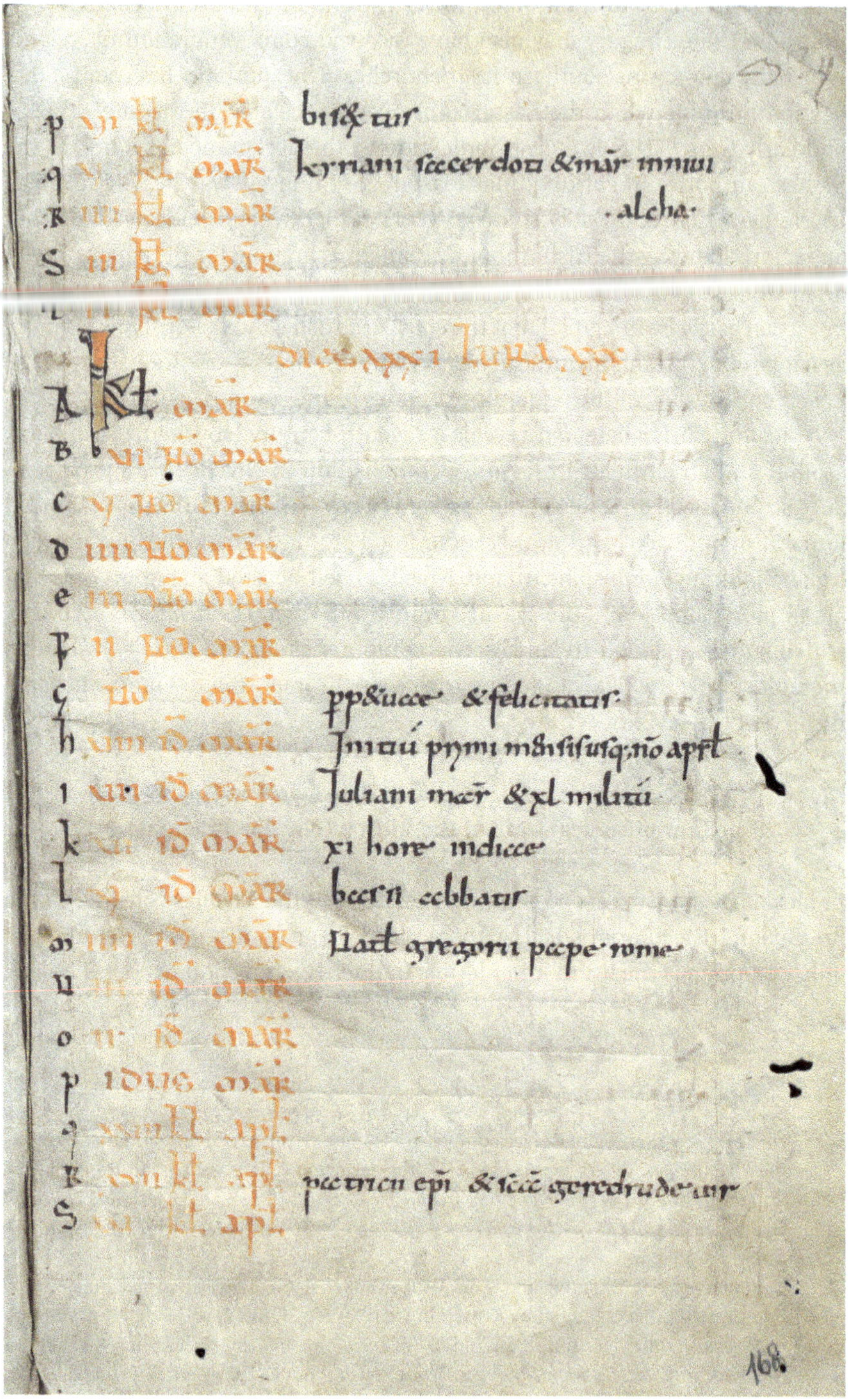

FIGURE 3.4 Zentralbibliothek Zürich, Ms. Rh. 30, fols. 168r–168v: March page
WITH PERMISSION OF THE ZENTRALBIBLIOTHEK ZÜRICH

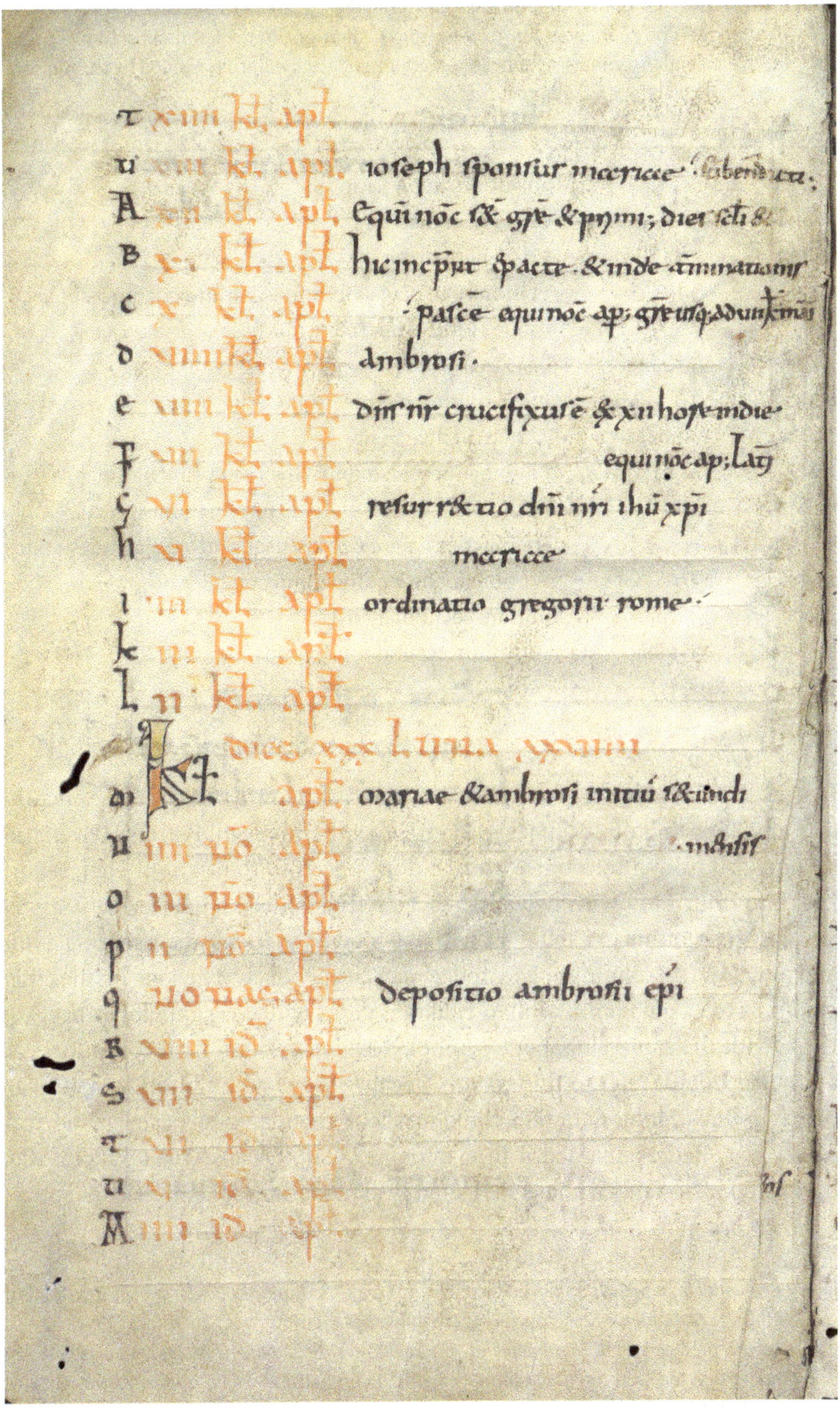

TABLE 3.3 Overview of the computistical entries in the Zürich / Nivelles calendar of c.AD 700 (Zürich ZB Rh. 30, 166v–169v). Additionally, the left-hand column of the calendar contains synodic lunar letters, and each month is headed by the number of calendar and lunar days.

25 Dec	VIII Kl Ian	Solstitium apud Latinum
9 Jan	V Id Ian	VII hore in diae
24 Jan	VIIII Kl Feb	VII(I) hore in die
8 Feb	VI Id Feb	[illegible] et terminus initium quadragesimae usque II Idus Mai- (recte Martii)
24 Feb	VI Kl Mar	bisextus
8 Mar	VIII Id Mar	Initium primi mensis usque Nonas Aprilis
10 Mar	VI Id Mar	XI hore in diae
21 Mar	XII Kl Apr	Equinoctium secundum Grecos et prim⟨i⟩us dies saeculi
22 Mar	XI Kl Apr	Hi incipient epacte et in die terminationis pasce, equinoctium apud Grecos, usque ad VII Kalendas Maii
25 Mar	VIII Kl Apr	XII hore in die, equinoctium apud Latinos
1 Apr	Kl Apr	Initium secundi mensis
20 June	XII Kl Iul	Solstitium secundum Egyptios

4.1 *Lunar Letters*

In the left-most column of the Nivelles calendar, there is a list of three successive alphabets (all omitting j; see Figure 3.4): the first simply from a to u; the second also from a to u but with two dots placed to the right of each letter; the third from a to t with two dots placed to the left of each letter (a–u, a:–u:, :a–:t). This amounts to a letter system of a total of 59 items, representing a full and a hollow lunar month of 30 and 29 days respectively. The idea behind this system is to easily identify Julian calendar dates with the same lunar age, i.e. those with the same alphabetic symbol.[94] The pre-Bedan Irish computistical textbooks do not refer to this system, because they had no interest in tabular oversimplification of data that could easily be calculated by basic calendrical algorithms. Bede explained it in exuberant detail, because his textbook was designed for students who needed lengthy explanations for simple concepts,[95]

94 See especially Sickel, 'Lunarbuchstaben', pp. 159–68; Borst, *Kalenderreform*, pp. 405–11; Nothaft, *Scandalous Error*, pp. 54–55 (with further literature).

95 In the preface to his *De temporum ratione* (Jones, *Bedae opera*, p. 175), Bede explicitly states that his brethren urged him to expand on his earlier *De temporibus*, as they considered that too concise.

this alphabetic system for the stupid among these ('for those who do not know how to calculate', who are 'lazy or bored').[96] He provided a table that assigns a specific lunar age to each of these letters for every year of the 19-year cycle (e.g., letter a has *luna* 9 in year 1, *luna* 20 in year 2, *luna* 1 in year 3, etc.). With this additional table, the exact lunar age of any given Julian calendar date could be inferred from the alphabetic symbol found in the calendar, provided the reader knew which year in the 19-year cycle one lived in. But, as mentioned, the system certainly also had its use without the additional table, indicating Julian calendar dates with the same lunar age. Thus, no calendar necessarily needed an additional table to translate the alphabetic symbols.

It is evident that Bede did not invent this system, either in form of the calendar column or the additional table. Willibrord's calendar already incorporated the alphabetic column, which, in this form, did not occur in the Roman models of 354 and Polemius Silvius. The additional lunar table also appears to be pre-Bedan (Bede himself ascribes its composition to *Antiquitas*; by the early ninth century it is attributed to Jerome). There certainly is no reason to doubt that this system circulated in the last three decades of the seventh century.[97] Interestingly, the three alphabets are rather sloppily noted in the Nivelles calendar, omitting most of the punctuation. This suggests that either the author or, more likely, the copyist did not understand the mechanics of the underlying system. But, to be sure, the author of this calendar would have had no problems in obtaining this alphabetic sequence from computistical circles in Ireland or in north-eastern France in the early eighth century.

4.2 *Number of Calendar and Lunar Days per Month*

In the Nivelles calendar, each month is headed by the number of Julian calendar days and lunar days (e.g., 31 and 30 respectively for March; see Figure 3.4). The length of each Julian calendar month could easily be deduced from Macrobius or the passages quoted above.[98] The number of lunar days per month is more interesting, because this entailed considerable computistical knowledge, at least in application. First, to use this information sensibly, the reader needed to know that the number of lunar days listed before each Julian calendar month referred to the lunar month *ending* in that specific month, not *beginning* in it. E.g., in the nineteenth year of the 19-year cycle, the lunar

96 Bede, *De temporum ratione* ch. 23 (Jones, *Bedae opera*, pp. 224–26): *quis computare non potest; quisque deses vel hebes est.*

97 See especially Jones, *Bedae pseudepigrapha*, pp. 65–68; and also Jones, *Bedae opera*, pp. 358–59; Borst, *Kalenderreform*, 406–8.

98 See above, section 2: 'Calendrical knowledge in seventh-century Ireland'.

age on 1 January was 27; thus, the January lunation ended with *luna* 30 on 4 January, and the lunation that began with *luna* 1 on 5 January belonged to February. There were exceptions to this rule, which made the system moderately complex.[99]

From a retrospective viewpoint, the common way of distributing full and hollow lunar months was in alternating sequence, as found in *all* early medieval calendars: January 30, February 29, March 30, April 29, etc. For a computist around AD 700, the situation was less clear-cut. The early Middle Ages knew three different [illegible] reckonings, the *latercus*, Victorius, and Dionysius.[100] None of the original authors of these systems was specific about the sequence of lunations employed. From the data available, computists (unanimously, as far as the surviving evidence goes) reconstructed the alternating sequence for Victorius. The *latercus* used an interesting alternative, which assigned hollow lunations (of 29 days) to short Julian calendar months (of 30 days), full lunations (of 30 days) to long Julian calendar months (of 31 days).[101] When faced with the Dionysiac Easter table, Irish computists had thus already experienced two different methods. The Dionysiac system appeared to represent a third option: the lunar data for years 8, 11, and 19 of the 19-year cycle suggested a full April lunation (of 30 days). Thus, some Irish computists reconstructed a sequence of January 30, February 29, March and April 30, May and June 29, July 30, August 29, September 30, October 29, November 30, December 29. The Munich Computus is explicit about this sequence of lunations.[102] The *Computus Einsidlensis* applies it, but then lists the alternating one for Dionysius;[103] I suspect that a later copyist replaced the older interpretation by his own custom. Certainly, the *Computus Einsidlensis* provides us with a flavour

99 These exceptions are best explained by Helperic, *Liber de computo* ch. 36 (PL vol. 137, cols. 46–47).

100 The *latercus* is edited by Mc Carthy, 'Easter principles', pp. 218–19; an English translation is provided by Blackburn and Holford-Strevens, *Oxford Companion to the Year*, pp. 870–75. Victorius' and Dionysius' computistica are edited by Krusch, *Studien*, pp. 4–52, 59–86 respectively. A quick overview over these reckonings and their differences is provided by Warntjes, 'Computus as scientific thought', pp. 159–68. For their technical details, see especially Holford-Strevens, 'Paschal lunar calendars', pp. 178–205.

101 See especially Warntjes, '84 (14)-year Easter reckoning', pp. 41–43 and the reference in the following note, which discuss all surviving evidence. Davis, 'Lunar year', now argues that this sequence of lunations also formed the basis of the famous second-century Coligny calendar.

102 Munich Computus ch. 50 (Warntjes, *Munich Computus*, p. 178, with further references).

103 Einsiedeln SB 321 (647), pp. 109–11.

of the shift from the earliest Irish interpretation of the Dionysiac sequence of lunations to the alternating one that was to become the standard from the AD 720s onwards. The third Irish computistical textbook, *De ratione conputandi*, stands at the end of this development, vehemently arguing for the alternating sequence, while explicitly condemning the older alternative:[104]

> Sciendum nobis quomodo menses lunares mensibus solis apud Grecos et Latinos secundum hanc regulam iunguntur. Ita etiam, id est .xxx. luna Ianuario, .xxviiii. Feburario, .xxx. Martio, .xxviiii. Apreli (etsi alii dicunt lunam Aprelis .xxx.mam esse semper apud Grecos, quod non sinit natura, nisi in tribus annis tantum, id est quando .xxvi. uel .xxvii. uel .xxviiii. in Kalendis Ianuarii fit. Nisi enim in his tribus annis .xxx.ma fuerit luna Aprelis, .xiiii. lunae primi mensis, quae contra Maium regulantur, non erunt in naturalibus mensis diebus); .xxx. Maio, .xxviiii. Iunio, .xxx. Iulio, .xxviiii. Agusto, .xxx. Septimbri, .xxviiii. Octimbri, .xxx. Nouimbri, .xxviiii. Decimbri.

> It should be known to us, how the Greek and the Latins link the lunar months to the solar months according to this rule: *luna* 30 in January, 29 in February, 30 in March, 29 in April (though some say that *luna* 30 always belongs to April according to the Greeks; this nature does not generally allow, but only in the three years with epact 26, 27, or 29 on 1 January; in fact, if there was not *luna* 30 in April in these three years, the full moons of the first month, which are ruled by May, would not be on the natural days of the month), 30 in May, 29 in June, 30 in July, 29 in August, 30 in September, 29 in October, 30 in November, 20 in December.

By the early eighth century, at the end of the debate outlined above and at the time when the genre of calendars took off, the alternating sequence seems to have been universally accepted for the Dionysiac reckoning. Any maker of a calendar could have implemented this basic information from any pre-Bedan computistical textbook.

104 *De ratione conputandi* ch. 72 (Walsh and Ó Cróinín, *Cummian's Letter*, pp. 179–80); cf. Warntjes, *Munich Computus*, pp. XCIV, CXCVIII–CC.

TABLE 3.4 The different sequences of lunations known in the seventh century

Month	*latercus*	Victorius	Dionysius 1	Dionysius 2
January	30	30	30	30
February	29	29	29	29
March	29	30	30	30
April	29	29	30	29
[illegible]	[illegible]	30	29	30
June	29	29	29	29
July	30	30	30	30
August	30	29	29	29
September	29	30	30	30
October	30	29	29	29
November	29	30	30	30
December	30	29	29	29

4.3 *Solstices and Equinoxes*

The dates of the solstices and equinoxes were crucial for early medieval computists. The Alexandrian reckoning set out the rule that Easter was supposed to fall on the first Sunday after the first full moon after the spring equinox, which Dionysius Exiguus linked to the council of Nicaea when he popularized this method in the Latin West.[105] Of the three known Easter reckonings of the early Middle Ages, only the Alexandrian / Dionysiac system followed this rule. The *latercus* and Victorius ensured that Easter Sunday would always occur after the spring equinox; the Easter full moon, however, could fall before it. These two systems are also interesting because they followed different dates for the spring equinox: the *latercus* set it on 25 March, while Victorius was following Alexandria in preferring 21 March.

This already illustrates that two different traditions for the dates of the solstices and equinoxes circulated in the seventh and eighth centuries. The first was principally transmitted through Isidore's *Etymologiae* and *De natura rerum*.[106] Here, the equinoxes and solstices fell on the eighth day before the

105 Dionysius, *Epistola ad Petronium* (Krusch, 'Studien', p. 65). For the spread of the myth that the Council of Nicaea in AD 325 codified this rule, see Jones, 'Legend of St. Pachomius'.

106 Isidore, *Etymologiae* ch. V 34.2–3 (edited and translated into Spanish by Yarza Urquiola and Andrés Santos, *Isidoro de Sevilla*, pp. 112–15); id., *De natura rerum* ch. VIII 1 (edited by Fontaine, *Isidore de Séville*, p. 205).

Calends of the respective month, i.e. *VIII Kal. Apr.* (25 March; spring equinox), *VIII Kal. Iul.* (24 June; summer solstice), *VIII Kal. Oct.* (24 September; autumn equinox), *VIII Kal. Ian.* (25 December; winter solstice). It came to be associated with the 'Latins' or 'Romans' in computistical literature. The alternative 'Greek' dates, on the other hand, were not readily outlined in any source available to mid-seventh-century intellectuals. Dionysius, in the prologue to his Easter table, explicitly mentioned 21 March as the spring equinox, and the letter of the Alexandrian patriarch Proterius translated by Dionysius from Greek into Latin did the same.[107] From this datum, the 12th day before the Calends of April for the spring equinox, the remaining three dates for the summer solstice, the autumn equinox, and the winter solstices were inferred as also occurring on the 12th day before the Calends of the respective month, mirroring 'Latin' practice: *XII Kal. Apr.* (21 March; spring equinox), *XII Kal. Iul.* (20 June; summer solstice), *XII Kal. Oct.* (20 September; autumn equinox), *XII Kal. Ian.* (21 December; winter solstice). All three Irish computistical textbooks refer to, in one form or another, both traditions,[108] and explain the difference between these.[109]

Because only parts of the Nivelles calendar survive (25 December–10 April, 28 May–13 July), it cannot be ascertained if all solstices or equinoxes were noted. The 'Greek' tradition is recorded in both places in which it was to be expected, 21 March and 20 June. The 'Latin' tradition is mentioned on 25 December and 25 March (see Figure 3.4), but not on 24 June (probably that entry was already considered overloaded with the birth of John the Baptist and the reception in Ephesus of John the Evangelist). Two aspects are noteworthy: First, the spring equinox on 21 March is ascribed to the Greeks (see Figure 3.4), as was common, while the terminology changed for the summer solstice on 20 June *secundum Egyptios*. This may well reflect a very early stage in the recording of the 'Greek' tradition, when only the spring equinox was known from the available sources, and the remaining equinox and solstices need to be inferred (and therefore received a different terminology). More interesting is the recording

107 *Epistola Proterii* ch. 8 (Krusch, *Studien*, p. 277); Dionysius, *Epistola ad Petronium* (Krusch, 'Studien', p. 65). See also the mysterious *Disputatio Morini* ll. 22–23 (edited by Graff, 'Recension of two Sirmond texts', p. 141).

108 Munich Computus ch. 38 (Warntjes, *Munich Computus*, pp. 102–5); *De ratione conputandi* chs. 47–50 (Walsh and Ó Cróinín, *Cummian's Letter*, pp. 157–62); for the *Computus Einsidlensis*, see the following note.

109 *Computus Einsidlensis* (Einsiedeln SB 321 (647), pp. 98–99); Munich Computus ch. 38 (Warntjes, *Munich Computus*, pp. 106–7). *De ratione conputandi* does not have a similar passage, but cf. the Irish *Dial. Langob.* discussed below (n. 111). See also Nothaft, *Scandalous Error*, pp. 29–33.

of an alternative 'Greek' date for the spring equinox, on 22 March (see Figure 3.4). The terminology changed slightly from the 21 March to the 22 March entry, from *secundum Grecos* to *apud Grecos*. *Apud* for *secundum* is relatively unusual in the calendar tradition (but not in computistical texts), but in the Nivelles calendar it recurs in the entries of the 'Latin' spring equinox on 25 March and summer solstice on 24 June, a tradition followed *apud Latinos*. This indicates that the unusual 22 March for the alternative 'Greek' spring equinox was taken from the same source as the 'Latin' entries. The Bobbio Computus transmits an interesting passage that shows the [illegible] (though not using *apud*): it discusses all 'Latin' equinoxes and solstices, but only one Greek alternative, the spring equinox on 22 March.[110] The reason for assigning 22 March to the spring equinox rather than the computistical 21 March is theological, and best explained by the eighth-century Irish tract *Dial. Langob.*:[111] If the celestial objects, including sun and moon, were created on the fourth day of creation, and if this coincided with the Latin spring equinox on 25 March, then 22 March was the first day of creation, which the Greeks considered as more appropriate for the spring equinox.

Thus, the Nivelles calendar's information for the equinoxes and solstices very well reflects the mindset of early eighth-century Irish computists. Bede, in one of his more dogmatic chapters, discusses the difference between 'Latin' and 'Greek' tradition, only to conclude that the 'Latin' dates violate the universal paschal doctrine of celebrating Easter after the equinox.[112] A calendar taking Bede as its authority for the computistical entries could not record these misleading dates.

4.4 *The Moveable Feasts Dependent on Easter*

The spring equinox leads directly into Easter calculations proper. For calendar makers, obviously the Julian calendar limits for Easter Sunday were the most significant information. Of the three Easter calculations of the early Middle Ages, each had a different range: *latercus* 26 March–23 April; Victorius 22 March–24 April; Dionysius 22 March–25 April. As mentioned earlier, *all* calendars that have survived are Dionysiac.[113] Thus, invariably (and frequently) the Julian calendar limits of 22 March–25 April are recorded in this genre of texts, at is the case in the Nivelles calendar under 22 March (see

110 Bobbio Computus chs. 29, 45 (*PL* vol. 129, cols. 1291, 1297; Milan BA H 150 inf., 14r, 19r–v). The first of these chapters can also be found in the Irish computus in Brussels KBR 8654–72, 203v–205v: 204r.

111 *Dial. Langob.* ch. 18 (Borst, *Schriften*, pp. 448–50).

112 Bede, *De temporum ratione* ch. 30 (Jones, *Bedae opera*, pp. 235–37).

113 Cf. n. 51 above, and also the following note.

Figure 3.4).[114] Easter Sunday regulates all other moveable feasts of the liturgical calendar, which principally are Lent, Rogation, and Pentecost. Historically, Lent entered the computistical discussion at a very early stage, as it already constituted a column in the *latercus* of the early fifth century.[115] Rogation and Pentecost become part of this discussion in the second half of the eighth century and the early ninth century at the latest, respectively.[116] Unsurprisingly,

---

114 Some calendars mark 24 April as the end of the Easter period, which could refer to Victorian practice. This, however, is simply a scribal error, a misplacement one line too early of the Dionysiac end of the Easter period on 25 April, as can be demonstrated by the only calendar of the 'major recensions' listed by Borst to have this feature (*Reichskalender*, pp. 809, 812), his calendar C6 = Melk SB 412, 44–55 (Auxerre, AD 836–854?): This calendar frequently lists computistical entries one or two days too early, which is best illustrated by the recording of the seventh and third embolisms on 4 and 5 March, which are then, realizing the mistake, connected to the following entries by drawing a line to 5 and 6 March respectively, the correct places (p. 46); in April, the last possible date for the beginning of the Easter lunation (5 April) and the last possible date of the Easter full moon (18 April) are correctly recorded for the Dionysiac reckoning, which implies that the last possible date for Easter Sunday was also supposed to follow this reckoning (i.e. to be recorded on 25 April) and was simply misplaced (one line too early, 24 April). More interesting is the note in calendars F2–3, 7, G4, H1–4 of Borst's edition for 23 April that this date constitutes the 'end of the Easter of the Latins' (Borst, *Reichskalender*, pp. 806, 808). The classification of 'the Latins' makes it immediately evident that the calendar makers here are referring to a tradition not their own, and at least F2 and G4 record the more appropriate custom of 'the Egyptians' on 25 April (Borst, *Reichskalender*, pp. 813–14, 816). Where this information originally came from, is an interesting question. The most obvious explanation is Victorius, *Epistola ad Hilarum* ch. 4 (Krusch, 'Studien', p. 19), where Victorius outlines the practice, before his reform, of the 'Latins' of Julian calendar limits for the Easter new moon of 5 March to 2 April (see also Borst, *Reichskalender*, pp. 651, 653, 751, 753, which Borst misunderstands), and Easter lunar limits of 16 to 22; this leads to the last possible Easter date being 23 April, *luna* 22. An interesting, though highly unlikely alternative possibility is that at least F2 (Karlsruhe BLB Aug. per. 167, 16v; the only calendar in Irish script; see nn. 128, 157–8) here originally refered to the *latercus* rather than the *Latini* (the term is abbr. as *lati-*). It certainly should be stressed again that all of these calendars followed the Dionysiac / Alexandrian reckoning.

115 See Mc Carthy, 'Easter principles', pp. 218–19 (edition); Blackburn and Holford-Strevens, *Oxford Companion to the Year*, pp. 870–75, esp. 873–75 (translation); Warntjes, '84 (14)-year Easter reckoning', pp. 80–82 (facsimile).

116 The Rogation features in the Victorian Easter table of AD 720–4 discussed in n. 120; in the Dionysiac context, the rogation is fully established by the time of *Lect. comp.* ch. I 12–14 of AD 760 (Borst, *Schriften*, pp. 556–60); the most substantial discussion can be found in the unpublished Computus of AD 789 (Fulda UB F III 15k, 36r–49r), which was the model for relevant chapters in Hrabanus Maurus' *De computo*; cf. Warntjes, 'Köln', pp. 54–56, 83–88. For Pentecost, see *Lib. calc.* ch. 7 (Borst, *Schriften*, pp. 1391–92); the earliest calendar to record this may be Berlin SBPK Phillipps 1831, 2v (Verona, AD 800–810); note that many calendars record *Primum Pentecosten* on 15 May (Borst, *Reichskalender*, pp. 891–92), which is not to be confused (as Meyvaert, 'Discovering the calendar', pp. 28–29 does) with the

then, the Nivelles calendar does not include any reference to Pentecost. Whether Rogation was noted anywhere must remain speculative, as its dates fall into the calendar's first lacuna.

The *initium quadragesimae*, the beginning of the Lenten fast, however, features prominently in the Nivelles calendar, which records the Julian calendar limits of the sixth Sunday before Easter, 8 February (*VI Id. Feb.*) to (and inclusive) 14 March (*II Id. Mar.*), under the first day of this range. It is important to remember that one of the main *desiderata* in Bede's accounts of time reckoning, his *De temporibus* of AD 703 and his length[illegible] AD 725, is a discussion of Lent. This demonstrates that some key computistical information could not be found in Bede's works by early medieval calendar makers. In Irish computistical textbooks of the early eighth century, on the other hand, Lent features prominently. The earliest two of these textbooks, the *Computus Einsidlensis* and the Munich Computus, explain in detail how to calculate the beginning of Lent.[117] *De ratione conputandi* of c.AD 720s, explicitly refers to the Julian calendar limits, here 7 February to 14 March (probably an exclusive count or a copying error).[118]

In Frankish computistics, a similar development can be reconstructed. The Anglo-Saxon missionary Willibrord introduced the Dionysiac Easter calculation to the Frankish kingdoms when he moved from Ireland to the Frankish-Frisian border in AD 690. None of the three texts with computistical information he brought with him, his calendar, Easter table, and computistical formulary, contain any information about Lent.[119] The recipient Frankish culture, at this time still firmly following Victorius, was more alert to the importance of Lent. A Victorian Easter table covering the years AD 700–770, and the Easter table in the Burgundian Dialogue of AD 727, contain a column for the beginning of Lent, probably following Irish examples.[120] When Dionysius got

first day of the 35-day Pentecostal range, which starts on 10 May (Borst, *Reichskalender*, pp. 875, 878); *Primum Pentecosten* on 15 May refers to the historic event of the Holy Spirit enlightening the apostles, corresponding to Christ's resurrection on 27 March; cf. also Borst, *Kalenderreform*, p. 426.

117 Einsiedeln SB 321 (647), pp. 119–122; Munich Computus ch. 58 (Warntjes, *Munich Computus*, pp. 230–41).

118 *De ratione conputandi* ch. 103 (Walsh and Ó Cróinín, *Cummian's Letter*, p. 207).

119 For Willibrord's calendar, see n. 13; for his Easter table, n. 19; for his formulary of calendrical algorithms, n. 20.

120 The Victorian Easter table of AD 700–770 is transmitted in Vatican BAV Reg. lat. 586, 9v–10v (Lake Constance region?, early ninth century?); Paris BnF Lat. 4860, 147v–148r (Reichenau, late ninth century); cf. Warntjes, *Munich Computus*, pp. LXXXIV–LXXXV. *Dial. Burg.* ch. 16 (Borst, *Schriften*, pp. 368–72); the Victorian Easter table for the years equivalent to AD 720–724 in Paris BnF Lat. 10756, 66v–67r not only records the beginning of

more widely accepted in the Frankish kingdoms in the AD 730s, Lent features more prominently. The earliest Frankish Dionysiac textbook of AD 737 contains a list of the 19 *termini quadragesimae* (the equivalent to the Easter full moon; it is generally *luna* 2 of the lunar month preceding the Easter lunation, i.e. exactly 6 weeks before the Easter full moon, but here *luna* 3, the earliest possible lunar age of the beginning of Lent), and explains that the beginning of Lent is to be set on the Sunday following these fixed Julian calendar dates.[121] The unpublished *Computus Rhenanus* of AD 776 contains rules of how to quickly calculate the beginning of Lent and its lunar age.[122] By the time of the Computus of AD 789, the moveable feasts in the Dionysiac reckoning are established as an integrated part of Frankish computistics. Here, the Julian calendar limits of the beginning of Lent are explicitly mentioned.[123]

This demonstrates that the Julian calendar limits for the Dionysiac beginning of Lent could easily have been calculated from information available to Irish computists at around AD 700, to Frankish computists around the AD 730s. By the AD 720s texts circulated in Ireland from which the Julian calendar limits of the Dionysiac beginning of Lent could simply be copied, and the same applied to the Frankish kingdoms a few decades later.

Of the moveable data related to Easter calculation, the Nivelles calendar also records the Julian calendar limits for the Easter new moon (8 March to 5 April, noted under the first day of this range; see Figure 3.4). This information was readily available to computists and calendar makers through Victorius of Aquitaine's Letter to Archdeacon (later Pope) Hilarus of AD 457, or Dionysius' Letter to Petronius, or Isidore of Seville's *Etymologiae*.[124] It can also be found

---

Lent, but also the Rogation; Paris BnF Lat. 10756 and Bern BB 611, which transmits the only surviving copy of *Dial. Burg.*, originally formed one manuscript from the first half of the eighth century; cf. https://www.e-codices.unifr.ch/de/description/bbb/0611/Mittenhuber, accessed 14 August 2020; the Paris codex is available online at: https://gallica.bnf.fr/ark:/12148/btv1b9065920c.r=10756?rk=42918;4, accessed 14 August 2020; the Victorian Easter table therein is transcribed and discussed in Krusch, 'Chronologisches', pp. 93–94.

121 *Dial. Neustr.* ch. 19 (Borst, *Schriften*, pp. 400–01).

122 Cologne DB 103, 186r–v (Cologne, AD 798; available online at: www.ceec.uni-koeln.de, accessed 14 August 2020); Wolfenbüttel HAB Cod. Guelf. 91 Weiss., 170r–v (Worms or Weissenburg, first half of ninth century; available online at: http://diglib.hab.de/?db=mss&list=ms&id=91-weiss, accessed 14 August 2020). For this text, see Warntjes, 'Argumenta', pp. 43, 70, 75–77, 101–02; id., 'Computus Cottonianus', pp. 189–90, 199–203, 207–12; id., 'Köln', 62–64.

123 Cf. n. 116. See also *Lect. comp.* ch. VI 1 (Borst, *Schriften*, pp. 613–14); this section of the text is datable to AD 790 (cf. Warntjes, 'Köln', pp. 61, 66–67).

124 Victorius of Aquitaine, *Epistola ad Hilarum* ch. 4 (Krusch, 'Studien', p. 20): *Sancte memoriae contra Theophilos quondam Alexandrinus antestis ad Theodosium imperatorem datis epistolis, ex primo ipsius et Gratiani quinto consulate conditum paschale direxit, in quo ab*

in all three Irish computistical textbooks of the early eighth century.[125] What is interesting here is the choice made by the calendar maker: Focusing on the Easter new and full moons, Easter Sunday, and Lent, but not on Rogation or Pentecost speaks for its early composition in the first half of the eighth century rather than the second.

4.5 *Bissextus*

In the Nivelles calendar, the additional day in leap-years (*bissextus*) is noted under 24 February; this information [illegible] Macrobian excerpt called *Disputatio Chori et Praetextati* discussed above, which circulated widely in the early Middle Ages.[126] Thus, the calendar maker here only states common knowledge, and it appears that no further comment is needed. Still, it is worth keeping in mind that this date became standard only in the eighth century. The earliest two Irish computistical textbooks, the *Computus Einsidlensis* and the Munich Computus, still experimented with two other options: 2 March as suggested by a rather obscure passage in Isidore's *Etymologiae*, and 21 March, understood as the first day of creation.[127] Interestingly, remnants of this seventh-century debate can still be found in some calendars with strong Irish connections.[128] The Nivelles calendar is

---

VIII. idus martias usque in diem nonarum aprilium, quolibet in medio eiusdem temporis spatio natam perhibit lunam, facere primi mensis exordium. Dionysius, *Epistola ad Petronium* (Krusch, 'Studien', p. 65): *Sed quia mensis hic, unde sumat exordium vel ubi terminetur, evidenter ibi non legitur, praefati venerabiles CCCXVIII pontifices antique moris observantiam et exinde a sancto Moyse traditam, sicut in septimo libro ecclesiasticae refertur historiae, sollertius investigantes, ab VIII. Idus Martii usque in diem Nonarum Aprilis natam lunam facere dixerunt primi mensis exordium.* Isidore of Seville, *Etymologiae* ch. VI 17.20 (edited and translated into Spanish by Chaparro Gómez, *Isidoro de Sevilla*, pp. 104–05): *Greci primi mensis lunam ab octauo Idus Martias usuqe in die Nonas Apriles obseruant.*

125 Einsiedeln SB 321 (647), p. 111; Munich Computus ch. 55 (Warntjes, *Munich Computus*, pp. 216–17); *De ratione conputandi* ch. 90 (Walsh and Ó Cróinín, *Cummian's Letter*, p. 197).

126 *Disputatio Chori et Praetextati* ch. 18 (Holford-Strevens, *Roman Calendar for Beginners*, pp. 80–81).

127 See especially Einsiedeln SB 321 (647), pp. 105–6; Munich Computus ch. 41 (Warntjes, *Munich Computus*, pp. 130–31). Cf. Warntjes, *Munich Computus*, pp. CXLIII–CXLIV, CXXXIX; Smyth, 'Once in four', pp. 230–36; Nothaft, *Scandalous Error*, pp. 25–26.

128 Of the calendars collated by Borst, 2 March according to Isidore as a potential place for the bissextile day is only recorded in the only calendar in Irish script, Karlsruhe BLB Aug. perg. 167, 16v (Soissons-Laon, c.AD 848; online at: https://digital.blb-karlsruhe.de/blbhs/content/titleinfo/20736, accessed 14 August 2020), and its copy St Gall SB 248, p. 73 (Soissons-Laon c.AD 850; online at: https://www.e-codices.unifr.ch/de/list/one/csg/0248, accessed 14 August 2020); cf. Schneiders, 'Irish calendar', pp. 46, 66. See also the two later Spanish calendars in El Escorial RB d.I.2, 6r–v (Albelda, 974–76) and Silos AM 4 (Albelda, AD 1052). The *bissextus* on 21 March is only recorded in Vatican BAV Ottob. Lat. 67, 23r

beyond that stage, and, as in the data for the *initium quadragesimae*, resembles the last of these three Irish textbooks (*De ratione conputandi*) more than the first two.[129]

4.6 *Epact*

Like the *bissextus* on 24 February, it was common knowledge that the *epactae* that constituted the second column in Dionysius' Easter table refer to the lunar age on 22 March. The first Latin author to explicitly mention this date, interestingly, was not Dionysius himself, but Isidore.[130] Through his *Etymologiae*, one of the most popular texts of the early Middle Ages, this datum circulated widely. The Irish computistical textbooks also note it,[131] though they preferred to calculate from the lunar age of 1 January as they had been accustomed to from the *latercus* and Victorius.[132] If noting the epact on 22 March thus is fairly non-indicative, it is worth pointing to an omission by the Nivelles calendar maker: A year is generally characterized by two data in combination, the lunar age and the weekday of a given day. In the *latercus* and the Victorian reckoning, 1 January was chosen for both. In Dionysius' Easter table, the lunar age on 22 March was matched by the weekday on 24 March, called the concurrent. That very term is nowhere to be found in the Irish computistical textbooks,[133] and it may be significant that the Nivelles calendar does not provide this information under 24 March.

(Angers?, c.AD 978?, based on a northern Italian exemplar of c.AD 810) and Karlsruhe BLB Karlsruhe 442, 40v (Angers?, saec. X/XI?; available online at: https://digital.blb-karlsruhe.de/blbhs/Handschriften/content/titleinfo/2589900, accessed 14 August 2020). Meyvaert, 'Discovering the calendar', pp. 33–34 ascribes the recording of this datum explicitly to Bede, because he believes that the Vatican calendar is the closest to Bede's original now lost; it is rather an Irish feature. The computistical contents of Karlsruhe 442 with a copy of *De computo dialogus* and a recension of *De divisionibus temporum* reveals and ultimately Irish background. Cf. also the calendar in Florence BML Plut.16.8, 2r (Florence, saec. XI^mid; available online at: http://mss.bmlonline.it/s.aspx?Id=AWOHy4AYI1A4r7GxMB2H&c=Martyrologium%20Bedae#/oro/13, accessed 14 August 2020). See Borst, *Reichskalender*, pp. 643, 645, 697, 700–01.

129 See especially *De ratione conputandi* ch. 57 (Walsh and Ó Cróinín, *Cummian's Letter*, pp. 168–69). Cf. Warntjes, *Munich Computus*, p. CXCVIII.

130 Isidore of Seville, *Etymologiae* ch. VI 17.31 (Chaparro Gómez, *Isidoro de Sevilla*, pp. 108–09).

131 See especially Einsiedeln SB 321 (647), p. 109; Munich Computus, ch. 49 (Warntjes, *Munich Computus*, pp. 168–69); *De ratione conputandi* ch. 103 (Walsh and Ó Cróinín, *Cummian's Letter*, p. 207).

132 Cf. Warntjes, *Munich Computus*, pp. LXXII–LXXIII, CLXII–CLXIII.

133 Cf. ibid., p. LXXIII n. 188.

4.7 *Second Month*

The Nivelles calendar provides unusual information for 1 April: *initium secundi mensis*. Generally, in the early medieval calendar tradition, April was considered the first, not the second month, based on Bede.[134] The entry in the Nivelles calendar reflects pre-Bedan practice, mirroring a passage preserved in the Munich Computus:[135] *Primus mensis apud Hebreos Martius, quia in eo mundus factus est, et primum phase factum est, conception et passio et resurrectio Christi*. The author is explicit why March is considered the first month: the world was created in this month (21 March), the first passover happened in it, as did the conception (25 March), passion and resurrection of Christ (25 and 27 March respectively). The ascription of this tradition to the Hebrews stems from Isidore.[136] It is important to note that the Munich Computist's statement here can be explained independent of the calendar tradition, which provided similar information. Polemius Silvius equated March with Nisan, the first month of the Hebrew lunar year.[137] Early in the eighth century at the latest, the correlation between Roman and Hebrew months was shifted by one, leading to the equation of Nisan with April. April thus became the first month.[138] The Munich Computist is still confused about the issue, clinging to the old tradition of March being the first month, but appearing irritated by the fact that this is supposed to be Hebrew Adar (the month before Nisan).[139] By the time of *De ratione conputandi* and Bede's *De temporum ratione*, i.e. by the AD 720s, the new theory of Nisan equaling April as the first month was fully developed, and through Bede it became popular in subsequent calendars.[140] Recording the older tradition of March being the first month (and April thus the second) is a clear indication of a pre-Bedan layer.

4.8 *Creation of the World*

Another, but more obvious and more indicative feature of pre-Bedan Irish computistical theory, is the information provided on the creation of the world. For 21 March, the Zürich / Nivelles calendar notes *equinoctium secundum Grecos et primius* [*recte primus*] *dies seculi*. Connecting the spring equinox with the first day of creation was a common feature in pre-Bedan computistical theory.

134 Borst, *Reichskalender*, pp. 743–44, 746.
135 Munich Computus ch. 24 (Warntjes, *Munich Computus*, p. 64).
136 Isidore, *De natura rerum* ch. 6.2 (Fontaine, *Traité de la nature*, p. 193).
137 Polemius Silvius, *Fasti* (Mommsen, *Corpus inscriptionum latinarum* 1, p. 339).
138 See especially Jones, 'Polemius Silvius'.
139 Munich Computus ch. 55 (Warntjes, *Munich Computus*, pp. 210–13).
140 *De ratione conputandi* ch. 41 (Walsh and Ó Cróinín, *Cummian's Letter*, p. 150); Bede, *De temporum ratione* ch. 11 (Jones, *Bedae opera*, p. 203).

The fifth-century text called the Acts of the Council of Caesarea argues that the world was created on a Sunday, 25 March, the spring equinox.[141] Victorius of Aquitaine, in his Prologue of AD 457, also mentions this older Roman tradition (carefully avoiding to refer to it as the spring equinox).[142] With the adoption of the Alexandrian / Dionysiac Easter reckoning with its 'Greek' spring equinox of 21 March, it was only natural to shift this theory to this earlier date, i.e. first day of creation on 21 March. Sixth- and seventh-century computists tried to justify this shift by arguing that the old Roman equinox of 25 March should be considered the fourth day of creation, when the heavenly bodies came into existence. Thus, the first day of creation was supposed to have happened three days earlier, on 22 March. This is how the late-fifth, early sixth-century text variously ascribed to Athanasius, Sulpicius Severus, and Martin of Braga portrays it.[143] This pushed the first day of creation to 22 March, but not 21 March. The final piece in the puzzle was the argument that the days of creation should be considered lunar days, starting at midday of the preceding Julian calendar day according to early medieval Irish computists.[144] Therefore, the first day of creation began at midday of 21 March. The Munich Computus explains this theory in detail,[145] lists both traditions (Greek and Roman) for the first day of creation (21 and 25 March respectively),[146] but insists on the Greek tradition as the binding one.[147] Shortly after the composition of the Munich Computus of AD 718/9, *De ratione conputandi* mentions only the Greek date,[148] and such Irish texts inform the Zürich / Nivelles calendar.

Bede, in AD 725, revisited the theory, and felt the need for further improvement.[149] The equinox is the date of equal periods of day and night (12 hours each). God made all things equal. Therefore, the heavenly bodies on

141 The *Acta Synodi Caesareae* exists in four recensions, labelled A to D; the B-recension, published in *PL* vol. 90, cols. 607–10, was the most widely used in the early Middle Ages. A new study of the complex history of this text by Leofranc Holford-Strevens is in preparation.

142 Victorius of Aquitaine, *Epistola ad Hilarum* ch. 7 (Krusch, 'Studien', p. 23): *octavo kalendarum aprilium, quo mundus traditur institutus.*

143 This work is certainly not from Martin of Braga's pen. It circulated in different recensions; a full-scale study with edition and translation by Alden Mosshammer is in preparation. For the passage in question see Barlow, *Martini episcopi Bracarensis Opera*, pp. 272–73; Krusch, *Studien*, pp. 331–32; Strobel, *Texte*, pp. 109, 113–14.

144 For this beginning of the lunar day, ascribed to the Hebrews in Irish sources, see Munich Computus ch. 8 (Warntjes, *Munich Computus*, pp. 26–27), with a discussion of the most relevant contemporary sources.

145 Ibid. ch. 44 (Warntjes, *Munich Computus*, pp. 146–49).

146 Ibid. ch. 11 (Warntjes, *Munich Computus*, pp. 42–43).

147 Ibid. ch. 36 (Warntjes, *Munich Computus*, pp. 98–99).

148 *De ratione conputandi* ch. 68 (Walsh and Ó Cróinín, *Cummian's Letter*, p. 177).

149 Bede, *De temporum ratione* ch. 6 (Jones, *Bedae opera*, p. 192).

the fourth day surely must have been created at the equinox. The correct spring equinox is that of the Greeks, 21 March. The first day of creation, accordingly, happened three days earlier, on 18 March (which coincides with the entry of the sun in the zodiac sign of Aries). Thus, Bede pushed the first day of creation forward by another 3 days to 18 March, and this is the information found in almost every early medieval calendar that records this datum.[150] The alternative of 21 March is very rare[151] but highly indicative characteristic of a pre-Bedan layer most prominently expressed in early eighth-century Irish texts.

4.9 *Increase and Decrease of Daylight*

Possibly the most indicative feature of Irish, pre-Bedan computistical influence in the Zürich / Nivelles calendar are the statements on the increase of daylight between 25 December and 25 March, the Roman winter solstice and spring equinox. The author of this calendar started on the assumption (not recorded) that there were 6 hours of daylight on the winter solstice (25 December), and then noted an increase of daylight of one hour per 15 days: on 9 January (*VII hore in diae*), 24 January (*VII⟨I⟩ hore in die*), 8 February (*VIIII hore in diae*), 10 March (*XI hore in diae*), and 25 March (*XII hore in die*), missing only the 10 hours on 23 February. This theory is based on Anatolius Latinus' *De ratione paschali*. Anatolius Latinus outlined the theory that daylight increased by 2 2/3 moments per day between the winter and the summer solstice, which he considered on the Roman dates (25 December, 24 June), like the author of the Zürich / Nivelles calendar; 40 moments complete an hour, and thus a daily increase of 2 2/3 moments leads to a full hour in 15 days (2 2/3 × 15 = 40), as recorded by the Zürich / Nivelles calendar. This theory of a daily increase in daylight between the winter and the summer solstice is taken up and discussed in much more detail in the Irish computistical textbooks. The *Computus Einsidlensis* and the Munich Computus of the early eighth century transferred it to the Greek dates (21 December, 21 March, 20 June, 23 September).[152] *De ratione conputandi* of a few years later proved more conservative, following the ultimate source (Anatolius Latinus) in its Roman dates.[153] This is also the practice of the mid-eighth century Irish text *Dial. Langob.*[154]

---

150 Borst, *Reichskalender*, pp. 687, 689–90.

151 None of the calendars used by Borst contains a reference to the first day of creation on 21 (or 22) March; cf. Borst, *Reichskalender*, pp. 699–706.

152 Einsiedeln SB 321 (647), p. 99; Munich Computus ch. 39 (Warntjes, *Munich Computus*, pp. 110–17).

153 *De ratione conputandi* ch. 50 (Walsh and Ó Cróinín, *Cummian's Letter*, pp. 159–62).

154 *Dial. Langob.* ch. 19C (Borst, *Schriften*, pp. 451–52).

It is interesting that the Zürich / Nivelles calendar records the one-hour increase (and decrease in the second half of the year) of daylight only in the first quarter of the year, between 25 December and 25 March. In the following months, the relevant entries remain empty. This may be a reaction to the earlier Irish debate, in which the first quarter of the year, between the winter solstice and the spring equinox, received special attention. According to these Irish texts, additional to the increase of daylight of 2 2/3 moments per day the bissextile day grew by 1 1/3 moment per day in this 90-day period, amounting to the 3 hours that constitute a quarter of a 12-hour day. But whether these three extra hours were to be accounted for as daylight or nighttime or dusk and dawn, the authors found difficult to explain. The solution was a rather vague theory of a cloud (*nubs*) symbolized by dusk and dawn. The message of the Zürich / Nivelles author was that only the regular increase of daylight of 2 2/3 moments per day was of relevance, the bissextile day should be accounted for by a different mechanism.

This theory of an increase of daylight between the winter and the summer solstice of 2 2/3 moments remained extraordinarily popular throughout the Carolingian period. Only a full assessment of all computistical manuscripts before c.AD 900 will provide a solid insight into which monastic centers followed this concept popularized by Irish computists. It remains noteworthy that Bede showed no interest in this theory. In the early medieval calendar tradition, noting the length of day and night never became a standard feature, ultimately because of Bede's disinterest. If this feature is included in a calendar, then usually only at the solstices and equinoxes[155] or, more frequently, at the end or beginning of each month,[156] not in 15-day intervals as in the Zürich / Nivelles calendar. A noteworthy exception is the only surviving calendar in Irish script, now Karlsruhe BLB Aug. perg. 167, and its copy now in St Gall.[157]

155 Cf. Borst, *Reichskalender*: spring equinox 21 March: pp. 697–98, 701; 24 March: pp. 710, 712; 25 March: pp. 713–14, 718; summer solstice: 19–20 June: pp. 998, 1000–1, 1003; 23–24 June: pp. 1011–13, 1016; autumn equinox: 20 September: pp. 1309, 1311; 23 September: pp. 1317, 1319; winter solstice: 20–21 December: pp. 1601, 1603–4, 1606; 24 December: pp. 1613–14.

156 Borst, *Reichskalender*, pp. 420, 422, 525, 531–33, 623, 632–33, 733, 741–42, 832, 839–40, 935, 942–43, 1035, 1041–42, 1144, 1149–50, 1244, 1250–51, 1341, 1348, 1350, 1445, 1449–50, 1538, 1546–47, 1633.

157 Karlsruhe BLB Aug. perg. 167, 16v–17v (Soissons-Laon, *c.*AD 848; online at: https://digital.blb-karlsruhe.de/blbhs/content/titleinfo/20736, accessed 14 August 2020); St Gall SB 248, pp. 72–76 (Soissons-Laon, *c.*AD 850; online at: https://www.e-codices.unifr.ch/de/list/one/csg/0248, accessed 14 August 2020). Cf. especially Schneiders, 'Irish calendar', with an edition of the Karlsruhe calendar; for the St Gall codex, see especially Cordoliani, 'Manuscrits', pp. 168–77; for the relationship between the Karlsruhe and the St Gall calendar, see Borst, *Reichskalender*, pp. 228–31.

Pre-Bedan Irish influence kept its appeal in monastic centres that resisted the Carolingian standardization process on Bede's model.

5 Conclusion: Towards a Generic Model

Arno Borst classified the entries in early medieval calendars in three categories:

a) liturgical / martyrological
b) computistical
c) astronomical

Of these, astronomical entries were an extension of computistical ones, and not essential for this genre of texts. In fact, they were introduced into calendars only with the reception of Pliny excerpts in the eighth century, and did not fulfill any role in Christian worship.

More important were the first two categories. A calendar was, essentially, a grid of 365 days, which were populated with entries important for the communities that produced them. The basic function was a liturgical reminder of key events, which comprised, principally, the feast days of saints considered most important for the communities in question (martyrological), and the basics of the moveable feasts of the liturgical calendar centred around Easter (computistical). It happened that either only martyrological, or only computistical data were noted. If it was only martyrological, it belonged to a different genre of texts, the martyrologies. If it was only computistical, it still classified as a calendar. Thus, the computistical entries are the defining criterion for an early medieval calendar.

If early medieval calendars with only computistical entries exist, they are extremely rare.[158] Typically, they contain at least both martyrological and computistical data. Both are equally important for modern scholars. As extensive research of martyrologies has shown, martyrological entries can often be classified in various chronological layers that reveal the transmission history of the text.[159] Surprisingly, computistical entries have not received similar attention.

---

158 Meyvaert, 'Discovering the calendar', pp. 12–16 hypothesized that Bede's calendar lacked martyrological entries, but besides computistical material also incorporated astronomical data. A good example mentioned by him is St Gall SB 248, pp. 72–76; note, however, that this is a copy of Karlsruhe BLB Aug. perg. 167, 16v–17v, which does contain martyrological entries. In other words, omitting martyrological entries here is the deliberate choice of a later copyist, not a marker of an early calendar close to Bede's time of writing (thus Borst, *Reichskalender*, pp. 228–31; *contra* Meyvaert, 'Discovering the calendar', pp. 40–42). Cf. previous note.

159 See, e.g., the excellent studies by Pádraig Ó Riain assembled in his *Feastdays of the saints*.

This may be due to the fact that there is a general, misguided perception that the computistical data recorded is, more or less exclusively, Bedan.[160] This perception was instrumental in the reconstruction of the origin of early medieval calendars: a Roman model similar to Polemius Silvius' calendar was brought to Northumbria in the second half of the seventh century. There, it led to the construction of native calendars with a more strongly Christian purview. From Anglo-Saxon England, this new calendar type (whether at that stage already *the* prototype of the genre or not) travelled in the bag of scholars, missionaries, or pilgrims into the Frankish heartland in the second third of the eighth century. In the Carolingian Renaissance, production of calendars then increased on a massive scale, with a wide distribution throughout Western Europe.

The computistical data provides the key to a more generic model. Computistical theory was not uniform, it certainly was not only Bedan. Before Bede, computus developed into a major subject of monastic education in seventh-century Ireland, and regional characteristic were also developed in Visigothic Spain and Francia. Traces of these, especially of Irish thought, can be found in the earliest calendars that survive from the eighth century. This article has proven that all the computistical information found in one test-case calendar (Zürich UB Rh. 30, 166v–169v) provides an almost exact match with Irish computistical knowledge of the early eighth century (and not, it will be noted, Bedan thought). No Roman model was needed for the construction of an early medieval calendar when this genre came to life towards the end of the seventh century. The 365-day grid could be reconstructed from the Macrobian excerpt known as *Disputatio Chori et Praetexti* or comparable and more compressed texts, and this grid was then populated not only by saints' feastdays of particular local relevance, but also by computistical characteristics developed in the same region. The origin of the early medieval calendar, therefore, is not only to be sought in the popular notion of Roman texts informing Anglo-Saxon scholars whose teaching laid the foundation for the Carolingian Renaissance. Calendars, no doubt, started to flourish independently in various places, without the need for a fixed model or centralized authority. The early medieval intellectual endeavour was more regional than modern master narratives would want to make us believe, and the calendar was no exception. In this context, it is worth reminding ourselves of Wallace-Hadrill's assessment of 1956:[161]

160 Meyvaert, 'Discovering the calendar', pp. 25–34 is only the most obvious example. Even Schneiders, 'Irish calendar', pp. 66–68 works from the premise that the computistical entries in the Carlsruhe calendar, the only one in Irish script, must have a Bedan origin.

161 Wallace-Hadrill, 'Rome and the early English church', pp. 547–48.

> I must leave the matter there. All I have done has been to hazard the suggestion that the devotion of the early English Church to Rome and to the name of Gregory the Great can a little blind us, even now, to the true nature of Rome's obligations towards England, some of which could only be fulfilled with the help of churches lying nearer to England than Rome herself lay. In England, at least, we need to be better informed about the state of the Frankish and the Visigothic churches, their teaching and learning and their ability to supply England with some of the books she needed. We need to learn not to be blinded by the highlights of insular calligraphy to the study of debased texts in humbler scripts and in barbarized Latin; and the completion of Lowe's Codices Latini will, some day, make a systematic appraisal of western manuscript resources in the sixth and seventh centuries a possibility at last. England is part of the picture. We have too long regarded her, as perhaps she liked to regard herself, as Rome's only peculiar child. In truth, do her difficulties, and the help she obtained from Rome in their solution, notably distinguish her from her sister-churches in the west? Everywhere we meet the same problems: of simony, of lay-seizure of church lands, of the foundation of false monasteries, of royal incursion on ecclesiastical privileges; and many more. These are the common problems of barbarian Christendom, as St. Boniface must only too well have known. What now needs investigation is not England's debt to Rome but Rome's activity in England as an integral part of her activity in Western Europe as a whole.

One of the common intellectual problems of Western Christianity after the Fall of Rome was to construct a calendar grid that would guide the faithful through the liturgical year. At around AD 700, this problem could be solved anywhere in Western Europe without the need for a Roman model, or the stimulus of central authority. The key to our understanding of the early medieval calendar tradition lies in acknowledging a high standard of calendrical learning throughout the Latin West before Bede and the Carolingians, with flourishing regional diversity fostering different intellectual traditions.

References

R. Abels, 'The council of Whitby: a study in early Anglo-Saxon politics', *Journal of British Studies* 23 (1983), pp. 1–25.

A. Arweiler, 'Zu Text und Überlieferung einer gekürzten Fassung von Macrobius Saturnalia I,12,2–I,15,20', *Zeitschrift für Papyrologie und Epigraphik* 131 (2000), pp. 45–57.

C.W. Barlow, *Martini episcopi Bracarensis* Opera omnia (New Haven: Yale University Press, 1950).

R. Bauerreiß, 'Ein angelsächsisches Kalendarfragment des bayrischen Hauptstaatsarchivs in München', *Studien und Mitteilungen zur Geschichte des Benediktiner-Ordens und seiner Zweige* 51 (1933), pp. 177–82.

J. Bisagni, 'The newly-discovered Irish and Breton computistica in Città del Vaticano, BAV, MS Reg. Lat. 123', *Peritia* 28 (2017), pp. 13–34.

J. Bisagni and I. Warntjes, 'Latin and Old Irish in the Munich Computus: a reassessment and further evidence', *Ériu* 57 (2007), pp. 1–33.

J. Bisagni and I. Warntjes, 'The Early Old Irish material in the newly discovered *Computus Einsidlensis* (c.AD 700)', *Ériu* 58 (2008), pp. 77–105.

B. Bischoff, *Die südostdeutschen Schreibschulen und Bibliotheken in der Karolingerzeit, Teil 1: Die bayerischen Diözesen* (3rd ed., Wiesbaden: Harrassowitz, 1974).

B. Bischoff, *Die südostdeutschen Schreibschulen und Bibliotheken in der Karolingerzeit, Teil 2: Die vorwiegend österreichischen Diözesen* (Wiesbaden: Harrassowitz, 1980).

B. Blackburn and L. Holford-Strevens, *The Oxford Companion to the Year: An Exploration of Calendar Customs and Time-Reckoning* (Oxford: Oxford University Press, 1999).

A. Borst, *Das Buch der Naturgeschichte: Plinius und seine Leser im Zeitalter des Pergaments* (Heidelberg: Universitätsverlag C. Winter, 1994).

A. Borst, *Die karolingische Kalenderreform* (Hannover: Hahn, 1998).

A. Borst, *Der karolingische Reichskalender und seine Überlieferung bis ins 12. Jahrhundert*, 3 vols (Hannover: Hahn, 2001).

A. Borst, *Der Streit um den karolingischen Kalender* (Hannover: Hahn, 2004).

A. Borst, *Schriften zur Komputistik im Frankenreich von 721 bis 818*, 3 vols (Hannover: Hahn, 2006).

V. Brown, 'Where have all the grammars gone? The survival of grammatical texts in Beneventan script', in P. de Paolis, M. de Nonno, and L. Holtz (eds), *Manuscripts and Tradition of Grammatical Texts from Antiquity to the Renaissance*, 2 vols (Cassino: Edizione dell'Università degli Studi di Cassino, 2000), pp. 389–414.

R.W. Burgess, 'The Chronograph of 354: its manuscripts, contents, and history', *Journal of Late Antiquity* 5 (2012), pp. 345–96.

R.W. Burgess, 'The new edition of the Chronograph of 354: a detailed critique', *Zeitschrift für antikes Christentum* 21 (2017), pp. 383–415.

C. Burnett, 'Translation and transmission of Greek and Islamic science to Latin Christendom', in D.C. Lindberg and M.H. Shank (eds), *The Cambridge History of Science, vol. 2: Medieval Science* (Cambridge: Cambridge University Press, 2015), pp. 341–64.

C. Burnett, 'The twelfth-century renaissance', in D.C. Lindberg and M.H. Shank (eds), *The Cambridge History of Science, vol. 2: Medieval Science* (Cambridge: Cambridge University Press, 2015), pp. 365–84.

M.J. Carton, *Three Unstudied Manuscripts of Macrobius' Saturnalia* (unpublished Ph.D. thesis, St. Louis, 1966).

C. Chaparro Gómez, *Isidoro de Sevilla, Etimologías, Libro VI: De las Sagradas Escrituras* (Paris: Les belles lettres, 2012).

T.M. Charles-Edwards, *Early Christian Ireland* (Cambridge: Cambridge University Press, 2000).

B. Colgrave, *The Life of Bishop Wilfrid by Eddius Stephanus* (Cambridge: Cambridge University Press, 1927, repr. 1985).

B. Colgrave, *The Earliest Life of Gregory the Great* (repr. Cambridge: Cambridge University Press, 1985).

J.J. Contreni, 'The Carolingian Renaissance: education and literary culture', in R. McKitterick (ed.), *The New Cambridge Medieval History, vol. 2: c.700–c.900* (Cambridge: Cambridge University Press, 1995), pp. 709–57, 1013–24.

J.J. Contreni, 'Learning for God: education in the Carolingian Age', *Journal of Medieval Latin* 24 (2014), pp. 89–130.

A. Cordoliani, 'Les manuscrits de comput ecclesiastique de l'Abbaye de Saint Gall du VIIIe au XIIe siècle', *Zeitschrift für Schweizerische Kirchengeschichte = Revue d'histoire ecclésiastique suisse* 49 (1955), pp. 161–200.

C. Corning, *The Celtic and Roman Traditions: Conflict and Consensus in the Early Medieval Church* (New York: Palgrave Macmillan, 2006).

E.T. Dailey, 'To choose one Easter from three: Oswiu's decision and the Northumbrian Synod of AD 664', *Peritia* 26 (2015), pp. 47–64.

B. Davis, 'The lunar year of the Coligny calendar as a precedent for the insular lunar year', *Talanta* 38–39 (2006–7), pp. 9–34.

G.B. de Rossi and L. Duchesne, *Acta Sanctorum Novembris 2.1* (Brussels: Bollandists, 1894).

A. Degrassi, *Inscriptiones Italiae 13: fasti et elogia, fasciculus 2: Fasti anni Numani et Iuliani, accedunt ferialia, menologia rustica, parapegmata* (Rome: Libreria delle Stato, 1963).

A. Dierkens, *Abbayes et chapitres entre Sambre et Meuse (VIIe–XIe siècles): contribution à l'histoire religieuse des campagnes du Haut Moyen Age* (Sigmaringen: Jan Thorbecke, 1985).

A. Dierkens, 'Saint Amand et la foundation de l'abbaye de Nivelles', *Revue du Nord* 69 (1986), pp. 225–35.

A. Dierkens, 'Prolégomènes à une histoire des relations culturelles entre les îles britanniques et le continent pendant le Haut Moyen Age: la diffusion du monachisme dit colombanien ou iro-franc dans quelques monastères de la région parisienne au VIIe siècle et la politique religieuse de la reine Bathilde', in H. Atsma (ed.), *La Neustrie: les pays au nord de la Loire de 650 à 850*, 2 vols (Sigmaringen: Jan Thorbecke, 1989), vol. 2, pp. 371–94.

A. Dierkens, 'Notes biographiques sur saint Amand, abbé d'Elnone et éphémère évêque de Maastricht († peu après 676)', in E. Bozoky (ed.), *Saints d'Aquitaine: missionaires et pélerins du haut Moyen Âge* (Rennes: Presses Universitaires de Rennes, 2010), pp. 63–80.

A. Dierkens, 'Maubeuge et Nivelles à la fin du VIIe siècle: note sur deux passages de la Vita sanctae Aldegundae prima (BHL 244)', in C. Depauw et al. (eds), *Hainault: la terre et les hommes* (Mons: Hannonia, 2016), pp. 23–47.

J. Divjak and W. Wischmeyer, *Das Kalenderbuch von 354: der Chronograph des Filocalus*, 2 vols (Wien: Holzhausen, 2014).

I. Dobcheva, 'The umbrella of Carolingian computus', in M.J. Muñoz Jiménez, P. Cañizares Ferriz, and C. Martin (eds), *La compilación del saber en la edad media* (Porto: Fédération Internationale des Instituts d'Études Médiévales, 2013), pp. 211–30.

B. Englisch, *Zeiterfassung und Kalenderprogrammatik in der frühen Karolingerzeit: das Kalendarium der Hs. Köln DB 83–2 und die Synode von Soissons 744* (Stuttgart: Thorbeke, 2002).

M. Esposito, 'On the pseudo-Augustinian treatise, De mirabilibus sanctae scripturae, written in Ireland in the year 655', *Proceedings of the Royal Irish Academy* 35C (1918–20), pp. 189–207.

A. Fingernagel, *Die illuminierten lat. Handschriften deutscher Provenienz der Staatsbibliothek PK Berlin, 8.-12. Jahrhundert* (Wiesbaden: Harrassowitz, 1991).

J. Fontaine, *Isidore de Séville: Traité de la nature* (Bordeaux: Féret et fils, 1960).

H. Frank, 'Die Briefe des heiligen Bonifatius und das von ihm benutzte Sakramentar', in *Sankt Bonifatius: Gedenkausgabe zum zwölfhundertsten Todestag* (Fulda: Parzeller, 1954), pp. 58–93.

K. Gamber, *Das Bonifatius-Sakramentar und weitere frühe Liturgiebücher aus Regensburg* (Regensburg: Friedrich Pustet, 1975).

E. Gibbon, *The History of the Decline and Fall of the Roman Empire*, 6 vols (London, 1776–89, repr. 2005).

L. Gougoud, *Christianity in Celtic Lands* (London: Sheed & Ward, 1932, repr. Dublin: Institute for Advanced Studies, 1992).

E. Graff, 'The recension of two Sirmond texts: *Disputatio Morini* and *De divisionibus temporum*', in I. Warntjes and D. Ó Cróinín (eds), *Computus and its Cultural Context* (Turnhout: Brepols, 2010), pp. 112–42.

M. Gretsch, 'Æthelthryth of Ely in a lost calendar from Munich', *Anglo-Saxon England* 35 (2006), pp. 159–77.

C. Grocock and I.N. Wood, *Abbots of Wearmouth and Jarrow* (Oxford: Clarendon Press, 2013).

P. Grosjean, 'Recherches sur les debuts de la controverse pascale chez les Celtes', *Analecta Bollandiana* 64 (1946), pp. 200–44.

P. Grosjean, 'Sur quelques exégètes irlandaise du VIIe siècle', *Sacris Erudiri* 7 (1955), pp. 67–98.

P. Grosjean, 'Un fragment d'obituaire Anglo-Saxon du VIIIe siècle naguère conservé à Munich', *Analecta Bollandiana* 79 (1961), pp. 320–45.

A. Hänggi and A. Schönherr, *Sacramentarium Rhenaugiense: Handschrift Rh 30 der Zentralbibliothek Zürich* (Freiburg i. d. Schweiz: Universitätsverlag, 1970).

C.H. Haskins, *Studies in the History of Mediaeval Science* (Cambridge: Harvard University Press, 1924).

C.H. Haskins, *The Renaissance of the Twelfth Century* (Cambridge: Harvard University Press, 1927).

A. Hauck, *Kirchengeschichte Deutschlands, vol. 1* (3rd and 4th ed., Leipzig: J.C. Hinrichs'sche Buchhandlung, 1904).

G. Henderson, *Vision and Image in Early Christian England* (Cambridge: Cambridge University Press, 1999).

L. Holford-Strevens, 'Paschal lunar calendars up to Bede', *Peritia* 20 (2008), pp. 165–208.

L. Holford-Strevens, 'Marital discord in Northumbria: Lent and Easter, his and hers', in I. Warntjes and D. Ó Cróinín (eds), *Computus and its Cultural Context* (Turnhout: Brepols, 2010), pp. 143–58.

L. Holford-Strevens, *The Disputatio Chori et Praetextati: The Roman Calendar for Beginners* (Turnhout: Brepols, 2019).

L. Holtz, 'Le Parisinus Latinus 7530, synthese cassinienne des arts libéraux', *Studi medievali*, 3rd series 16 (1975), pp. 97–152.

E. Honée, 'St Willibrord in recent historiography', in P.N. Holtrop and H. McLeod, *Missions and Missionaries* (Woodbridge: Boydell, 2000), pp. 16–31.

C.W. Jones, 'Polemius Silvius, Bede, and the names of the months', *Speculum* 9 (1934), pp. 50–56.

C.W. Jones, *Bedae pseudepigrapha: Scientific Writings Falsely Attributed to Bede* (Ithaca: Cornell University Press, 1939).

C.W. Jones, 'A legend of St. Pachomius', *Speculum* 18 (1943), pp. 198–210, repr. in W.M. Stevens (ed.), *Bede, the Schools and the Computus* (Aldershot: Variorum, 1994), article VII.

C.W. Jones, *Bedae opera de temporibus* (Cambridge: The Medieval Academy of America, 1943).

C. Kendall and F. Wallis, *Bede: On the Nature of Things and On Times* (Liverpool: Liverpool University Press, 2010).

J.F. Kenney, *The Sources for the Early History of Ireland: Ecclesiastical* (New York: Columbia University Press, 1929, repr. Dublin: Four Courts, 1997).

B. Krusch, *Studien zur christlich-mittelalterlichen Chronologie: der 84jährige Ostercyclus und seine Quellen* (Leipzig: Veit, 1880).

B. Krusch, 'Die Einführung des griechischen Paschalritus im Abendlande', *Neues Archiv der Gesellschaft für ältere deutsche Geschichtskunde* 9 (1884), pp. 99–169.

B. Krusch, 'Chronologisches aus Handschriften', *Neues Archiv der Gesellschaft für ältere deutsche Geschichtskunde* 10 (1885), pp. 81–94.

B. Krusch, 'Das Datum des Concils von Soissons 744 März 3.', *Neues Archiv der Gesellschaft für ältere deutsche Geschichtsforschung* 30 (1905), pp. 708–09.

B. Krusch, 'Studien zur christlich-mittelalterlichen Chronologie: die Entstehung unserer heutigen Zeitrechnung', *Abhandlungen der Preußischen Akademie der Wissenschaften Jahrgang 1937, phil.-hist. Klasse* 8 (1938).

W. Levison, 'St. Willibrord and his place in history', *Durham University Journal* 32 (1940), pp. 23–41, repr. in idem, *Aus rheinischer und fränkischer Frühzeit: ausgewählte Aufsätze* (Düsseldorf: Schwann, 1948), pp. 314–29.

E.A. Loew (later Lowe), *Die ältesten Kalendarien aus Monte Cassino* (Munich: Beck, 1908).

M. Mac Carron, 'Bede, Irish computistica and Annus Mundi', *Early Medieval Europe* 23 (2015), pp. 290–307.

M. Mac Carron, *Bede and Time: Computus, Theology and History in the Early Medieval World* (London: Routledge, 2019).

G. MacGinty, 'The Irish Augustine: De mirabilibus sacrae scripturae', in P. Ní Chatháin and M. Richter (eds), *Ireland und die Christenheit: Bibelstudien und Mission* (Stuttgart: Klett-Cotta, 1987), pp. 70–83.

D. Mc Carthy, 'Easter principles and a fifth-century lunar cycle used in the British Isles', *Journal for the History of Astronomy* 24 (1993), pp. 204–24.

P. Meyvaert, 'Discovering the calendar (*annalis libellus*) attached to Bede's own copy of *De temporum ratione*', *Analecta Bollandiana* 120 (2002), pp. 5–64.

T. Mommsen, *Corpus Inscriptionum Latinarum*, vol. 1 (Berlin, 1863, 2nd ed. 1893).

G. Morin, 'Les quatre plus anciens calendriers du Monte-Cassin (VIIIe–IXe siècles)', *Revue Bénédictine* 25 (1908), pp. 486–97.

A.A. Mosshammer, *Easter Computus and the Origins of the Christian Era* (Oxford: Oxford University Press, 2008).

C.P.E. Nothaft, *Walcher of Malvern: De lunationibus and De dracone* (Turnhout: Brepols, 2017).

C.P.E. Nothaft, *Scandalous Error: Calendar Reform and Calendrical Astronomy in Medieval Europe* (Oxford: Oxford University Press, 2018).

D. Ó Corráin, *Clavis litterarum Hibernensium: Medieval Irish Books and Texts (c.400–c.1600)* (Turnhout: Brepols, 2017).

D. Ó Cróinín, 'The Irish provenance of Bede's computus', *Peritia* 2 (1983), pp. 229–47, repr. in idem, *Early Irish History*, pp. 173–90.

D. Ó Cróinín, 'Rath Melsigi, Willibrord, and the earliest Echternach manuscripts', *Peritia* 3 (1984), pp. 17–49, repr. in idem, *Early Irish History*, pp. 145–72 (with an appendix by T. Fanning).

D. Ó Cróinín, 'Merovingian politics and insular calligraphy: the historical background to the Book of Durrow and related manuscripts', in M. Ryan (ed.), *Ireland and Insular Art, A.D. 500–1200* (Dublin: Royal Irish Academy, 1987), pp. 40–43.

D. Ó Cróinín, 'Bischoff's Wendepunkte fifty years on', *Revue Bénédictine* 110 (2000), pp. 204–37.

D. Ó Cróinín, 'Bede's Irish computus', in idem, *Early Irish History*, pp. 201–12.

D. Ó Cróinín, *Early Irish History and Chronology* (Dublin: Four Courts, 2003).

P. Ó Riain, 'Les Vies de saint Fursy: les sources irlandaises', *Revue du nord* 68 (1986), pp. 405–13.

P. Ó Riain, *Feastdays of the Saints: A History of Irish Martyrologies* (Bruxelles: Société des Bollandistes, 2006).

P. Ó Riain, *A Dictionary of Irish Saints* (Dublin: Four Courts, 2011).

M. Ohashi, *The Impact of the Paschal Controversy: Computus, Exegesis and Church History in Early Britain and Ireland* (unpublished Ph.D. thesis, Nagoya, 1999).

J.T. Palmer, 'Computus after the paschal controversy of AD 740', in I. Warntjes and D. Ó Cróinín (eds), *The Easter Controversy of Late Antiquity and the Early Middle Ages: Its Manuscripts, Texts, and Tables* (Turnhout: Brepols, 2011), pp. 213–41.

J.T. Palmer, 'Calculating time and the end of time in the Carolingian world, c. 740–820', *English Historical Review* 126 (2011), pp. 1307–31.

J.-M. Picard, 'Church and politics in the seventh century: the Irish exile of King Dagobert II', in idem (ed.), *Ireland and Northern France, AD 600–850* (Dublin: Four Courts, 1991), pp. 27–52.

F. Piper, *Karls des Grossen Kalendarium und Ostertafel* (Berlin: Decker, 1858).

H. Pirenne, *Mahomet et Charlemagne* (Paris, 1937).

C. Plummer, *Baedae opera historica*, 2 vols (Oxford: Oxford University Press, 1896).

W. Pohl, 'The transformation of the Roman world revisited', in J. Kreiner and H. Reimitz (eds), *Motions of Late Antiquity: Essays on Religion, Politics, and Society in Honour of Peter Brown* (Turnhout: Brepols, 2016), pp. 45–61.

F. Prinz, *Frühes Mönchtum im Frankenreich* (2nd ed., Darmstadt: Wissenschaftliche Buchgesellschaft, 1988).

O. Rackham, *Transitus beati Fursei: a Translation of the 8th-Century Manuscript Life of Saint Fursey* (Norwich: Fursey Pilgrims, 2001, repr. 2007).

P. Riché, *Les écoles et l'enseignement dans l'Occident chrétien: de la fin du Ve siècle au milieu du XIe siècle* (Paris: Aubier Montaigne, 1979).

P. Salmon, 'Le martyrologe-calendrier conservé dans le manuscript latin 14086 de Paris et ses origins', *Revue Bénédictine* 56 (1945–6), pp. 42–57.

M.R. Salzmann, *On Roman Time: The Codex-Calendar of 354 and the Rhythms of Urban Life in Late Antiquity* (Berkeley: University of California Press, 1990).

J. Schmid, *Die Osterfestberechnung auf den britischen Inseln vom Anfang des vierten bis zum Ende des achten Jahrhunderts* (Regensburg: Verlagsanstalt, 1904).

M. Schneiders, 'The Irish calendar in the Karlsruhe Bede (Karlsruhe, Badische Landesbibliothek, Cod. Aug. CLXVII, ff. 16^v–17^v)', *Archiv für Liturgiewissenschaft* 31 (1989), pp. 33–78.

H. Schüling, 'Die Handbibliothek des Bonifatius', *Archiv für Geschichte des Buchwesens* 4 (1961–3), pp. 286–349.

E. Schwartz, 'Christliche und jüdische Ostertafeln', *Abhandlungen der königlichen Gesellschaft der Wissenschaften zu Göttingen, philologisch-historische Klasse, Band 8, Nr. 6* (Berlin: Weidmannsche Buchhandlung, 1905).

T. Sickel, 'Die Lunarbuchstaben in den Kalendarien des Mittelalters', *Sitzungsberichte der Akademie der Wissenschaften in Wien, phil.-hist. Klasse*, 38 (1862), pp. 153–201.

P. Siffrin, 'Das Walderdorffer Kalenderfragment saec. VIII und die Blätter eines Sakramentars aus Regensburg', *Ephemerides Liturgicae* 47 (1933), pp. 201–24.

M. Smyth, 'Once in four: the leap year in early medieval thought', in I. Warntjes and D. Ó Cróinín (eds), *Late Antique Calendrical Thought and its Reception in the Early Middle Ages* (Turnhout: Brepols, 2017), pp. 229–64.

K. Springsfeld, 'Eine Beschreibung der Handschrift St. Gallen, Stiftsbibliothek, 225', in I. Warntjes and D. Ó Cróinín (eds), *Computus and its Cultural Context in the Latin West, AD 300–1200* (Turnhout: Brepols, 2010), pp. 204–37.

S. Stern, *Calendars in Antiquity: Empires, States, and Societies* (Oxford: Oxford University Press, 2012).

A. Strobel, *Texte zur Geschichte des frühchristlichen Osterkalenders* (Münster: Aschendorffsche Verlagsbuchhandlung, 1984).

A. van Berkum, 'Willibrord en Wilfrid: een oderzoek naar hun wederzijdse betrekkingen,' *Sacris Erudiri* 23 (1978), pp. 347–417.

G.S.M. Walker, *Sancti Columbani Opera* (Dublin: Dublin Institute for Advanced Studies, repr. 1997).

J.M. Wallace-Hadrill, 'Rome and the early English church: some questions of transmission', in *La chiese nei regni dell'Europa occidentale e i loro rapport con Roma sino all'800* (Spoleto: Centro italiano di studi sull'alto medioevo, 1960), pp. 519–48 (article), pp. 575–82 (discussion).

M. Walsh and D. Ó Cróinín, *Cummian's Letter* De controversia paschali *and the* De ratione conputandi (Toronto: Pontifical Institute of Mediaeval Studies, 1988).

C. Wampach, *Geschichte der Grundherrschaft Echternach im Frühmittelalter*, 2 vols (Luxemburg: Luxemburger Kunstdruckerei, 1929–30).

I. Warntjes, 'The Munich Computus and the 84 (14)-year Easter reckoning', *Proceedings of the Royal Irish Academy* 107C (2007), pp. 31–85.

I. Warntjes, *The Munich Computus: Text and Translation. Irish Computistics between Isidore of Seville and the Venerable Bede and its Reception in Carolingian Times* (Stuttgart: Franz Steiner Verlag, 2010).

I. Warntjes, 'The Computus Cottonianus of AD 689: a computistical formulary written for Willibrord's Frisian mission', in I. Warntjes and D. Ó Cróinín (eds), *The Easter Controversy of Late Antiquity and the Early Middle Ages: Its Manuscripts, Texts, and Tables* (Turnhout: Brepols, 2011), pp. 173–212.

I. Warntjes, 'Irische Komputistik zwischen Isidor von Sevilla und Beda Venerabilis: Ursprung, karolingische Rezeption und Forschungsperspektiven', *Viator* 42 Multilingual (2011), pp. 1–31.

I. Warntjes, 'Köln als naturwissenschaftliches Zentrum in der Karolingerzeit: die frühmittelalterliche Kölner Schule und der Beginn der fränkischen Komputistik', in H. Finger and H. Horst (eds), *Mittelalterliche Handschriften der Kölner Dombibliothek, Viertes Symposion* (Köln: Erzbischöfliche Diözesan- und Dombibliothek, 2012), pp. 41–96.

I. Warntjes, 'Seventh-century Ireland: the cradle of medieval science?', in M. Kelly and C. Doherty (eds), *Music and the Stars: Mathematics in Medieval Ireland* (Dublin: Four Courts, 2013), pp. 44–72.

I. Warntjes, 'An Irish eclipse prediction of AD 754: the earliest in the Latin West', *Peritia* 24–25 (2013–14), pp. 108–115.

I. Warntjes, 'Victorius vs Dionysius: the Irish Easter controversy of AD 689,' in P. Moran and I. Warntjes (eds), *Early Medieval Ireland and Europe: Chronology, Contacts, Scholarship* (Turnhout: Brepols, 2015), pp. 33–97.

I. Warntjes, 'Computus as scientific thought in Ireland and the early medieval West', in R. Flechner and S. Meeder (eds), *The Irish in Early Medieval Europe: Identity, Culture and Religion* (New York: Macmillan, 2016), pp. 158–78.

I. Warntjes, 'The continuation of the Alexandrian Easter table in seventh-century Iberia and its transmission to ninth-century Francia (Isidore, *Etymologiae* 6.17)', *Revue d'histoire des textes* n.s. 13 (2018), pp. 185–94.

I. Warntjes and D. Ó Cróinín (eds), *Computus and its Cultural Context in the Latin West, AD 300–1200* (Turnhout: Brepols, 2010).

I. Warntjes and D. Ó Cróinín (eds), *The Easter Controversy of Late Antiquity and the Early Middle Ages: Its Manuscripts, Texts, and Tables* (Turnhout: Brepols, 2011).

L.T. White (ed.), *The Transformation of the Roman World: Gibbon's Problem after Two Centuries* (Berkley: University of California Press, 1966).

C. Wickham, *The Inheritance of Rome: A History of Europe from 400 to 1000* (London: Allan Lane, 2009).

H.A. Wilson, *The Calendar of St. Willibrord* (Woodbridge: Henry Bradshaw Society, 1918).

I.N. Wood, *The Transformation of the Roman West* (Leeds: Arc Humanities Press, 2018).

V. Yarza Urquiola and F.J. Andrés Santos, *Isidoro de Sevilla, Etimologías, Libro V: De legibus – De temporibus* (Paris: Les belles lettres, 2013).

M. Ziegler, 'The Ripon connection? Willibrord, Wilfrid, and the mission to Frisia', *The Heroic Age* 6 (2003).

H. Zimmer, 'Keltische Kirche in Britannien und Irland', *Realencyklopädie für protestantische Theologie und Kirche, 3rd ed., vol. 10* (1901), pp. 204–43, trans. by A. Meyer as *The Celtic Church in Britain and Ireland* (London: David Nutt, 1902).

CHAPTER 4

The Chronology of Early Islam

The Ancient Calendar at Mecca and the Origin of the Islamic Calendar

François de Blois

The early history of Islam, the life of the Prophet and of his companions and successors, is the subject of a very detailed, chronologically explicit traditional Islamic narrative. This narrative takes the form of a large body of reports (*ʾaxbār*) preserved, allegedly, by an uninterrupted chain of oral informants stretching back to the time of the events themselves. These traditions were codified in a number of different literary genres: the biography of the Prophet (*sīra*), collections of his sayings (*ḥadīϑ*), commentary on the Qurʾān (*tafsīr*), and so forth, differing for the most part only in the organisation of the common material, in the *sīra* chronologically, in the *ḥadīϑ* according to subject matter, in the *tafsīr* according to the explicated Qurʾānic verses.[1]

This traditional narrative is (for the most part) internally coherent and chronologically plausible. The difficulty for historical researchers is that the proto-Islamic narrative, very much like the proto-Christian narrative, is not explicitly confirmed by any contemporary sources. The Prophet Muḥammad is not in fact mentioned by contemporary Byzantine or Syriac historians, nor is there any contemporary epigraphic evidence. The emergence of Islam occurs essentially in a historic vacuum.[2]

In this chapter I will focus on three key data for the chronology of early Islam.

First there is the *hijra*, the emigration of the Prophet from his native town Makka to the Jewish settlement of Yaϑrib, later renamed the City of the Prophet, *madīnatu n-nabiyy*, or al-Madīna (anglicised as Medina) for short,

1 Shorter versions of this paper were presented at a workshop organised by our ERC-financed project on 'Calendars in Antiquity and the Middle Ages' at University College London on 18 February 2015, and again as the William Wright Memorial Lecture at the Faculty of Asian and Middle Eastern Studies, University of Cambridge, 7 May 2015. – The older literature on this topic is surveyed in my previous articles, 'Taʾrīkh' I, 1 and 'Qurʾān 9:37 and CIH 547'. Some recent contributions are: Bonner, 'Time has come full circle'; Rink and Hansen, 'Der altarabische Kalender'; Hideyuki, 'The calendar in pre-Islamic Mecca'. I regret to have to say that I have not profited from these contributions, mainly because they operate with methodologies different from my own.

2 I have developed this line of inquiry in 'Islam in its Arabian context'.

 | DOI:10.1163/9789004459694_006

where – according to the traditional narrative – the Prophet exploited the dissention amongst the local Jews to establish himself as ruler of the town. This event was later to become the point of departure of the Islamic era, which has its notional epoch on either Thursday 15th or Friday 16th of July 622.[3] The actual *hijra*, the Prophet's physical departure from Mecca, is said to have been on a Monday (though it is debated on which Monday) in the month of Rabīʿ al-ʾawwal,[4] which would put it in September 622 (on the assumption that no intercalation took place after this date; if there were one or more subsequent intercalations the Julian date of the *hijra* would have to be put one or more months earlier, but still firmly in the year 622).

The second datum is the Prophet's Farewell Pilgrimage. In the year 10 of the hijra the Prophet, having defeated his old enemies in Mecca and established his rule of that city, returned one last time to his native place, took part in the pilgrimage, and addressed his followers in a final sermon, in which, among other things, he announced the abolition of *nasīʾ* ('postponement'). We will discuss in a moment what exactly this means; for the present it will suffice to say that it has something to do with the calendar. This was in the pilgrimage month (δū l-Ḥijja) of the mentioned year 10 of the hijra, or March 632. The death of the Prophet is put in the following year, the year 11 of the hijra, whereby there is much disagreement about the exact date.

The third point of reference is the institution of the era of the hijra by the second caliph, ʿUmar, according to some sources in the year of hijra 17, which begins in January 638, according to others one year earlier or later.[5] This becomes the official Islamic calendar.

These events are recorded in the traditional Islamic narrative but not (as I have mentioned) in any contemporary sources. Modern historians have thus three options. Some authors accept them as valid historical facts. Others reject them as being part of an invented narrative from a much later time. Others again accept some of them and reject others, or more precisely, they accept those reports that fit in with their own pet theories, and reject those that do not fit in; this seems to be the prevalent methodology in current research on the early history of Islam. Here I am attempting something different and more

3 As I have emphasised elsewhere, if every month begins with the local sighting of the new moon then the question of whether the first day of the first year of the era was a Thursday or a Friday is purely theoretical.

4 al-Bayrūnī, *Chron.*, p. 330, discusses this and comes out in favour of Monday 8 Rabīʿ al-ʾawwal, which would correspond to 20 September 622 CE.

5 al-Bayrūnī, *Chron.*, pp. 29–31, gives an elaborate account of the introduction of the hijrī era in AH 17, citing aš-Šaʿbī; aṭ-Ṭabarī, p. 1253, puts the introduction of the era in the year 18.

modest, namely to try to measure their historical and arithmetic plausibility. This is, I think, as far as we can venture.

1 The Institution of the Islamic Calendar

My point of departure is the 'standard' or 'classic' form of the Islamic calendar. This is very well attested in the mediaeval period and down to the present. It is a purely lunar calendar without intercalation. The months begin with the local observation of the new crescent, a day or two after conjunction. Although the precise date of the sighting of the moon cannot be predicted it can be estimated within a day or two and tables can be constructed to indicate the approximate equivalents of hijrī dates in other calendars; tables are fine as long as one does not take them too seriously. The year consists always of 12 months, with an average of 354 days, about 11 days short of a solar year, but no attempt is made to compensate for this difference. The months are thus not tied to any particular season, but move fairly quickly through the whole year. Years are counted according to the hijrī era with its epoch at the sighting of the moon in July 622.

Calendars based on the phases of the moon are widespread throughout the whole world, but the Islamic calendar seems to be unique[6] in not attempting in any way to coordinate the months with the solar year, notably through the intercalation of a 13th month at regular intervals. It will be necessary to find some explanation for this anomaly.

At present the oldest unambiguous evidence for the 'standard' Islamic calendar is a papyrus from Egypt, now in Vienna.[7] It is a bilingual (Greek and Arabic) tax receipt and is dated in two calendars. The Arabic text is dated to the month Jumādā I of the year 22, without specifying the day of the month; the Greek text has a date according to the well-known Coptic (or Alexandrian)

---

6 It has on occasion been suggested that the Assyrians in the Middle Assyrian period had a purely lunar calendar, without intercalation (differently from their Babylonian contemporaries, and from the Old Assyrian period calendar), but the evidence for this is not conclusive. See most recently Jeffers, 'The Nonintercalated Lunar Calendar'.

7 The first published description of the papyrus is by Josef Karabacek in the exhibition catalogue Krall, Wessely, and Karabacek, *Papyrus Erzherzog Rainer*, p. 128, no. 508; Karabacek summarises the content of the papyrus and calls it 'die älteste Schriftdenkmal des Islâm'. It was subsequently published several times by Grohmann, for example in his article 'Aperçu de papyrologie arabe', esp. pp. 40–43, with planche IX, where it is cited (wrongly?) as '*PERF* 558', and, with the same siglum, in several later publications. See also Jones, 'The strange neglect of PERF 558', who refers to a publication of the papyrus in Grohmann, *Allgemeine Einführung in die arabischen Papyri* (not accessible to me). Jones's article contains a fairly large number of factual errors.

calendar, namely the 30th day of the month Pharmouthi of the 1st indiction. The two dating formulae complement each other and give us an equivalent Julian date of 25 April 643.[8] But this synchronism will only work if we assume that the Arabic text is in fact dated according to the hijrī era and the 'standard' Islamic calendar.

There is at least one other Arabic papyrus (P Berol 15002)[9] dated to this same year 22, but without indication of the month or day and without a synchronism to any other calendar, and then an increasing number dated to the subsequent years. In theory, one could argue – if one wanted to be really stubborn – that these are possibly dated to some other calendar or other era, but in the case of the Vienna papyrus this is not possible. Thus, we can say with certainty that the 'standard' Islamic calendar was in place by the year 22 of the hijra at the latest.

As mentioned before, the Islamic tradition maintains that the hijrī era was introduced by the caliph 'Umar in or about the year 17. This is just five or so years before the earliest texts dated according to this era. Although the papyri do not necessarily confirm the traditional date for the introduction of this calendar, they certainly do not contradict it. In any case, they show that the 'standard' Islamic calendar was in operation during what the tradition defines as 'Umar's lifetime; he is said namely to have died in the year 23 (644).

2 Intercalation and 'Postponement'

But what was the situation before 'Umar's reform, and, more to the point, before the rise of Islam? We have a fairly broad range of reports, both in specifically Islamic texts, and in astronomical works from the mediaeval period, suggesting that the ancient Arabs practiced some form of intercalation to keep their months in line with the seasons, but that this was abolished by the Prophet of Islam. This procedure is linked with the concept of *nasīʾ* (postponement), which is mentioned once in the Qur'ān. I have discussed this whole complex twice already: first in 1998,[10] and then, in greater detail, and with some modification of my previous argument, in 2004.[11] I will reiterate only very briefly the results of these two contributions, and then move on to something new.

8 Here, and throughout this paper, all conversions have been carried out or verified using KAIROS Version 4.0, copyright Raymond Mercier 2007–8.

9 See Grohmann, 'Aperçu de papyrologie arabe', p. 44.

10 'Ta'rīkh' I, 1.

11 'Qur'ān 9:37 and CIH 547', pp. 101–4.

Qur'ān 9:36 (*at-tawba*) states:

> Verily, the number of the months according to God is twelve months in God's book on the day that he created the heavens and the earth, of which four are sacred/forbidden (…).

إِنَّ عِدَّةَ ٱلشُّهُورِ عِندَ ٱللَّهِ ٱثۡنَا عَشَرَ شَهۡرٗا فِي كِتَٰبِ ٱللَّهِ يَوۡمَ خَلَقَ
السَّمٰوَاتِ وَالأرْضَ مِنْهَا اربَعَةٌ حُرُمٌ

And the next verse continues:

> Postponement (*an-nasīʾ*) is merely an increase in disbelief; those who disbelieved have been misled in (or: by) it, in that they permit it one year and forbid it another year so as to equalise the number of what God forbade and to permit what God forbade (…).

إِنَّمَا ٱلنَّسِيٓءُ زِيَادَةٞ فِي ٱلۡكُفۡرِۖ يُضَلُّ بِهِ ٱلَّذِينَ كَفَرُواْ يُحِلُّونَهُۥ , عَامٗا
وَيُحَرِّمُونَهُۥ , عَامٗا لِّيُوَاطِـُٔواْ عِدَّةَ مَا حَرَّمَ ٱللَّهُ فَيُحِلُّواْ مَا حَرَّمَ ٱللَّهُ

The key words here are *ḥarām* and *ḥalāl*, and the corresponding verbs *ḥarrama* and *ʾaḥalla*. At one level *ḥarām* means 'forbidden' and *ḥalāl* means 'permitted', but at the same time *ḥarām* means 'sacred' in the sense 'sacrosanct, inviolable, too sacred to touch', and *ḥalāl* means 'not sacred, profane'. The cited verses indicate clearly that there were four 'sacred' or 'forbidden' months, and the commentators explain that during these months the Arabs were not allowed to engage in warfare or raiding; the other months were 'profane', 'permitted', in the sense that fighting was allowed in them. The four sacred months (still according to the tradition) were Rajab, and then the three consecutive months centred on the pilgrimage: ðū l-Qaʿda, ðū l-Ḥijja, and Muḥarram. The illicit activity of *nasīʾ* consists in altering the distribution of sacred and profane months by de-sanctifying one or more of the sacred months in a given year and then 'equalising the number of what God forbade (or sanctified)' by sanctifying one or more otherwise profane months in the following year. These verses were revealed at the time of the farewell pilgrimage and were underlined by the Prophet's sermon on that occasion.

All of the astronomical authors, and at least some of the Qur'ānic scholars, maintain that this dislocation of the sacred months was the result of intercalation. Normally (we are told) the Arabs had twelve months in their year, but from time to time they inserted a thirteenth month, which resulted in the postponement of the sacred months in the following year. However, the description of this procedure in the Islamic sources is highly confused and contradictory to the extent that it is virtually useless for historical purposes.[12]

In my two mentioned papers I investigated the problem of *nasīʾ* primarily from a linguistic point of view, arguing that *nasīʾ* means literally 'postponement', not 'intercalation'. I drew attention to an Ancient South Arabian (Sabaic) inscription from perhaps some time between the 2nd century BCE and the 1st century CE, in which the cognate verb n-s_1-ʾ is used to describe the illicit postponement of a ritual event by two months. It is however very clear from the inscription that this was not the result of intercalation, but rather of the fact that the tribesmen were away at war at the time when they should have performed the ritual in question. But this does not rule out the possibility that in Mecca, five centuries or more later, the term *nasīʾ* had come to be used specifically to designate a postponement of festivals as the result of intercalation. This is perhaps supported by the quoted passage from the Qurʾān: By insisting that 'the number of the months according to God is twelve months in God's book on the day that he created the heavens and the earth' the creator states clearly that any intercalation is unacceptable as it would alter the established number of months in each year (twelve); the prohibition of intercalation goes thus hand in hand with the banning of 'postponement'.

3 The Reconstruction of the Ancient Arab Calendar

Let us for the moment give the traditional interpretation the benefit of the doubt: the Arabs at Mecca practiced intercalation up until the time when this was banned by the Prophet of Islam. We will now examine the consequences of this hypothesis, but first it is necessary to outline two basic assumptions.

First this: The ancient calendar at Mecca, like the Islamic calendar, was a lunar calendar with each month beginning at about the time of the New Moon. This means that each month would begin at roughly the same time as some month in all other lunar calendars which begin their months at the New Moon, for example the Babylonian calendar, the Jewish calendar, the ancient

12 The material is collected, and its essentially problematic nature elucidated, by Moberg in his article 'An-Nasi (Koran 9,37) in der islamischen Tradition'. The confused and (I suggest) mendacious account of the matter by Abū Maʿšar al-Balxī is discussed in my paper 'Qurʾān 9:37 and CIH 547'.

Greek calendars, the Indian calendars[13] and even the Chinese calendar. The exact time of the beginning of the month can differ between calendars by perhaps one or two days, but the approximate time of the beginning of the month will be the same. This means that any month in any lunar calendar will overlap almost entirely with a single month in any other lunar calendar. This is what I mean when I say that the month X of the Arab calendar 'corresponds to' the month Y of some other calendar.

Secondly, if the ancient Arabs practised intercalation, and if they (if we may pursue for a moment this hypothesis) inserted the extra month at exactly the same time as (for example) the Jewish calendar, then every Arabian month would correspond always with one and the same Jewish month. If, on the other hand, the Arabs practiced intercalation, but with a different cycle of intercalation than (for example) the Jews, then every Arab month would correspond to a given Jewish month, or the month just before, or the month just after, depending on the rhythm of intercalation in each of the two calendars. In this case we can say that an Arab month 'corresponds on average' to one and the same Jewish month; it will not deviate from the latter by more than one month in either direction.

If the ancient Arabian calendar was intercalated, and if the intercalation was carried out correctly (whether according to a cycle, or empirically by observing astronomical phenomena) then its months would of course have been fixed to the seasons. Some hints as to how this might have worked can be deduced from the names of the Islamic months, even if this evidence is limited.[14] A few only of the month names are actually Arabic common nouns: Muḥarram (or al-Muḥarram, as both forms are used) means 'sacred' or 'forbidden', δū l-Ḥijja means '(month) of the pilgrimage', but these do not tell us anything about the time of year. Then there are the two months called Rabīʿ (first and second Rabīʿ). The common noun *rabīʿ* means 'spring' in modern Arabic, but the classic lexica are unanimous in stating that it is also (or even primarily) used for 'autumn' and is synonymous with *xarīf*. The verb *rabaʿa* is supposed to mean 'it rained' and there is a likelihood that *rabīʿ* actually means 'rainy season', which of course can be at different times in different regions. This is

13 Some Indian calendars begin the month at the New Moon, some at the Full Moon. Here I am referring to the former.

14 The (supposed) etymologies of the Arabic month names are discussed by the lexicographers, beginning (as far as extant texts are concerned) with Quṭrub and al-Farrāʾ in their treatises specifically on the vocabulary of time-keeping (which, curiously, contain absolutely no information about the mechanics of the calendar), then also by the astronomer al-Bayrūnī, *Chron.*, p. 60 and (with in part different explanations) p. 325, and the historian al-Masʿūdī, *Murūj* III, pp. 416–19, and others.

consistent with the Semitic parallels:[15] Syriac *rḇīʿā* generally means 'autumn rain', and it occurs in the same meaning also in Jewish Palestinian Aramaic and Mishnaic Hebrew.

Some of the other month names do not occur as common nouns in Arabic, but can be attached more or less plausibly to Arabic roots: the name Ṣafar has been connected with the colour adjective *ʾaṣfar* 'yellow', which might suggest a localisation in the late summer or early autumn. Ramaḍān has been connected with *ramaḍ* 'parchedness, scorchedness', and the two months with the name Jumādā with the common verb *jamada* 'to freeze'; there are also a fair number of references in pre-Islamic poetry to cold damp weather in Jumādā. It has been objected (among others, by me) that it seems strange that the 'freezing' months of Jumādā should be followed only three months later by the 'scorched' month of Ramaḍān, but it could also be noted that in Central Arabia the weather does suddenly get very hot in the early spring. This is about as far as anyone has got with possible weather connections for the Arabic month names; the subject has been discussed in much detail, notably in classic works by Wellhausen[16] and Nöldeke[17] already in the nineteenth century.

There is however another side to the examination of the Arabic month names. A very unusual feature of the Arab calendar is the fact that, although there are twelve months, there are only ten different month names, with two of the names covering two months each: two months called Rabīʿ followed by two called Jumādā.

An obviously similar situation can be observed in the Syrian variant of the Babylonian calendar, which has two consecutive months called Tešrī followed by two called Kānōn.[18] The ancient Babylonian calendar had twelve different

15 Nöldeke, *Neue Beiträge*, p. 81.

16 Wellhausen, *Reste*, pp. 94–101 ('Die ursprüngliche Zeit der Feste und der Kalender').

17 See in particular his review of Barth's *Dịwan des ʿUmeir ibn Schujeim al-Quṭâmî herausgegeben und erläutert*, esp. pp. 278–79.

18 I digress here with a linguistic point. Jumādā is a feminine proper noun and is followed in the regular manner by a determined feminine adjective (e.g. *jumādā l-ʾāxiratu*, 'the last Jumādā'); the Arabic (masculine) forms of the Syrian paired month names work in exactly the same way (e.g. the diptotic *tišrīnu l-ʾawwalu*), but concerning the two Rabīʿ months the information in the lexica and grammars (native and European) is ambiguous and in part contradictory. In actual texts three different forms can be observed. **First:** *rabīʿun-i l-ʾawwalu, šahru rabīʿin-i l-ʾawwali* etc., which imply that the month-name Rabīʿ is declined exactly like the triptotic common noun *rabīʿun*. These names are consistently vocalised thus (i.e., with *tanwīn*) in the early fourteenth-century Edinburgh Ms. of al-Bayrūnī's *Chronology* (Edinburgh, University Library 161), and thus also in Sachau's careful edition of that book (e.g., in the tables on pp. 20, 29 and often). **Second:** *rabīʿu l-ʾawwalu, šahru rabīʿa l-ʾawwali* (thus, with *fatḥ*, for example in the codex optimus of Quṭrub's *Kitābu l-ʾazmina*; see the facsimile in Ḥaddād's edition p. 80, 3rd line of the text);

names for the twelve months, but in the Syrian version the Babylonian Tišrītu becomes the 'first Tešrī' and the immediately following month of Araḫsamna is renamed 'second Tešrī'. The months with two pairs of identical names are very well known as the Syrian names for the Julian months: Tešrī I is October, Tešrī II is November, Kānōn I is December, Kānōn II is January, and so forth. These were the usual names for the Julian months in Western Asia (Iraq, Syria etc.), and are still used in that region as the Arabic names of the Gregorian months. However, there is evidence for the use of these names for lunar months as well. al-Bayrūnī[19] writes that the pagans in Ḥarrān (the so-called Sabians) use the 'Syrian names' but apply them 'in the manner of the Jews',[20] with each month beginning on the second day after conjunction. The counting of the months begins with Hilāl Tišrīn al-ʾawwal, followed by Hilāl Tišrīn al-ʾāxir. The intercalary month is inserted after Hilāl Ṣubāṭ and is called Hilāl ʾĀδār al-ʾawwal. It is thus evident that the 'Syrian' forms of the month names were used not only for the Julian months, but also for lunar months. We can accordingly distinguish a 'Syro-Babylonian' (lunisolar) calendar and a 'Syro-Julian' (solar) calendar, both using the same month names.

The fact that both the Arab and the Syrian calendar have two consecutive sets of paired month-names would be a remarkable coincidence. I propose

---

these imply that the month name *rabīʿu* is diptotic and thus different from the name of the season *rabīʿun*. **Third:** *rabīʿu l-ʾawwali, šahru rabīʿi l-ʾawwali*, analysed as a possessive construction of the type styled *ʾiḍāfatu t-tafsīr* or *ʾiḍāfatu l-bayān*. Although this interpretation has been favoured by such experts as de Goeje, apud Wright, *Grammar of the Arabic Language*, vol. II, pp. 232–33, and Reckendorf, *Arabische Syntax*, p. 140, it seems very bizarre, especially as it is difficult to understand why *rabīʿ* and *jumādā* should be construed differently. Moreover, there do not seem to be any really convincing examples for the use of '*muḍāf* + genitive adjective' in the meaning 'noun + attributive adjective' (the Qurʾanic *dāru l-ʾāxirati* [16:30], mentioned by de Goeje in this context, can be explained differently). This interpretation seems to be based on a false analysis of شهر ربيع الاول, where in fact *šahr*, not *rabīʿ*, is the headword of the *ʾiḍāfatu t-tafsīr*. I add that although the lexicographers are unanimous in claiming that the names of the two Rabīʿ months must always be preceded by the word *šahr*, this 'rule' is not universally followed in literary texts, and certainly not in dated documents. Of the three mentioned options the first seems the most straightforward. I add further that for 'Modern Standard' Arabic, Wehr's *Arabisches Wörterbuch* (1st German ed.) has '*r. al-awwali*', (5th German ed.) '*r. al-ʾawwal*', Schregle's *Arabisch-deutsches Wörterbuch* has رَبِيعُ الأَوَّل etc. Incidentally, the Syriac month names *tešrī qḏem* (or *qḏīm*), *kānōn ḥrāy* etc., with an attributive adjective in the absolute state, also seem unusual, or at least unexpectedly archaic (Nöldeke, *Kurzgefasste syrische Grammatik*, p. 145).

19 *Chron.*, pp. 318–22, collated by me with Beyazıt 4667, fol. 180b–181a (= B.).

20 *wa qad yusammūna š-šuhūra bi l-ʾasmāʾi s-suryāniyyati wa yaslukūna fīhā šabīha* (B.: *sabīla*) *ṭarīqati l-yahūd*.

that it is not a coincidence.[21] The most straightforward explanation for the similarity of the two calendars would be that the Northern Arabs simply adopted the 'Syro-Babylonian' lunisolar calendar from their neighbours in the Fertile Crescent, with new month names, but retaining not only the arithmetic principles of the Babylonian calendar, but also the Syro-Babylonian convention of having two consecutive sets of paired months. This Arabicised variant of the Syro-Babylonian calendar seems possibly to have replaced the (in North Arabia) previously prevalent calendrical system, which had solar years counted according the era of the Roman Provincia Arabia with Coptic-style months, that is: twelve (non-lunar) months of exactly 30 days, and five (in a leap year: six) epagomenal days.[22] The abandonment of the Coptic-style solar year and the adoption of a Babylonian-style lunisolar year would have reflected a political reorientation of at least some of the North Arabian tribes from the Roman to the Sasanian world.

If this is correct, we can propose the configuration of the Arab and Syro-Babylonian calendars as in Table 4.1 (where the names of the sacred months of the Arabs in bold).

This configuration is consistent with the conclusions suggested by the analysis of the Arabic month names. Ṣafar is in the late summer, when the grass turns yellow. The two Rabīʿ months are in the autumn. The 'freezing' months – the two Jumādās – are in the winter, and the 'parched' month of Ramaḍān falls in the hot Arabian spring. But more importantly it fits broadly with the contemporary testimony of two Byzantine historians, Procopius and Nonnosus.

Procopius[23] discusses the relations with the Arabs in the year 541 CE. Some people warn the Byzantine commander Belisarius that the Lakhmid king al-Munδir is about to attack Byzantine territory, but Belisarius rejects this information, because 'it was the season of the summer solstice,[24] and at this season the Saracens always dedicated about two months to their god, and during this time never undertook any inroad into the land of others'.

---

21 A connection of some sort between the paired months of the Islamic calendar and those of the Syro-Julian (solar) calendar was suggested already in an interesting article by the Assyriologist Winckler, 'Zur altarabischen Zeitrechnung', with wide-ranging conclusions that I am not able to accept.

22 For the calendar of Provincia Arabia see: Meimaris, Kritikakou, and Bougia, *Chronological Systems*, supplemented by Mercier, 'Intercalation'. This calendar was used both with Macedonian month names (in the Greek inscriptions) and with Babylonian month names (in the Aramaic inscriptions).

23 Procopius Bel. II 16, 18: Βελισάριος δὲ τὼ ἄνδρε τούτω ὡς ἥκιστα ἀληθῆ οἴεσθαι ἔλεγε. τοῦ γὰρ καιροῦ τροπὰς θερινὰς εἶναι. ταύτης δὲ τῆς ὥρας δύο μάλιστα μῆνας ἀνάθημα τῷ σφετέρῳ θεῷ Σαρακηνοὺς ἐς ἀεὶ φέροντας ἐν ταύτῃ ἐπιδρομῇ τινι οὔποτε χρῆσθαι ἐς γῆν ἀλλοτρίαν.

24 Dewing, *Procopius*, has wrongly 'vernal equinox'.

TABLE 4.1 The Arab months with their Syro-Babylonian equivalents

Arab months (lunisolar)	Syro-Babylonian months (lunisolar)	Approximate Julian equivalents
1. **Muḥarram**	Āḇ	July/August
2. Ṣafar	Ēlūl	August/September
3. Rabīʿ I	Tešrī qdīm	September/October
4. Rabīʿ II	Tešrī ḥrāy	October/November
5. Jumādā I	Kānōn qḏīm	November/December
6. Jumādā II	Kānōn ḥrāy	December/January
7. **Rajab**	Šḇāṭ	January/February
8. Šaʿbān	Āḏār	February/March
9. Ramaḍān	Nīsān	March/April
10. Šawwāl	Iyyār	April/May
11. **ḏū l-Qaʿda**	Ḥzīrān	May/June
12. **ḏū l-Ḥijja**	Tāmūz	June/July

And a fragment of the Histories of Nonnosus, preserved by Photius,[25] says that the Saracens assemble at a sacred meeting-place twice a year, once for a whole month, 'almost down to the middle of spring, when the sun enters Taurus', and once for two months 'after the summer solstice'. 'During these meetings complete peace prevails, not only amongst themselves but also with all the local population; even the animals are at peace with themselves and with human beings.'

These reports obviously need to be connected with the Islamic traditions concerning the sacred months during which no fighting is permitted: the single month of Rajab and a cluster of months later in the year. The Islamic tradition speaks in the latter case of three sacred months, two at the end of the year followed by one at the beginning of the next year. Procopius says that the armistice lasts for 'about two months'. 'About two' could conceivably means 'three'.

25 Nonnosus, quoted in Photius Bibl. Cod. 3: ὅτι τῶν Σαρακηνῶν οἱ πλεῖστοι, οἵ τε ἐν τῷ Φοινικῶνι καὶ οἱ τοῦ Φοινικῶνος καὶ τῶν ὀνομαζομένων Ταυρηνῶν ὀρῶν ἐπέκεινα, ἱερόν τι χωρίον νομίζουσιν ὁτῳδὴ θεῶν ἀνειμένον, καὶ ἐνταῦθα συλλέγονται κατ' ἐνιαυτὸν ἕκαστον δίς· ὧν τὴν μὲν τῶν πανηγύρεων αὐτῶν μὴν ὅλος μετρεῖ παρατείνων, σχεδόν που τοῦ ἔαρος κατὰ τὸ μέσον τελουμένην, ὅτε τὸν ταῦρον ὁ ἥλιος ἐπιπορεύεται· ἡ δὲ ἑτέρα πανήγυρις δυσὶ μησὶ παρα τείνεται. μετὰ θερινὰς τροπὰς ἄγουσι ταύτην. ἐν ταύταις, φησί, ταῖς πανηγύρεσιν πᾶσαν ἄγουσιν εἰρήνην, οὐ πρὸς ἀλλήλους μόνον, ἀλλὰ καὶ πρὸς ἅπαντας τοὺς ἐνδημοῦντας ἀνθρώπους· φασὶ δὲ ὅτι καὶ τὰ θηρία πρὸς τοὺς ἀνθρώπους, οὐ μόνον δὲ ἀλλὰ καὶ αὐτὰ πρὸς ἄλληλα.

Another possibility is that if the Arabs inserted the intercalary month at the end of the year, after ðū l-Ḥijja, and if the intercalated month was a profane (non-sacred) month, then in the intercalated year there would have been only two successive sacred months (ðū l-Qaʿda and ðū l-Ḥijja), followed by a profane intercalated month and then the sacred month Muḥarram in the next year. The Byzantines would not necessarily have known when the Arabs inserted the extra month; Procopius' statement to the effect that they renounced fighting for 'about' two months would be a cautious way of saying that one could count on an armistice of at least two months. But Procopius says that this was in 'the season of the summer solstice', which is exactly the time of the sacred months ðū l-Qaʿda and ðū l-Ḥijja according to the suggested configuration of the Arab and Syro-Babylonian calendars.[26]

Nonnosus puts the longer armistice 'after the summer solstice', which would be correct in the Arab year following an intercalation,[27] when the beginning of ðū l-Qaʿda would have been shifted to around the middle of June. But his statement that the one-month armistice was 'almost down to the middle of spring, when the sun enters Taurus' seems about one month too late for Rajab, even in a post-intercalary year. It is possible that there is an error in his report, or possibly in Photius's transcript of the same. Some scholars have raised the possibility that the sacred months were arranged differently in various regions, but the fact that these months marked a universal armistice between the warring tribes means that the system only makes sense if it is applied the same way everywhere.

4 The *nasīʾ* Formula

In Ibn Hišām's redaction of the life of Prophet by Ibn Isḥāq there is a short but invaluable account of the 'postponement' procedure in Mecca.[28] We read here that if the following year was to be affected by *nasīʾ* the *nāsiʾ* would address

26 The two months' truce at the time of the summer solstice has, of course, been discussed by everyone who has written about the ancient Arabian calendar, with wildly differing interpretations (for Mahmoud Effendi, 'Mémoire', Procopius was referring to Rajab; for Sprenger, 'Ueber den Kalender', he had the Persian Nawrōz in mind, etc., etc.). I do not think that it is now useful to review all of these suggestions.

27 This is the case regardless of whether the intercalary month is put after ðū l-Ḥijjah, or after Muḥarram, as it would be according to the hypothesis outlined below.

28 Ibn Hišām, ed. Wüstenfeld, p. 30. The same narrative, but without citing the formula, is found in aṭ-Ṭabarī, p. 934. In the *Tāju l-ʿarūs*, s.v. قلمس, the formula is cited in a slightly different form ('version 2'), namely:
اللهم إنّي قد أحللت أحد الصفرين وحرمت صفر المؤخر

the crowds assembled for the pilgrimage at Mecca with an obviously ritualised formula that went like this:

اللهم أحْلَلْتُ لهم أحد الصفرين الصفر الأول ونَسَأْتُ الآخر للعام المقبل

> O God! I have de-sanctified for them one of the two Ṣafars – the first Ṣafar – and I have postponed the latter (Ṣafar) for the following year.

The main problem here is the implication of the dual form *aṣ-ṣafarāni* 'the two Ṣafars'. The Arabic dictionaries are consistent in stating that this refers to the two consecutive months Muḥarram and Ṣafar and it has been widely argued (at least since Wellhausen[29]) that the first two months of the Arab year were 'originally' called Ṣafar I and Ṣafar II and that only later were they renamed Muḥarram and Ṣafar respectively; this would mean that the calendar had not two, but three consecutive pairs of identically named months (two Ṣafars, two Rabīʿs, two Jumādās), followed by six unpaired months. Although this is possible, it is not the only conceivable explanation. Another option would be to regard *aṣ-ṣafarāni* as an instance of what the grammarians call *al-muϑannā ʿalā t-taγlīb* (alias *dualis a potiori*),[30] whereby a pair of closely related terms can be expressed by the dual of one of them. Thus, one says *al-qamarāni* 'the two moons' for 'sun and moon', or *al-ʾabawāni* 'the two fathers' for 'father and mother', or *al-furatāni* 'the two Euphrates' for 'Tigris and Euphrates', or *al-ʿumarāni* 'the two ʿUmars' for the first two caliphs, Abū Bakr and ʿUmar. The same phenomenon can be observed also in Indo-European, for example in Sanskrit *pitarau* 'the two fathers' for 'father and mother', or *kr̥ṣṇau* 'the two Krishnas', that is 'Rama and Krishna'; in both instances the dual represents in effect a truncated *dvandva*. In this case *aṣ-ṣafarāni* would mean 'the two months of which Ṣafar is one', that is: 'the two consecutive months Muḥarram

---

that is: 'Oh God! I have de-sanctified one of the two Ṣafars and sanctified Ṣafar of the following (year).' In this version the pair of verbs *ʾaḥlaltu* and *nasaʾtu* has been replaced by the more obviously opposing pair *ʾaḥlaltu* and *ḥarramtu*, evidently in allusion to Qurʾān 9:37 (*yuḥillūnahū … yuḥarrimūnahū*). The version in Ibn Hišām ('version 1') has the benefit of greater antiquity, besides being clearly the *lectio difficilior*.

29 Wellhausen, *Reste*, p. 95, n. 2.

30 Copiously collected in Grünert, 'Begriffs-Präponderanz'; see also Reckendorf, *Die syntaktischen Verhältnisse*, p. 31; Wright et al., *Grammar*, vol. I, p. 190.

and Ṣafar'. The month name *ṣafarun*[31] is a proper noun, determined by nature, but without the article, like *muḥammadun*. If there were two months of that name they would surely be called *ṣafarun-i-l-ʾawwalu* and *ṣafarun-i-l-ʾāxiru*, like *jumādā* (not '*al-jumādā*') *l-ʾūlā* and *jumādā l-ʾāxiratu*. But in this case the form *aṣ-ṣafaru l-ʾawwalu* in our *nasīʾ* formula would not be a month name; rather the article would function to transform the proper name into a common noun ('the first of the two months of which Ṣafar is one'[32]), as it does in the mentioned *al-ʿumarāni*, or again in the same way that the proper name *muḥammadun* is turned into a common noun in *al-muḥammadūna fī t-taʾrīx* ('the various men in history named Muḥammad'). A third possible interpretation will be discussed in just a moment.

To understand the *nasīʾ* formula we need – I think – to make two assumptions. First: the year begins (as in the Islamic calendar) with Muḥarram and ends with the pilgrimage month ðū l-Ḥijja. The *nāsiʾ* declared the character of the following year in the presence of the assembled pilgrims. And second: the intercalary month was inserted at the end of the year, after ðū l-Ḥijja, and was a non-sacred (profane) month. In a common year the two sacred months ðū l-Qaʿda and ðū l-Ḥijja are followed immediately by the sacred first month of the next year, Muḥarram. But in an intercalated year the last month is followed by the profane intercalated month; the *nāsiʾ* declares thus that he has desacralized (*ʾaḥlaltu*) the month after ðū l-Ḥijja, which is now no longer called 'sacred' (i.e., *muḥarram*), but merely 'one of the two months of which Ṣafar is one' (*ʾaḥada ṣ-ṣafarayni*), and to be more precise 'the first of the two' (*aṣ-ṣafara l-ʾawwala*), and that he has postponed the latter of those two months (*wa nasaʾtu l-ʾāxira*), namely the profane month Ṣafar, 'belonging to the following year' (*li l-ʿāmi l-muqbili*),[33] by inserting the sacred month (Muḥarram) between the two. This manipulation of the calendar is expressed not in terms of intercalation, but in terms of 'desacralizing' one of the *ṣafarān* and 'postponing' the other. A calendrical operation is thus expressed in purely cultic terms.[34]

---

31 There is some disagreement as to whether the name of the second month of the year is triptotic (*ṣafarun*) or diptotic (*ṣafaru*), though the former seems to be favoured, for example by al-Farrāʾ, *al-ʾAyyām*, p. 17 (accusative *ṣafaran*).

32 This interpretation seems to be shared by *Lisānu l-ʿarab* and *Tāju l-ʿarūs*, s.v. *ṣafar*, which state that 'when they combine it (sc. Ṣafar) with al-Muḥarram they say: two Ṣafars' (*fa ʾiðā jamaʿūhu maʿa l-muḥarrami qālū ṣafarāni*).

33 In theory, *nasaʾtu l-ʾāxira li l-ʿāmi l-muqbil* could mean 'I postponed the latter until the next year', but I do not see how this would make any sense.

34 In 'version 2' of the formula (see n. 28), the *nāsiʾ* 'de-sanctifies' the first of the two Ṣafars and 'sanctifies' the second, in effect reversing Muḥarram and Ṣafar. The result is very much the same.

A comparable phenomenon can be observed in the Jewish calendar, where, in a common year, each of the twelve months has its apportioned festivals and fasts. But in the intercalated year the month of Adar is repeated and the holidays that are otherwise observed in Adar are transferred to the second Adar; the first Adar is ritually empty.[35] I am proposing a similar situation in the Arab calendar before Islam: in the intercalated year the month following the pilgrimage was stripped of its sanctity and the cultic events associated with it were transferred to the next following month, which assumes its name 'Sacred' (*muḥarram*). In both cases, the intercalation expresses itself in as a postponement (*nasīʾ*) of the sacred festivals by one month.

This interpretation assumes – as mentioned – that the year began with Muḥarram and that the intercalary month was put after δū l-Ḥijja. But we need to consider the possibility that this was not always the case. The ancient lunisolar calendars in the Near East – and elsewhere – begin their year, as a rule, either at the spring equinox or at the autumn equinox. The Babylonians began their year with Nisannu, in the spring, and put the intercalary month either just before the spring equinox (doubling Addaru/Āḏār) or just before the autumn equinox (doubling Ulūlu/Ēlūl), but most of the Near Eastern calendars in late antiquity begin the year in the autumn. Before the Achaemenid introduction of the regular 19-year cycle of intercalations, the New Year was presumably governed by the observation of some recurring astronomical (or agricultural) phenomenon, for example the heliacal rising or setting of some star, or the appearance of one of the lunar mansions, or the entry of the sun into one of the zodiacal signs, or something similar. If the phenomenon associated with the equinox had not occurred at the completion of twelve months the year would have been extended, ad hoc, by one month. If we assume now that the ancient Arab year (and presumably also that of its Syro-Babylonian prototype) began in the autumn, with Rabīʿ I (= Tešrī I) as its first month, and that the intercalary month was inserted before the autumn equinox by repeating the month of Ṣafar (= Ēlūl), then the 'two Ṣafars' (*aṣ-ṣafarān*) would have been the ordinary Ṣafar and the intercalary month of the same name, like the two Ulūlus of the Babylonian calendar. (I have previously adduced circumstantial evidence that the Sabaic calendar in South Arabia also had an intercalary month in the autumn.[36]) We would then have to think that the beginning of the year was later transferred to Muḥarram, that is: the month immediately following the *ḥajj*, the ceremony at which the intercalation was publicly promulgated, thus minimising the delay between the promulgation and the

35 This is explained neatly by al-Bayrūnī, *Chron.*, pp. 279–80.

36 De Blois, 'Qur'ān 9:37 and CIH 547', p. 103.

TABLE 4.2 The older form of the Central Arabian calendar (hypothetical reconstruction)

Arab months (lunisolar)	Syro-Babylonian months (lunisolar)	Approximate Julian equivalents
1. Rabīʿ I	1. Tešrī qḏīm	September/October
2. Rabīʿ II	2. Tešrī ḥrāy	October/November
3. Jumādā I	3. Kānōn qḏīm	November/December
4. Jumādā II	4. Kānōn ḥrāy	December/January
5. **Rajab**	5. Šḇāṭ	January/February
6. Šaʿbān	6. Āḏār	February/March
7. Ramaḍān	7. Nīsān	March/April
8. Šawwāl	8. Iyyār	April/May
9. **ðū l-Qaʿda**	9. Ḥzīrān	May/June
10. **ðū l-Ḥijja**	10. Tāmūz	June/July
11. **Muḥarram**	11. Āḇ	July/August
12. Ṣafar	12. Ēlūl	August/September
and in an intercalated year:		
13. ⟨Ṣafar II⟩	13. ⟨Ēlūl II⟩	⟨September/October⟩

actualisation of the intercalation. The intercalary month, still called by its old name, the 'first *ṣafar*', would have been moved forward to the position just before the new beginning of the year, while the second *ṣafar* is (as before) the ordinary month Ṣafar before Rabīʿ al-ʾawwal, but now separated from its namesake by the sacred month Muḥarram. The situation in the older form of the calendar would then have been as in Table 4.2.

I add that the two Ṣafars are mentioned also in a verse by the celebrated poet Abū Δuʾayb al-Huðalī, who is supposed to have been a contemporary of the Prophet of Islam, where he says of his mistress Umm ar-Rahīn:[37]

ʾaqāmat bihī ka muqāmi l-ḥanī- -fi šahray jumādā wa šahray ṣafar

that is: 'She remained there the way that the *ḥanīf* remains (in one place), for the two months of Jumādā and the two months of Ṣafar'. In this context[38] *ḥanīf* evidently means 'hermit, anchorite'; the implication is that Umm ar-Rahīn stayed in her tent in the desert, eschewing the pleasures of urban life, for an

37 *Der Diwan des Abu Ḏuʾaib*, poem no. 9, vs. 3.
38 See in general: De Blois, 'Naṣrānī'.

extended period of time. I do not think that the verse implies that 'the two months of Ṣafar' followed immediately after the two Jumādās[39] nor can I see any plausible calendrical model that would permit such an interpretation. Rather, the poet seems to be saying that his lady resided in the desert for the two months of Jumādā (in the winter) and was still there in the two months of Ṣafar (in the following late summer). This is consistent with his statement that she was there 'in the summer' (vs. 4: *fī ṣ-ṣayfi*), and that she referred (evidently during this sojourn) to the poet's uncharacteristically pious behaviour at the time of 'our pilgrimage' (vs. 7: *fī ḥajjinā*). This would make sense best if we assume that the poet is referring to 'the two months of Ṣafar' in an intercalated year, here perhaps as a poetic formula inherited from ancient times, rather than as a reference to actual calendrical practice at the time of Abū Δu'ayb.

5 When Was Intercalation Abandonned?

If then the first and second Rabīʿ were synchronised with the first and second Tešrī, and the first and second Jumādā with the first and second Kānōn, and the Arab New Year with the beginning of Āḇ, and if the intercalation was carried out regularly and correctly, this synchronism would have remained valid down to the time when intercalation was abandoned. After that, the Islamic and Syro-Babylonian months drift apart. This means that if we want to find the year in which intercalation could have been discontinued, we need to project the non-intercalating Islamic calendar back to a year where there is an 'average correspondence' (in the sense that I defined above) between the two Rabīʿs and the two Tešrīs. This average correspondence will be valid for two or three consecutive years (depending on the intercalation cycle) about every 34 years. This gives us lots of options. But, we have seen the documentary evidence that the standard Islamic calendar was in place already in the year 22 of the hijra. The abolition of intercalation must consequently have been before that date. If, however, we calculate the last time before AH 22 when there is an average correspondence of first and second Rabīʿ with first and second Tešrī we will make a very surprising discovery: this correspondence manifests itself precisely at the time of the hijra, as defined by the Islamic tradition, that is: in the year before the hijra (AH 0), the year of the hijra (AH 1), and the year after the hijra (AH 2).

---

39 As implied in Hell's translation (p. 25 of the German section): 'Dort hat sie geweilt (einsam) wie der Ḥanife in den beiden Ǧumādā- und Reǧeb-Monaten', i.e. 'the two months of Ṣafar' are here (apparently) Rajab and Šaʿbān. In a footnote, Hell refers to Wellhausen, *Reste*, pp. 95 and 239, but, as I understand him, Wellhausen says nothing of the sort.

TABLE 4.3 Synchronic table[a]

	Jewish (back-projected Rabbinic) equivalents of:	
	1st of Muḥarram	**1st of Rabīʿ al-ʾawwal**
AH 0	2 Av	2 Tishri
AH 1	2 Av	2 Tishri
AH 2	1 Av	1 Tishri
AH 3	30 Sivan	30 Av
AH 4	30 Sivan	30 Av
AH 5	1 Tammuz	1 Elul
AH 6	1 Sivan	1 Av
AH 7	1 Sivan	1 Av
AH 8	2 Iyar	2 Tammuz
AH 9	1 Iyar	1 Tammuz
AH 10	1 Iyar	1 Tammuz
AH 11	1 Nisan	1 Sivan
AH 12	2 Nisan	2 Sivan
AH 13	1 Nisan	1 Sivan
AH 14	1 Second Adar	1 Iyar
AH 15	30 Shevat	30 Nisan
AH 16	1 Adar	1 Nisan
AH 17	2 Shevat	2 Nisan
AH 18	1 Shevat	1 Nisan
AH 19	3 Shevat	2 Second Adar
AH 20	1 Tevet	1 Adar
AH 21	1 Tevet	1 Adar
AH 22	1 Tevet	1 Adar

a All conversions are thanks to KAIROS (see above, n. 8). For the Muslim dates of 1st of Muḥarram and 1st of Rabīʿ al-ʾawwal, the Thursday epoch and tabular calendar are assumed.

For the purpose of calculating the approximate luni-solar equivalents of the Muslim dates I have applied in a synchronic table the well-established arithmetic parameters of the Rabbinic version of the Jewish calendar,[40] but any other version of the luni-solar calendar will give broadly identical results (Table 4.3).

40 The earliest evidence for the Rabbinic calendar in its present form is the treatise on Jewish chronology by the Muslim astronomer Muḥammad b. Mūsā al-Xuwārizmī, written in or about 1135 Seleucid (AD 823–4), for which see provisionally my paper 'Some early

The observation that in the year of the hijra the Arab month Rabīʿ I corresponded to the Syro-Babylonian Tešrī I, and thus also with the Jewish Tishri, is in agreement with the Islamic tradition according to which when the Prophet arrived in Medina on the 8th of Rabīʿ al-ʾawwal AH 1 he found the local Jews to be fasting for Yōm Kippūr, alias *ʿāśōr*, Arabic *ʿāšūrāʾ*, which falls on the 10th of Tishri.[41] But this synchronism works only if there are no intercalations after that date.

The Islamic tradition does not — as we know — put the abolition of intercalation at the time of the hijra in AH 1, but rather at the time of the Farewell Pilgrimage in AH 10. As one can see from the synchronic table, the average correspondence between 1 Muḥarram and 1 Āḇ is not valid at that time. Within the framework of Islamic sacred history, there is good reason to assume that the nascent Muslim community abandoned intercalation already at the time of the hijra. The flight to Medina meant that the Prophet broke all ties with the temporal and spiritual rulers in Mecca; he would have seen no reason to abide by the calendrical ordinances of the intercalators in Mecca, nor indeed with those of the Jewish religious authorities in his new home. The new state in Medina warranted a new calendar. An obvious historical parallel is Julius Caesar, who is also reported to have abolished intercalation of the Roman calendar in an attempt (presumably) to establish himself as the sole arbiter of the calendar, as of everything else.[42] Later, when Muḥammad returned in triumph to the conquered Mecca he would have imposed the new calendar on his native city, announcing it publicly in the farewell sermon.

I would maintain that this in essence purely mathematical hypothesis vindicates the historicity of the hijra, this key element of Islamic sacred chronology. And it brings the early history of Islam just a bit closer to documented world history.

Islamic and Christian sources'. (My edition/translation of the treatise is forthcoming.) It is however possible and legitimate to project the parameters of this calendar back to the seventh century, in the same way that it would be possible and legitimate to project forward the parameters of the ancient Babylonian calendar. The results would be the same, with only slight variation. The Jewish month names are cited here in their usual anglicised spellings.

41 As is discussed, together with its calendrical implications, by al-Bayrūnī, *Chron.*, p. 330, though without observing that this story is irreconcilable with al-Bayrūnī's own claim that intercalation continued until the year of the Farewell Pilgrimage. The historical relationship between the Muslim *ʿāšūrāʾ* on 10 Muḥarram and the Jewish *ʿāśōr* on 10 Tishri will be discussed on another occasion. The recent article by Mazuz, 'Day of Atonement', unfortunately does not address any of the chronological issues involved.

42 Cf. Suetonius 1,40.

6 Recapitulation

The hypothesis that I have presented here, namely that intercalation was abolished at the time of Muḥammad's emigration from Mecca to Medina and the establishment of the Islamic polity in the latter city is dependent on the assumption that the Arabs in Central Arabia had an effective lunisolar calendar and that they carried out intercalation regularly and correctly and that they continued to do so up until the moment when intercalation was abolished and the months were set free to wander through all the seasons. This assumption cannot, of course, be taken for granted. It is possible that the intercalations were not carried out regularly and that the calendar fell into disarray years or even centuries before the banning of intercalation, or indeed that the abolition of intercalation was the direct response to the neglect of the same. All this is possible. In this case, the fact that what seems to be the original configuration of the Arab months (the two Rabīʿs corresponding to the first and second Tešrī in the autumn; the two Jumādās corresponding to the first and second Kānōn in the winter) recurs precisely at the time of the hijra (as defined by the Islamic tradition) would have to be pure coincidence. I suggest that it would be a very astonishing coincidence.

References

Primary Sources

Abū Δu'ayb (Xuwaylid b. Xālid), *Der Dīwan des Abu Ḏu'aib* (ed. and trans. J. Hell; Hannover: Lafaire, 1926).

al-Bayrūnī (*vulgo* al-Bīrūnī, Abū Rayḥān Muḥammad b. Aḥmad), *al-ʾĀϑāru l-bāqiyatu ʿani l-qurūni l-xāliya*, (*Chronologie orientalischer Völker* [ed. E. Sachau; Leipzig: Brockhaus, 1878]); (ed. Parwīz Azkāʾī; Tehran: Mīrās-I Maktūb, 2001); new edition in preparation by the present author. Cited here by the page numbers in Sachau's edition, which are also indicated in the margins of his translation *The Chronology of Ancient nations* (London: W.H. Allen & Co., 1879), and in those of the Tehran edition. The citations here have been collated with the oldest Mss.

al-Farrāʾ (Abū Zakariyyāʾ Yaḥyā b. Ziyād), *al-ʾAyyāmu wa l-layālī wa š-šuhūr* (ed. Ibrāhīm al-Abyārī; Cairo, 1956).

Ibn Manẓūr (Abū l-Faḍl Muḥammad b. Mukarram), *Lisānu l-ʿarab* (Būlāq 1300/1883 to 1308/1891).

al-Masʿūdī (Abū l-Ḥasan ʿAlī b. al-Ḥusayn) *Kitābu murūji δ-δahabi wa maʿādini l-jawhar* (*Les prairies d'or* [ed. and trans. C. Barbier de Meynard and P. de Courteille; Paris: Impr. Impériale/Nationale, 1861–77]; revised edition by Ch. Pellat (Beirut, 1965–79);

revised translation by Ch. Pellat (Paris: Société asiatique, 1962–89). Cited by pages of the first edition.

Quṭrub (Abū ʿAlī Muḥammad b. al-Mustanīr), *Kitābu l-ʾazminati wa talbiyati l-jāhiliyya* (ed. Ḥannā Jamīl Ḥaddād; Zarqāʾ [Jordan], 1405/1985).

aṭ-Ṭabarī (Abū Jaʿfar Muḥammad b. Jarīr), *Taʾrīxu r-rusuli wa l-mulūk* (ed. M.J. de Goeje et al.; Leiden: Brill, 1879–1901).

az-Zabīdī (Muḥammad Murtaḍā), *Tāju l-ʿarūsi min jawāhiri l-qāmūs* (Būlāq 1306/1889 to 1307/1890).

Secondary Literature

F.C. de Blois, 'Taʾrīkh' I, 1, in *Encyclopaedia of Islam* (2nd ed.; Leiden: Brill, 1998), pp. 257–64.

F.C. de Blois, 'Naṣrānī (Nazoraios) and hanīf (ethnikos): Studies on the religious vocabulary of Christianity and of Islam', *Bulletin of the School of Oriental and African Studies* 65 (2002), pp. 1–30.

F.C. de Blois, 'Qurʾān 9:37 and CIH 547', *Proceedings of the Seminar for Arabian Studies* 34 (2004), pp. 101–4.

F.C. de Blois, 'Islam in its Arabian context', in A. Neuwirth, N. Sinai, and M. Marx (eds), *The Qur'an in context* (Leiden: Brill, 2010), pp. 615–23.

F.C. de Blois, 'Some early Islamic and Christian sources regarding the Jewish calendar (9th to 11th centuries)', in S. Stern and Ch. Burnett (eds), *Time, Astronomy, and Calendars in the Jewish Tradition* (Leiden: Brill, 2014), pp. 65–78.

M. Bonner, 'Time has come full circle: Markets, fairs and the calendar in Arabia before Islam', in A. Ahmed, B. Sadeghi, and M. Bonner (eds), *Scholars and Scholarship of the Islamic World* (Leiden: Brill, 2011), pp. 15–47.

H.B. Dewing, *Procopius, History of the Wars, Books I and II* (Loeb Classical Library; London: Heinemann, 1914).

A. Grohmann, *Allgemeine Einführung in die arabischen Papyri* (*Corpus Papyrorum Archiducis Raineri* III, Series Arabica I, Pt. 2, XXI-XXVI; Vienna: Burgverlag, Ferdinand Zöllner, 1923–24).

A. Grohmann, 'Aperçu de papyrologie arabe', *Études de papyrologie* (Société royale égyptienne de papyrologie), I (Cairo, 1932), pp. 23–95.

M. Grünert, 'Die Begriffs-Präponderanz und die Duale a potiori im Altarabischen', *Sitzungsberichte der Philosophisch-Historischen Classe der Kaiserlichen Akademie der Wissenschaften* (Bd. 110/11; Vienna: Verlag der Österreichischen Akademie der Wissenschaften, 1886), pp. 559–96.

I. Hideyuki, 'The calendar in pre-Islamic Mecca', *Arabica* 61 (2014), pp. 471–513.

J. Jeffers, 'The Nonintercalated Lunar Calendar of the Middle Assyrian Period', *Journal of Cuneiform Studies* 69 (2017), pp. 151–91.

A. Jones, 'The dotting of a script and the dating of an era: The strange neglect of PERF 558', *Islamic Culture* LXXII/4 (1998), pp. 95–103.

J. Krall, C. Wessely, and J. Karabacek, *Papyrus Erzherzog Rainer: Führer durch die Ausstellung*, part 1 (Vienna: A. Hölder, 1892).

Mahmoud Effendi, 'Mémoire sur le calendrier arabe avant l'islamisme', *Journal Asiatique*, série v, 11 (1858), pp. 109–92.

H. Mazuz, 'The day of Atonement and Yawm 'Āshūrā': From assimilation to differentiation', *Ancient Near Eastern Studies* 50 (2013), pp. 255–61.

Y.E. Meimaris, K. Kritikakou, and P. Bougia, *Chronological Systems in Roman-Byzantine Palestine and Arabia: The Evidence of Dated Inscriptions* (Melethmata 17; Athens: Research Centre for Greek and Roman Antiquity, National Hellenic Research Foundation, 1992).

R. Mercier, 'Intercalation in the Era of the Province of Arabia', *Revue Biblique*, 108 (2001), pp. 101–8.

A. Moberg, 'An-Nasi (Koran 9,37) in der islamischen Tradition', *Acta Univ. Lundensis* 27 (1931), No. 1.

Th. Nöldeke, *Kurzgefasste syrische Grammatik* (Lepizig: Tauchnitz, 1898).

Th. Nöldeke, review of J. Barth's *Dịwan des 'Umeir ibn Schujeim al-Quṭâmî herausgegeben und erläutert, Wiener Zeitschrift für die Kunde des Morgenlandes* 16 (1902), pp. 275–285.

Th. Nöldeke, *Neue Beiträge zur semitischen Sprachwissenschaft* (Strasbourg: Trübner, 1910).

H. Reckendorf, *Die syntaktischen Verhältnisse des Arabischen* (Leiden: Brill, 1895).

H. Reckendorf, *Arabische Syntax* (Heidelberg: Carl Winter's Universitätsbuchhandlung, 1921).

C. Rink and R. Hansen, 'Der altarabische Kalender', in G. Wolfschmidt (ed.), *Sonne, Mond und Sterne – Meilensteine der Astronomiegeschichte. Zum 100jährigen Jubiläum der Hamburger Sternwarte in Bergedorf* (Nuncius Hamburgensis – Beiträge zur Geschichte der Naturwissenschaften 29; Hamburg: Tredition Gmbh, 2013), pp. 199–249.

G. Schregle, *Arabisch-deutsches Wörterbuch* (Wiesbaden: F. Steiner, 1981).

A. Sprenger, 'Ueber den Kalender der Araber vor Moḥammed', *Zeitschrift der Deutschen Morgenländischen Gesellschaft*, 12 (1859), pp. 134–75.

H. Wehr, *Arabisches Wörterbuch für die Schriftsprache der Gegenwart* (Wiesbaden: Otto Harrassowitz, 1st ed 1952, 5th ed. 1985).

J. Wellhausen, *Reste arabischen Heidentums gesammelt und erläutert* (zweite Ausgabe; Berlin: Reimer, 1897).

H. Winckler, 'Zur altarabischen Zeitrechnung', *Altorientalische Forschungen*, Reihe 2, Band 1 (1898), pp. 324–50.

W. Wright, *A Grammar of the Arabic Language* (ed. M.J. de Goeje and W.R. Smith; 3rd ed.; Cambridge: Cambridge University Press, 1896).

CHAPTER 5

The Institution of the Jalālī Calendar in 1079 CE and Its Cohabitation with the Older Persian Calendar

Johannes Thomann

1 Introduction

One of the universal problems in the history of calendars is the relationship between the description of a particular calendrical system in literary texts and its presence in documentary sources. In general, the first kind of source provides information on its technical details and its historical contexts, while the second kind of source allows us to draw conclusions on its use in everyday life, which in German is called 'Sitz im Leben'. Highly specialized studies may well concentrate on one of the two kinds of sources, but general accounts should always have an eye on both sides. In the present article an exercise in this double view will be presented, and hopefully it will be shown how misleading it can be if one of the two sides is neglected. The case represents an example of opposing characteristics in literary and documentary traditions.

The calendrical system in question is a modernized version of the Persian calendar, introduced during the reign of the Seljuq Sultan Malikshāh (1072–1092 CE).[1] The calendar is named after his honorific title Jalāl al-Dawla, therefore 'Jalālī'. Alternatively, the terms *malikshāhī* and *malikī* were also used.

Malikshāh was the son of Alp Arslān, who defeated the Byzantine army at Manzikert in 1071 CE. In the following years, Malikshāh conquered almost the entirety of Anatolia, and for the first time in history, this territory became populated by Turks.[2]

One of the first accounts of Malikshāh's calendrical reform mentions his famous vizier Niẓām al-Mulk as its leading initiator. Niẓām al-Mulk is well known as the author of the *Siyāsatnāma*, a Persian book on political governance. The secular character of the book has branded its author as the Islamic Machiavelli.[3]

1 De Blois, 'Ta'rīkh', p. 262; Ginzel, *Handbuch*, vol. I pp. 300–305; Barani, 'Jalali calendar'; Abdullāhī, *Tārīkh-i*, pp. 279–299; Naba'ī, *Gāhshumārī dar tārīkh*.

2 Vryonis, *Decline*; Beihammer, *Byzantium*.

3 Meri (ed.), *Medieval Islamic Civilization*.

 | DOI:10.1163/9789004459694_007

2 The Persian Calendar in Achaemenid, Sasanian and Early Islamic Times[4]

Originally, the Persian calendar was a Babylonian type lunisolar calendar, the so-called Old Persian calendar. In 525 BCE, Egypt was conquered and became a province of the Achaemenid Empire. Sometime later, probably between 481 BCE and 479 BCE, the Egyptian calendar was adapted by the emperor for the organization of the Zoroastrian cult; we might call it the Egypto-Zoroastrian calendar, or the Persian-Zoroastrian calendar. The Achaemenid rulers were proud of Egypt, their richest province, and therefore they might have chosen it for establishing uniformity of calendrical practice in the northern and eastern satrapies.[5] For non-religious purposes the Old Persian calendar remained in use: in a document written in 11 February 356 BCE, the dates are in the Old Persian calendar as well as in the Persian-Zoroastrian calendar.[6]

The Egyptian and Persian-Zoroastrian year consists of twelve months of 30 days and five additional days at the end of the year, the so-called epagomenal days. With its 365 days, the calendrical year is a quarter of a day shorter than the tropical solar year; every four years, its beginning falls one day behind in the seasons. Nevertheless, this calendar remained unchanged in Persia until the end of the fifth century CE.

At the beginning of the sixth century, a reform of the calendar was implemented in the Sasanian Empire. Two changes were made: the New Year was transferred from the beginning of the first month to the beginning of the ninth; and consequently, the epagomenal days were transferred from after the end of the twelfth month to after the end of the eighth. This shift of the epagomenal days led to a loss of five days and caused an earlier start of the year by five days. In the history of the Persian calendar from its beginning until early Islamic times, this was the only reform that really happened; all alleged reports of intercalated months are only legend.[7]

After the end of the Sasanian Empire in 651 CE, the years were counted according to an era based on the regnal years of the last Emperor Yazdgird III, with the epoch of 16 June 632 CE.

After 1007 CE, the five epagomenal days were shifted back from after the eighth month (Ābān) to the end of the year.

---

4 This section is based on De Blois, 'Persian Calendar', and Stern, *Calendars in Antiquity*, pp. 169–91.

5 Stern, *Calendars in Antiquity*, p. 190.

6 De Blois, 'Zoroastrian Calendar', p. 111.

7 Id., 'Persian Calendar', p. 50.

The months are all 30-days long, and their names are:

1. Farwardīn
2. Urdībihisht
3. Khurdādh
4. Tīr
5. Murdādh
6. Shahrīwar
7. Mihr
8. Ābān
9. Adhar
10. Day
11. Bahman
12. Isfandārmudh

3 Attempts to Reform the Persian Calendar before the Jalalī Reform

The oldest source on Malikshāh's calendar reform is the *Nawrūznāma*, attributed to ʿUmar Khayyām, but most probably not written by him.[8] In a section on various attempts to reform the Persian calendar, he reports the inconvenience of the beginning of the year having shifted from summer to spring, due to the Persian calendar having a fixed year of 365 days: the people had to pay their taxes earlier in the year, when the crop was still short. The Abbasid caliph al-Mutawakkil (r. 847–861) tried to reform the calendar, since in his time, the beginning of the year had shifted to April.[9] His reform did not last. In later documents, Persian dates are still based on the era of Yazdgird beginning from 16 June 632 CE.

The *Nawrūznāma* reports a second attempt to reform the calendar by Khalaf ibn Aḥmad, the Saffarid Amīr of Sīstān (r. 963–1009 CE).[10] In his time, Nawrūz had shifted from June to March. But recorded dates show that the Era of Yazdgird was still used. This is evident from inscriptions.

A particular case is the Arabic inscription on the Waruh Gorge, which has three dates: a complete Hijra date, and the corresponding months of the Persian and Syrian calendars. The Hijra date is fixed by the day of the week and corresponds to 19 December 1041 CE, in the Syrian calendar the month Kanūn I. In the Persian calendar based on the Yazdgird era, this corresponds to day 29

8 De Blois, *Persian Literature*, p. 359.
9 [ps.-]ʿUmar Khayyām, *Nawrūznāma*, p. 12; De Blois, 'Persian Calendar', pp. 45–46.
10 [ps.-]ʿUmar al-Khayyām, *Nawrūznāma*, p. 12.

of the Persian month Day if the five epagomenal days came after the month Isfandarmudh, and to day 24 of the same month if they came after Ābān. At the beginning of the 11th century CE, the epagomenal days were shifted back to their original place after the twelfth month (Isfandarmudh), but it remains to be investigated when this was implemented in practice in the different regions of the Eastern Islamic world. But if a reformed calendar as indicated in the *Nawrūznāma* had been used, another month would have been written in this inscription.

However, the reform was not entirely forgotten. Nāṣir-i Khusraw (1004–1060 CE) wrote a travel book on his Journey. He lived in Marw, and after a dream vision, he asked for leave from his administrative duties and travelled to the West, to Iran, Iraq, Syria, Egypt and the Arab Peninsula. What concerns us here is that in four cases he supplemented the usual Hijra date with the corresponding Persian month and the day in that month:[11]

1. [Tuesday] 20 Ṣafar 438 AH = 25 Shahrīwar [415 AY] [= 26 August 1046 CE]
2. Sunday 6 Ṣafar 439 AH = Urmuzd [1] Shahrīwar [416 AY] [= 2 August 1047 CE]
3. 1 Rajab – 20 Dhū l-Ḥijja 442 AH [= 19 Nov. 1050–5 May 1051 CE] ~ 15 Farwardīn [420 AY] [= 18 March 1051 CE]
4. Friday 19 Dhū l-Ḥijja 442 AH = 1 Khurdādh [420 CE] [= 3 May 1051 CE]

Cases 1, 2 and 4 show an accurate match of the dates in the two calendars, provided the era of Yazdgird is used; Nāṣir-i Khusraw thus still used the traditional Persian calendar (case 3 is inconclusive). However, he must have been aware of the existence of a reformed calendar. In all cases, he has added the word *qadīm* ('old'): *Shahriwar māh-i qadīm, Farwardīn-i qadīm, Khurdād mah-i qadīm*. We can only guess which new calendar these dates were meant to be distinguished from. Marw, the home of Nāṣir-i Khusraw, was under Ghaznavid rule.

There was indeed a calendar reform during the rule of Masʿūd the son of Maḥmūd of Ghazna (r. 1030–1040 CE).[12] It was praised in panegyrical poems. The main point appears to have been the readjustment of two holidays: *sāda*, the 'feast of the return of the light', and *nawrūz*, the celebration of New Year. Both were celebrated with traditional rituals, and in their symbolism, they were connected to the noticeable increase of day length in January and to the spring equinox in March. The reform seems to have had the purpose of putting the two holidays back to their appropriate place in the tropical year. It might have been restricted to the calendar of Persian festivals, and had no significance for other areas of life.

11 Nāṣir Ḫusrau, *Book of travels = Safarnāmah*, pp. 6, 51, 92, and 105.

12 Cristoforetti, *Il natale della luce in Iran*, p. 167.

4 The Reform of Malikshāh

The next reform reported in the *Nawrūznāma* was the one by the Seljuq ruler Malikshāh. He had been told about the inconvenience of the early Nawrūz, and decided to fix the beginning of the year at the spring equinox. For this purpose, he commissioned the best astronomers of the time from Khurasan. They built an observatory and began with observation. But according to the report in the *Nawrūznāma*, the task was not completed; as is explicitly stated, the intercalation did not take place. The text is somewhat unclear when it comes to determining the reason. Either Malikshāh lost patience and stopped the project, or he died and his successor was no longer interested.[13] If this report can be trusted, one may ask how the era that bears his name came into being.

The first historian who gave an account of this reform was Ibn al-Athīr (1160–1233 CE). He separates the reform of the calendar from the construction of an observatory and the astronomical observations. According to him, the observations ceased after Malikshāh's death and the observatory was demolished.[14] In the context of the calendar reform, in which *nawrūz* was set to the spring equinox, he mentions an anonymous group of astronomers. But in the context of the astronomical observations, he reports the names of three astronomers: the famous ʿUmar Khayyām, the well-known Abū l-Muẓaffar al-Isfizārī, and the unknown Maymūn ibn al-Najīb al-Wāsiṭī. ʿUmar Khayyām may well have joined the group of astronomers at a later time, or he might have been involved in the observations only and not in the calendar reform. From Ḥajjī Khalīfa's Lexicon we know that ʿUmar Khayyām produced astronomical tables dedicated to Malikshāh, the *Zīj-i Malikshāhī*, which is generally believed to be lost.[15] However, the Union Catalogue of Manuscripts in Iran refers to a copy of a *Zīj Malikshāhī* (Tehran, Millī 26211).[16] These tables might have been based on the observations at the court of Malikshāh in Isfahan.

Al-Khāzinī (active ca. 1120–1130 CE), a younger contemporary of ʿUmar Khayyām, prepared astronomical tables for Malikshāh's son Sanjar, *al-Zīj al-Sanjarī*. In the section on the different eras in history, he reports on Malikshāh's reform.[17] According to him, the Sultan ordered first to determine the vernal equinox, and after that the times when the sun enters the next zodiacal sign.

---

13 [ps.-]ʿUmar al-Khayyām, *Nawrūznāma*, p. 89.

14 Ibid., p. 89; on these observations, see in general Rozenfeld and Juškevič, *Omar Chajjam*, pp. 66–76.

15 Ḥajjī Khalīfa, *Keşf-el-zunun*, vol. 2, 966.

16 Dirāyatī, *Fihristgān*, vol. 17, p. 809.

17 Leichter, *The Zīj as-Sanjarī of Gregory Chioniades*, p. 29.

Two different dates of the new era are found in the sources. In the *Zīj* of Ulugh Beg (d. 1449 CE), the problem of these two different eras is briefly discussed; the dates are Sunday, 5 Shaʿbān 468 AH (13 March 1076), and Friday, 10 Ramaḍān 471 AH (15 March 1079); the later one is declared as the preferred.[18] The earlier date seems to have been the original one; at some time it was replaced by the later date.[19] Furthermore, an intercalation scheme was introduced. It was based on the Julian intercalation scheme, but every 25 years, another intercalary day was added. Both the epoch and the intercalation are erroneous. The true equinox was on 15 March at that epoch, and the correction of the Julian intercalation scheme is in the wrong direction. Instead of increasing the intercalary days, they should have been reduced. Other accounts report exactly such a correct improvement of the Julian intercalation scheme.[20]

Al-Khāzinī's account is the first technical description of the Jalālī calendar. The beginning of its epoch is set to Sunday, 13 March 1076, three years earlier than in later reports. In the text, a table and a description follow on how to convert dates of the Jalālī calendar.

The Jalālī calendar is used in the *Dustūr al-munajjimīn*, but no chronological parts of the work have been preserved.[21]

In later astronomical tables, the Jalālī calendar is regularly dealt with. Naṣīr al-Dīn al-Ṭūsī (1201–1274 CE) devoted a longer chapter to it in his *Zīj-i Īlkhānī* for the Mongol emperor Hülägü in 1272 CE.[22] The great success of these tables might have reinforced the inclusion of the Jalālī calendar as a topic in the chronological parts of astronomical tables produced in the Eastern parts of the Islamic World, and later in the Ottoman Empire.

The most influential of all later tables were those produced under the direction of Ulugh Beg (1394–1449 CE) in Samarqand.[23]

5 Documentary Evidence for the Jalālī Calendar

The historical and astronomical sources describe Malikshāh's reform of the Persian calendar as a prominent event and leave the impression that the older Persian calendar was abandoned. But how far can this be supported by documentary evidence?

18 Sédillot, *Prolégomènes*, pp. 25–26 (Persian text) and 27 (translation).
19 Giahi Yazdi, '*Jalālī* Calendar', p. 180.
20 See text and translation in Appendix I.2.
21 Van Dalen, 'Malikī Calendar'.
22 See text and translation in Appendix II.2.
23 Sédillot, *Prolégomènes*; Uluğ Beg, *Zidž*; id., *Uluğ Beg'in astronomi.*

The first place to direct one's attention to are inscriptions. The only inscription recorded in the *Thesaurus d'Épigraphie Islamique* containing a Jalālī year is found on an astrolabe which used to be part of a private collection in Benares.[24] Its present location is unknown, and no reproduction was published. According to the description of Morley in 1856 CE, years in four calendars were inscribed on both sides of the astrolabe.

Recto: 'year 190 of Malikshāh'
Verso: '640 of Yazdgird, 1582 of Iskandar Rūmī'
'Maḥmūd ibn 'Alī ibn Yūsha … year 669 of the Hijra'

If the standard eras are used, the dates are close together but do not match perfectly. However, the Jalālī era of 1079 is closer to the other dates than the era of 1076 assumed in *al-Zīj al-Sanjarī.* Moreley concluded from the owner's inscription that the astrolabe was made for the chief Vizier of the Mamluk Sultan Baybars. This would be an important clue for the use of the Jalālī era in the Western part of the Islamic world. But the fact that it appears on an astronomical instrument limits this evidence. The astrolabe maker might have converted the Hijra date, which appears together with his signature, into all calendars described in his astronomical tables.

In official monumental inscriptions from the time of Malikshāh, the era named after him does not occur. In the Friday mosque of Isfahan, two domes were built during his reign. The southern dome was founded by Malikshāh's chancellor Niẓām al-Mulk, and the northern dome by his rival Tāj al-Mulk. The inscription in the northern dome has a date: 'in the months of the year four-hundred eighty-one'. This date must be a Hijra date. If it were a Persian year based on the Yazdgird era, this year would be long after the death of Malikshāh. The inscription of the southern dome contains a long praise of Malikshāh and the name of Niẓām al-Mulk, but no date. From evidence provided through the titles mentioned, it can be concluded that the construction must have taken place no earlier than 1086 CE. This was definitely after the calendar reform, but the era was not used for official purposes. The same holds true for later inscriptions.

Iranian dates with years in the Yazdgird era do appear in some inscriptions. The earliest is on an epitaph in Sri Lanka.[25] Next in time is a foundation

24 http://www.epigraphie-islamique.org, accessed 4 November 2017, no. 2263.

25 For this and the following inscriptions see Appendix I.5.

inscription for the year 1203 CE.[26] Later inscriptions with years in the era of Yazdgird are restricted to astronomical instruments produced in Iran.[27]

Another place to look at would be numismatic evidence of the use of the Jalālī era. It seems, however, that years in the Jalālī era do not occur in coins.[28] The same seems to be true for official and private documents. Generally, archives in Iran do not contain documents earlier than 1500 CE. A singular exception is the archive of the Ṣafawid order in Ardabil, which was discovered in 1970.[29] The few extant documents of the early Seljuq period are all dated with Hijra years only.[30] The earliest of them, a deed from Khotan, bears the date 501 without further specification, but it is most likely a Hijra year, and certainly not a Jalālī year.[31] In an Arabic court order from Yārkand (Central Asia) the year 503 and the Arabic Month Shaʿbān are mentioned, definitely a Hijra date.[32]

These examples suggest that the Jalālī era was not used for administrative purposes. But then the question arises, why the astronomical tables provide descriptions and conversion tables for converting Jalālī dates at all.

The only documentary (or semi-literary) sources containing such dates are astronomical-astrological documents. The earliest example appears in two fragments in Berlin, found in a book binding.[33] They belonged to two consecutive leaves of a yearbook containing astronomical and astrological information which can be dated astronomically to the year 1182/1183 CE. This type of yearbook was first produced in the 11th century CE.[34] Two fragments containing data for the Persian year 413 AY (= 1044/1045 CE) are the first example of this type.[35] An almost entirely preserved leaf of a yearbook for the Hijra year 544 AH (= 1149/1150 CE) gives a clear picture of the layout of the spread with information for one month.[36] On the right side, the traditional layout of an ephemeris with the calendarium and the daily positions of sun, moon,

26 See text and translation in Appendix I.6.

27 See texts and translations in Appendix I.7–12. For exclusive use of the hijra in other documents, see Herrmann, *Persische Urkunden*; id., 'Urkundenfunde in Āẕerbāyǧān'; Gronke, *Derwische*, pp. 54–5.

28 Personal information by Lutz Ilisch, Tübingen (e-mail 26 June 2017).

29 Herrmann, *Persische Urkunden.*

30 For a survey of early Persian documents see Gronke, *Derwische*, p. 14.

31 Minorsky, 'Documents I', p. 186.

32 Gronke, 'Arabic Yārkand Documents', p. 489; id., *Arabische und persische Privaturkunden*, pp. 54–5.

33 See Appendix I.13.

34 Thomann, 'From Katarchai to Ikhtiyārāt', pp. 342–54.

35 Id., 'Ephemeride'.

36 Id., 'Arabic Ephemeris', pp. 216–217, figures 1–2.

and planets is depicted. The calendarium contains the days of the Arabic, Coptic, Syriac and Persian months. The Persian dates are based on the era of Yazdgird, despite the fact that it was created more than half a century after Malikshāh's reform.

In contrast to these examples, which were produced in Egypt, the yearbook of 1182/1183 was made in northern Iraq or Syria, perhaps in Aleppo. For this reason, the Coptic calendar is not included in the calendarium: it only contains the Persian, Arabic and Syriac calendars. The two headlines in larger writing indicate the Arabic and the Persian months. But the corresponding columns of the calendarium, of the Arabic and Persian calendars, as well as of the Syriac calendar, do not begin with day one, but start on the first line with a day numbered higher than one. This means that none of them served as the basis for the month contained in any one page. The dates in the Persian calendar are based on the era of Yazdgird. Thus, the only plausible assumption is that the months contained in each page are based on the Jalālī calendar. The position of the Sun in the first line of the page, indicating the first day of the month, is not at the beginning of a zodiacal sign. Therefore, the months were not regulated according to the entry of the Sun into new signs, as it is stated in the Sanjarī Zīj. The alternative would have been to keep the old system of twelve months of thirty days, with additional epagomenal days filling the gap before the New Year. Indeed, if one counts backwards from the months of the two fragments to the beginning of the first month, one ends up at the day of the vernal equinox. This was the main point of Malikshāh's reform of the calendar, to put *nawrūz* back in its proper place in time. This was accomplished in the yearbook of 1182/1183. Since the beginning of the yearbook is missing, we cannot tell if the year based on the era of Malikshāh was even mentioned.

The next example of a Jalālī date is a deluxe horoscope. It is the famous horoscope of Iskandar Sultan, a grandson of Tīmūr. It is one of the most beautiful illuminated Persian manuscripts of its age, kept in the Wellcome Library.[37] The birth date of Iskandar Sultan is given in five calendars:

> Monday, 3 Rabīʿ I 786 of the Hijra,
> corresponding to 15 Urdibihisht, the Jalālī month, year of Malikshāh 306,
> conforming to 17 old month Murdādh, Yazdgird year 753
> identical with 25 Rūmī month Nīsān, year of Iskandar 1695,
> calendar of the Chinese people [...] xin wei (辛未) [day]

37 Keshavarz, 'Horoscope'; Keshavarz, *Catalogue*, pp. 396–99 no. 224; Caiozzo, 'Horoscope'; Tourkin, 'Astrological Images'; id., 'L'Oroscopo di Shāh Tahmāsp'; id., 'The Horoscope for Shah Tahmāsp'.

The dates in the three traditional calendars all correspond to 25 April 1384. The year 306 of Malikshāh leads to an epoch of 1079 CE. The Chinese date given agrees with the date in the Imperial Chinese calendar.[38]

In this context, variation seems to be a value in and of itself. It can be seen in the varying colours at the beginning of each dating phrase, and the synonyms used for 'corresponds to'. The manuscript contains 86 lavishly illuminated folios. The five redundant calendrical dates can be interpreted as a form of chronological eloquence.

The Iskandar horoscope is singular both in the length of the text and the quality of the decoration. The closest in length to the horoscope of Iskandar Sultan is the horoscope of Shāh Tahmasp, with 50 folios.[39] It also contains the date of the horoscope in the Jalālī calendar: Wednesday 19 Isfandarmuz 435. Further calendars used for the date are the Hijrī calendar, the Alexandrian calendar with the Seleucid era, the Persian calendar with the era of Yazdgird III, and the Chinese calendar:

26 Dhū l-Ḥijja 919 AH
22 Shubāt 1825 AS
17 Tīr 883 AY
28 I 88.639.930 AC.

They all correspond to 22 February 1514 AD in the Julian calendar, which is a Wednesday.

Other deluxe horoscopes typically contain only a few pages.[40] However, their history is still barely explored and more examples of the type of the horoscope like Iskandar Sultan's may come to light in the future.

In the fifteenth century, a new type of deluxe yearbooks came into use.[41] Most are in Persian and Turkish, and only few exist in Arabic. A relatively early example can be found in a codex in the Biblioteca Medicea Laurenziana (MS Or.027).[42] It contains a number of leaves of one yearbook for the year 1450,

---

38 Fang, *Zhongguo shi liri he Zhong xi liri duizhao biao*, p. 585. For the Chinese calendar in Islamic context, see van Dalen, Kennedy, and Saiyid, 'The Chinese-Uyghur Calendar in Ṭūsī's Zīj-i Īlkhānī'; Kennedy, 'The Chinese-Uyghur Calendar as Described in the Islamic Sources'; Melville, 'The Chinese-Uyghur Animal Calendar'.

39 Tourkin, 'L'Oroscopo di Shāh Tahmāsp'.

40 For an example see Adamova and Bayani, *Persian Painting*, pp. 28–40 no. 2.

41 For a list with 78 items see Şen, *Astrology in the Service of the Empire*, p. 353ff.

42 Not mentioned by Şen, *Astrology in the Service of the Empire*. Images: http://teca.bmlonline.it/ImageViewer/servlet/ImageViewer?idr=TECA0001485178#page/1/mode/1up, accessed 5 November 2017.

mixed up with leaves of a later yearbook. The text is in Ottoman Turkish. Here, the Jalālī dates are visible in a separate column.[43] It is striking that it is not added to the calendarium with the traditional calendars, but rather placed to the left of the columns of the holidays, as if the scribe did not want to disrupt the traditional order of the columns. On the first day of the first month the sun is close to the beginning of Aries. But on the first day of the tenth month, the sun is in 24° of Capricornus. As in the yearbook of 1182 CE, the months have a fixed length of 30 days, and generally do not coincide with the entry of the sun into a zodiacal sign.

The first known Arabic case of this type of yearbook is an almanac for the year 1601/1602 CE (MS Venice, Biblioteca Marciana, Or. 84 (= 14)). This calendarium contains three traditional calendars, the Arabic, the Roman, and the Coptic. Separated by two columns is the column with the Jalālī dates. The script is unclear, but it is most likely called *al-sulṭānī*. In the top left field, year 524 is written in green and red. It is close to the column of the Jalālī dates and, indeed, if subtracted from the Julian year 1603, we result with the era of 1079 CE. The older Persian calendar is absent in this almanac. One could speculate that it had become obsolete by then, but the evidence disproves this claim. In the numerous yearbooks from later years, the older Persian calendar is a regular encounter. An astronomical ephemeris without an astrological part (Berlin Staatsbibliothek, or. quart. 211, fol. 1r) provides a good example.[44] The calendarium contains seven calendars. Among them are the four traditional calendars: the Arabic, Coptic, Greek (*rūmī*) and older Persian calendars. Additionally, the Latin calendar is included and labeled as 'gentile' (*ʿajamī*). Also included is the Hebrew calendar, a rare occurrence in an ephemeris. Finally, the Jalālī calendar is included in the calendarium. The first leaf is missing, and on the remaining pages there is no dating year. The first page contains the data for the second month Ṣafar of the Arabic calendar. The value for the first day and its equivalences in the other calendars are as follows:

WD	Ṣafar	Pachons	Greek Nīsān	April	Persian Ādhar	Hebrew Sivan	Jalālī Urdhībihisht
Fr	1	4	29	29	1	1	22

43 For text and translation see Appendix III.1.

44 For text and translation see Appendix III.2.

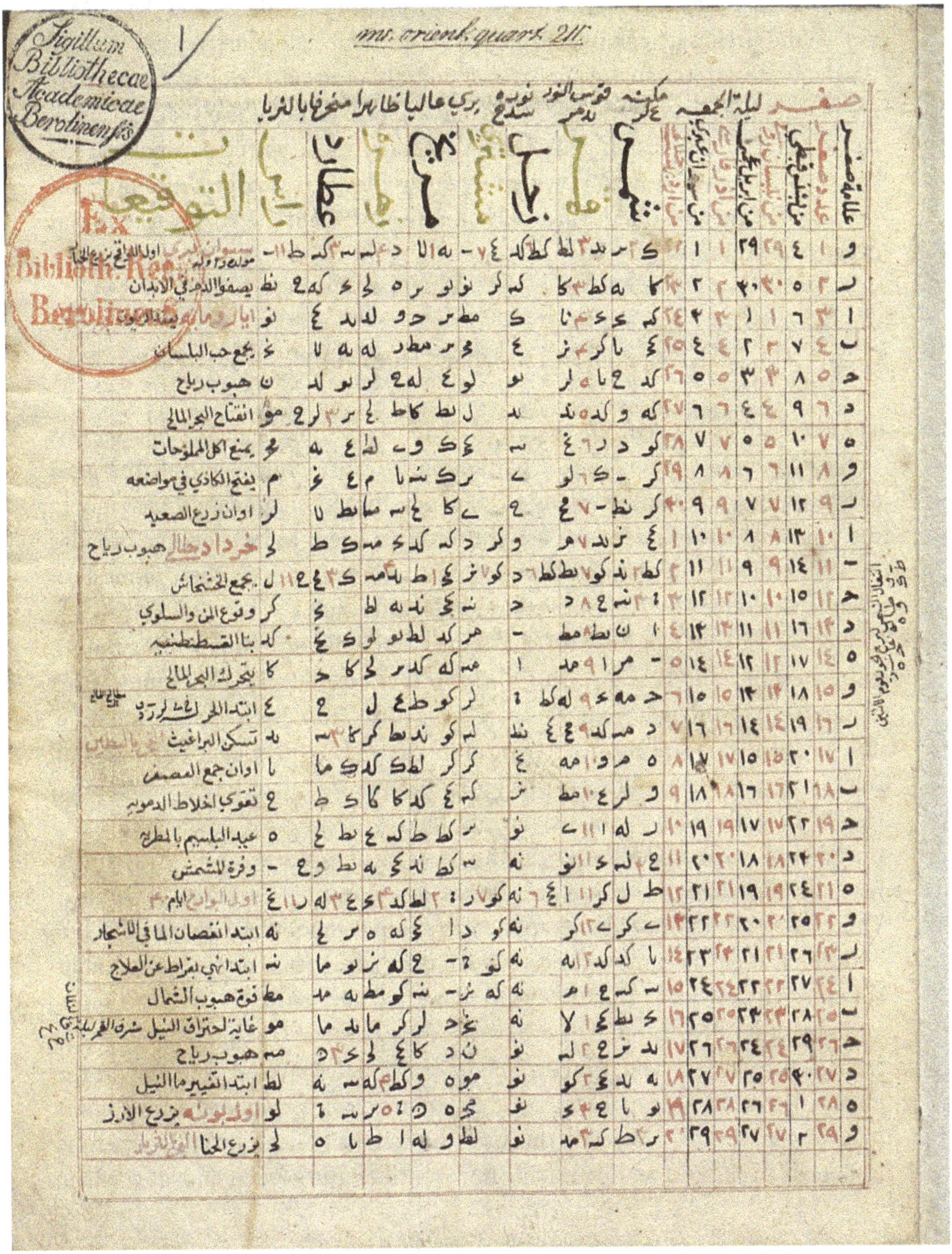

FIGURE 5.1 Berlin Staatsbibliothek, or. quart. 211, fol. 1r. Arabic Ephemeris for the year 1804 CE with dates in the Arabic, Coptic, Syriac, Julian, Persian (Yazdgird), Hebrew, and Jalālī calendars
PUBLIC DOMAIN MARK 1.0; IMAGE: STAATSBIBLIOTHEK ZU BERLIN – PK, https://digital.staatsbibliothek-berlin.de/werkansicht?PPN=PPN752162500&PHYSID=PHYS_0005

The solar, lunar and planetary positions point to the date of Friday 11 May 1804 in the Gregorian calendar (29 April in the Julian calendar; JD 2380088). All the dates are correct. In the Persian calendar that is used, the epagomenal days are placed after the 8th month Ābān, and not at the end of the year. This indicates that very old chronological tables were used. The spring equinox in 1804 was on 21 March (Gregorian), which is 51 days before 11 May. 22 Urdhībihisht is 51 days later then 1st of Farwardīn. Therefore Nawrūz in the Jalālī year 726 was at the true equinox.[45]

6 The Jalālī Calendar in a Broader Context of Time Practices[46]

These examples of occurrences of Jalālī dates show that the Jalālī calendar was exclusively used in documents pertaining to astronomical and astrological activities. But it would be wrong to conclude that its significance was only in the minds of astronomers and astrologers. This needs to be seen in a broader social context. More general thoughts on time concepts in the pre-modern Islamic world may help to understand its significance.

Jacques Le Goff's famous article 'Temps de l'église et temps du marchand', published in 1960, is still a classic in medieval studies. Even if some of his claims were disproved by later research, his distinction between 'church time' and 'civil time' is still regarded as useful.[47] The city tower of Aire-sur-la-Lys, built in 1355 CE, exemplified the beginning of a new kind of time, professional time.[48] Its bells did not ring for prayers, but for commercial transactions and the labor of workers. For the merchant, the time of business gave a structure to his daily life, while the bells of the church served another horizon of existence. This conception can easily be applied to the pre-modern Islamic world, but with the more generic categories of 'civil time' and 'ritual time'. In the public sphere, civil time and ritual time existed both in parallel.

As an example, a sundial inscribed with the name of Nūr al-Dīn al-Zangī in Syria as its maker is said, in the same inscription, to have been made 'for the knowledge of the seasonal hours and the hours of the prayers'; seasonal hours refer to civil time.[49] However, this dual model of time can be supplemented from texts in early paper documents from tenth-century Egypt and later, which

45 Spuler, *Wüstenfeld-Mahler'sche Vergleichungs-Tabellen*, p. 39.

46 This final section is an updated version of the conclusions in Thomann, 'Tools of Time'.

47 Dohrn-Van Rossum, *History of the Hour*, pp. 13–14, 226–31; Champion, *The Fullness of Time*, pp. 8–9; Kiening, 'Mediating the Passion', p. 140.

48 Le Goff, 'Au Moyen Age', p. 35.

49 Casanova, 'Le montre du sutan Noûr ad Dîn'; Thomann, 'Tools of Time'.

attest a third type of time, cosmic time, and show how it was managed in everyday life. These are the so-called almanacs, which contain daily entries for a particular year with information of an astronomical and astrological nature. Central to the almanacs were predictions based on the aspects of the moon with the sun and the planets. The theory behind these predictions was a part of astrology called *katarchai* in Greek, *ikhtiyārāt* in Arabic and *electiones* in Latin. The earliest available example is a paper fragment of an almanac for the year 297 of the Hijra.[50] A similar almanac, made for the year 334 of the Hijra, contains additional information, by adding the hour of day or night when the aspects take place.[51] This seems to allow for even more precise scheduling of actions, or avoidance of actions, at particular hours. More concrete descriptions can be found in an almanac for the Coptic year 707 (990/991 CE),[52] where besides more general prescriptions, e.g. 'contacts with the authorities' and 'to accomplish messages', are found.

These documentary examples have no archaeological context. All we know is that they were probably found in the region of the Fayyūm or further to the south. However, a number of almanacs were found among the documents from the Cairo Geniza.[53] They are very similar in layout and content to the almanacs found elsewhere in Egypt. In one single case, an astrological document was found in a regular excavation. Some hundred Arabic documents were excavated in al-Fustat, among them an astrological responsum.[54] The building where it was found belonged to a group of houses which have been characterized as 'the workers' quarter'.[55] The presence of an astrological document at such a place seems to indicate that astrology of an average level was not limited to the elite or the rich, but found its way to the middle class.

The evidence thus suggests that a third type of time has to be taken into consideration: cosmic time. It had its public appearance too. The Talisman gate in Baghdad presents the divine person of the moon as the dominant ruler of cosmic time, between the divine beings of the lunar nodes.[56] In this conceptual model, civil time serves interaction within society, ritual time is directed towards the transcendental aspect, and cosmic time keeps contact with the heavenly rulers, but is directed to the individual itself. Concerning rulership, civil time is implicitly atheistic, ritual time is decidedly monotheistic, and

50 Thomann, 'Fragment'.
51 Id., 'From Katarchai to Ikhtiyārāt', pp. 344–45, fig. 3.
52 Id., 'Almanach'.
53 Goldstein and Pingree, 'Astrological Almanacs'; id., 'Additional Astrological Almanacs'.
54 Richards, 'Written documents', p. 68.
55 Scanlon, 'Fustat', p. 367.
56 Sarre and Herzfeld, *Archäologische Reise*, pl. X–XI.

cosmic time is glaringly polytheistic. In structure, civil time is homogeneous, ritual time is dichotomous, and cosmic time is highly heterogeneous. Civil time manages interpersonal activity, ritual time manages otherworldliness, and cosmic time promises to manage self-fulfillment.[57] Moreover, the threefold concept of time can be applied to the three groups of calendars considered here. The lunar calendar based on the Hijra era was indispensable for the organization of rituals in the course of the year. Various, different simple solar calendars were used for different practical purposes in civil life. Finally, the Jalālī calendar was used exclusively in connection to astrological practices, and served to plan personal time choices in order to maximize individual success.[58] For a large part of society, this seems to have been guaranteed by the stars that were thought to govern all aspects of human life. To stay in synchronism with the heavens was for many the preferred way of life.[59] The Jalālī calendar was designed to achieve this.

Appendix 1: Arabic Texts

1.1 *Inscription, 1041 CE, of Waruh*[60]

كتابت (كتبت ؟) يوم الثلثاء من أيّام شهر جمادى الأولى سنة ثلث وثلثين وأربعمائة
أمر بها / تشريفا للتكين الأجلّ السيّد العالم العادل معزّ الدولة ارسلان تكين أبو
الفضل العبّاس من / وُيّد العدل ايلك ابن الأمير نصر بن علىّ سعيد خان مولى أمير
المؤمنين والى اسفره رودبار / رم (؟) أعزّ الله شرفه و غفر الله له و لوالديه / وألحقه
برسوله محمّد صلّى الله عليه وآله و أصحابه و أهل بيته / فى دى ماه الفرس سنة عشر
و أربعمائة فى كانون الأوّل الروميّة

Inscription of Tuesday/the third of the month of Jumādā I of the year four hundred thirty-three [December 29, 1041]. It was ordered to honor the most exalted prince, the wise and just lord, Muʿizz al-Dawla Arslān Tikīn Abū l-Faḍl al-ʿAbbās ibn Mu'ayyad al-ʿAdl Ilik, son of the amir Naṣr ibn ʿAlī Saʿīd Khān, client of the Commander of the Faithful, by the governor of Isfara, Rūdbār and Rūm [?], may God glorify his nobility

57 More extensive arguments and fuller documentation of this conceptual model are found in Thomann, 'Ritual Time'.

58 For the 'egocentric perspective' of astrology see Mozaffari, 'Astrological Opinions', p. 360.

59 The expression is borrowed from the book title of David King's work *In Synchrony with the Heavens*.

60 Text edition and translation by Blair, *Monumental Inscriptions*, p. 115 no. 42.

and pardon him and his parents and join him to His prophet Muḥammad, Gods's blessing be upon him and his family and his companions and the people of his house, in the Persian month Day in the year four hundred and ten and Kanūn I of the Greeks.

1.2 *Al-Khāzinī*, al-Zīj al-Sanjarī[61]

والخامس من التواريخ المشهورة سنو الكبائس السلطانية الملكشاهية
ولمّا كان أمر السنة الشمسية استيفاء الفصول الأربعة وعرد النسو خرج
الأمر العالي السلطاني الملكشاهي أنار الله برهانه بافتتاح التقويم من لدن بلوغ مركز
النير الأعظم نقطة الاعتدال الربيعي وكانت سنو التواريخ غير مطابقة لها فأردنا
الحاقها

لها فأردنا لحاقها إليها ليسفعك؟ على المقوم افتتاحها لاستخراج
أوساط حركات الكواكب بأي تاريخ أراده هو على محاذاة السنة
الشمسية وأولها يوم الأحد من شعبان سنة ثامن وستّين وأربعماة
وافتتاحها اليوم الّذي ينزل مركز الشمس نقطة الاعتدال
الربيعي وأسماء شهورها فارسيّة وأيامها بحسب دخول
الشمس أوائل البروج بخلاف الروم والفرس ودورها
٢٢٥ سنة شمسية يقع فيه نجـ يوما كبيسة منها
خمس وأربعون رابونا وثمانية خاموسا ومجموعها نجى . :

Translation:[62]

The fifth famous calendar is the leap year of Sultan Malikshāh. When he had ordered [to determine] the solar year, to treat exhaustively the four seasons and the return of growth, the order of Sultan Malikshāh – God may enlighten his inspiration – was issued 'to begin an astronomical table from [the time] onwards when the center of the great light [i. e. the sun] had reached the point of the spring equinox'. The years of the calendars were not adequate for that. We want to add [days] to them, in order to make it easy for the assessor to deduce the mean motion of the planets for any date he wants parallel to the solar year. Its beginning is Sunday of Shaʿbān in the year 468. Its beginning was the day at which the center of the sun reaches the point of the spring

61 Text established by the author, based on MS Istanbul, Süleymaniye Kütüphanesi, Hamidiye 859.

62 Translation by the author.

equinox. The names of their months are Persian. Their days are [determined] according to the entering of the sun into the beginnings of the zodiacal signs, as opposed to [systems of] the Greeks and the [ancient] Persians. The cycle is 225 solar years, into which 53 leap days fall, 45 of which are in the fourth [year after the last leap year], and 8 are in the fifth [year after the last leap year].

1.3 *Ibn al-Athīr*[63]

سنة ٤٦٧

وفيها جمع نظام الملك والسلطان ملكشاه جماعة من أعيان المنجّمين وجعلوا النيروز أوّل نقطة من الحمل ، وكان النيروز قبل ذلك عند حلول الشمس نصف الحوت وصار ما فعله السلطرن مبدأ التقواويم .

وفيها أيضًا عمل الرصد للسلطان ملكشاه واجتمع جماعة من أعيان المنجمين في عمله منهم عمر بن إلباهيم الخيامي وأبو المظفر الاسفزاري وميمون بن النجيب الواسطي وغيرهم وخرج عليه من الأموال شيء عظيم وبقي الرصد دائرًا إلى أن مات السلطان سنة خمس وثمانين وأربعمائة فبطل بعد موته .

Translation:[64]
Year 467 [= 1074 CE]
In this year Niẓām al-Mulk and Sultan Malikshāh gathered together a group of the most distinguished astronomers, and they set the first day of the year (*nayrūz*) at the beginning of Aries, whereas that day had come to coincide with the sun's entry into the middle of Pisces. This act of the Sultan established the beginning of the ephemerides (*takwīm*). Likewise, in this year the observatory was built (*ʿumila l-raṣadu*) for Sultan Malikshāh, and a group of outstanding astronomers came together for its foundation, among whom were ʿUmar ibn Ibrāhīm al-Khayyāmī, Abū l-Muẓaffar al-Asfizārī, and Maymūn ibn Najīb al-Wāsiṭī, and others. An enormous amount of money was spent for this purpose. The observatory continued functioning until the death of the Sultan in 485, and it came to an end after his death.

1.4 *Astrolabe, 1271 CE*[65]

أ - سنة قسج ملكشاهى / خم يزدجرديّة غثفب اسكندر روميّة / صنعة محمود بن علىّ بن يوشع الـ ...سرى سنة خسط هجريّة / برسم خزانة الصدر المعظّم الصاحب

63 Text: Ibn al-Athīr, ʿAlī ibn Muḥammad, *al-Kāmil fī l-taʾrīkh*, vol. 8, pp. 408–9.
64 Translation by Sayılı, *Observatory in Islam*, p. 161.
65 Text edition and translation by Morley, *Description*, pp. 33 and 35.

الأعظم المؤيّد العالم العادل ملك الأمراء خسرو الآفاق بهاء الدولة و الدين
شمس الإسلام و المسلمين غياث الملوك و الخواقين الحسن بن علىّ الشديد أعزّ
الله أنصاره و ضاعف (؟) جلاله

Translation
The year 193 Malikshāhī / [the year] 649 of the Yazdjirdiyya (era), 1583 of the Iskandar Rūmiyya (era) / Constructed by Maḥmūd ibn ʿAlī ibn Yūsha' al-...ri in the year 669 Hijriyya (era) / For the Museum of the honored Prime Minister the supreme lord, the assisted [by God], the wise, the just, the king of the amirs, the Khusraw of the quarters of the world, the splendor of the state, and of the faith, the sun of the Islam and of Muslims, the succor of princes and lords, al-Ḥasan ibn ʿAlī al-Shadīd: may God glorify his friends, and double his dignity.

1.5 *Inscription, 948 CE*

Text (excerpt):[66]

(٢١) [...] وكتب يوم الإثنين لخمس حه ا (كذا) [= خلون ؟] من
(١٣) رجب سنة سبع وثلثين وثلثمائة و بالفارسيّة (؟)
(١٤) ماين دين و رحر (؟) سنة سبعة اعشر (كذا)
(١٥) وثلثمائة و صلّى الله على نبيّه محمّد وسلّم

Translation:
(12) Written Monday 5 (or 6?)
(13) Rajab of the year 337, in the Persian [calendar]
(14) ... of the year 317. May God bless His prophet Muḥammad and salute him.

1.6 *Inscription, 1203 CE*

Text:[67]

هذا بناء القلعة [...] فى تأريخ المرداذ ماه سنة ستّمائة

Translation:[68]
This is the construction of the fortress [...] on the date Murdād māh of the year 600

66 Text edition by Kalus and Guillot, 'Réinterprétation' (with French translation); Combe, Sauvaget, and Wiet, *Répertoire chronologique d'épigraphie arabe*. Here: *RCEA* 4 (1933), pp. 113–14 no. 1435.

67 Text edition: Combe, Sauvaget, and Wiet, *RCEA* 9 (1937), pp. 257–58 no. 3580 (with French translation).

68 Translation by the author.

1.7 *Inscription on an Astrolabe, 1388 CE*

Text:[69]

أ - (١) صنعة

(٢) جعفر بن عمر الكرمانى

-ب - (١) فى سنة

(٢) ذص الهجرّية و ذنح (؟) اليزدجر [دّ]ية (اليزدجرديّة)

Translation:[70]

Work of Jaʿfar ibn ʿUmar al-Kirmānī in the year 790 of the Hijra, and 758 of Yazdjird

1.8 *Inscription, 1393 CE*

Text:[71]

فى سنة
(٢) ذصو الهجرّية
(٣) و ذسد اليزدجرديّة

Translation:[72]

In the year 796 of the Hijra, and 764 of Yazdjird

1.9 *Inscription, 1423 CE*

Text:[73]

أ - فى سنة ضكز الهجرّية
-ب - و ذصح اليزدرجرديّة

Translation:[74]

In the year 827 of the Hijra, and 798 of Yazdjird

---

69 Text edition: Combe, Sauvaget, and Wiet, *RCEA* 18 (1991), pp. 107–8 no. 790 011 (with French translation).

70 Translation by the author.

71 Text edition: Kalus, *Thesaurus d'Épigraphie Islamique*.

72 Translation by the author.

73 Text edition: Kalus, ibid.

74 Translation by the author.

1.10 *Inscription, 1430 CE*

Text:[75]

في سنة ضلد
(٤) الهجرية و ض ه اليزدرجدية
(٥) و غذلج الاسكمدرية

Translation:[76]
In the year 834 of the Hijra, and 800 of Yazdjird, and of 1633(?) of Alexander

1.11 *Inscription, 1522 CE*

Text:[77]

في سنة
(٢) ظكط الهجريّة
(٣) ضصب اليزدجريّة

Translation:[78]
In the year 929 of the Hijra, and 892 of Yazdjird

1.12 *Inscription, 1573 CE*

Text:[79]

(٧) في سنة ظ فا هجريّة
(٨) ٩٨١ و ظ نب
(٩) يزدجرديّة

Translation:[80]
In the year 981 of the Hijra, and 952 of Yazdjird

---

75 Text edition: Kalus, ibid.
76 Translation by the author.
77 Text edition: Kalus, ibid.
78 Translation by the author.
79 Savage-Smith, *Islamicate Celestial Globes*, p. 223, no. 9. Text edition: ibid., pp. 286–87.
80 Translation: ibid., p. 223.

1.13 *Ephemeris of 1182 CE*

Text:

4	3	2	1	
المتهلّ			اجتماع الأوّل [...]	i
ه		الأيّام		ii
ك		من ربيع الاوّل		iii
يا		من آب		iv
يد		من مهر ماه		v
د كد مز	الشمس	مهر ماه أوّل يوم السبت		vi
ا يه يا	القمر			vii
د كز نب	زحل			viii
ج يا مج	المشتري			ix
يا كه يح	المرّيخ			x
ما ه ح	الزهرة			xi
ن د ط	عطارد			xii
[جـ] يجـ مد	جوزهر			xiii
[يجـ] يط	الساعات			xiv
[...]	الارتفاعات			xv

Translation:

	i	ii	iii	iv	v	vi	vii	viii	ix	x	xi	xii	xiii	xiv	xv
1	Conjunction of Jumādā I [...]														
2		Days	of	of	of	Month Mihr, begin on Saturday									
3		[of the week]	Rabīʿ I	Āb	month Mihr	Sun	Moon	Saturn	Jupiter	Mars	Venus	Mercury	Node	Hours	Altitude
4		5	20	11	14	4 24 47	1 15 11	4 27 52	3 11 13	11 25 18	5 8 41	4 9 50	[3] 13 44	[13] 19	[...]

1.14 *Ephemeris of 1603 CE*

Text:[81]

[..]رافقه اليوم السادس والعشرين من مرداد⟨د⟩ ماه الفارسي اليزدجردي القديم لاسمه اليوم الأوّل من فروردين ماه الجلالي الملكي الشاهي السلجوقي سنة ٥٢٤ جلالية سلجوقيّة ملك ساه بن اولب قسلان [...]

xi	x	ix	viii	vii	vi	v	iv	iii	ii	i	
سنة ٥٢٤ الإختيارات القمريّة	فروردين ماه «سلجقي»	أحوال الأيّام	أحوال الجوّ	مواضع القمر	العدد من برمهات قبطي	العدد من آدار الرومي	العدد من رمضان المعظم	أيّام الأسبوع	المواسم اليوميّة	الأحكام الجزئيّة	1
اعتدال الزمان في سائر المعور	١	كلاب	تغيير في الجوّ	حوت	١٥	١١	٢٧	خميس ١٢	نوروز الملوك	الأحكام	2

Translation:[82]

[...] in company with it is day 26 of the old Persian month Murdād [in the era of] Yazdgird. In touch with it is day 1 of the Jalālī, Shāhī, and Salǧuqī month Farwardīn in the year 524 Jalālī, Salǧuqī year of Malik Shāh ibn Ūlp Qaslān [sic!] [...]

	i	ii	iii	iv	v	vi	vii	viii	ix	x	xi
1	Particular judgments	Daily feasts	Days of the week	Number of Ramaḍān	Number of Ādār of the Greeks	Number of Barmahāt	Places of the moon	Qualities of the air	Qualities of the days	Saljuqī month Farwardīn	Year 523. Choices of the moon
2	Prescriptions	Nawrūz of the kings	Thursday 12	27	11	15	Pisces	Change of the air	Dogs [?]	1	Balanced time in the rest of the world

81 Text established by the author, based on MS Venice, Bibliotheca nazionale Marciana, or. 84 (= 14), ff. 8r and 9r.

82 Translated by the author.

1.15 *Ephemeris of 1804* CE *(Header and First Day Only)*

Text (table turned here 90°anticlockwise):[83]

2	1	
و	علامة صفر	i
١	عدد صفر	ii
٤	من دشنس قبطي	iii
٢٩	من نيسان رومي	iv
٢٩	من أدريل عجمي	v
١	من آذر فابسي	vi
١	من سيوان عبري	vii
٢٢	من أردبهشت جلالي	viii
ك ٢ يز	شمس	ix
يد ٣ ٣٩	قمر	x
٢٩ ٦ كد	زحل	xi
كح ٧ ب	مشتري	xii
يه ١ لا	مرّيخ	xiii
د ٤ لب	زهرة	xiv
يب ٣ كب	عطارد	xv
ط ١١ ب	رأس	xvi
سيوان عبري أوّل اللواقح بزرع الخيّار مولد داؤد	التوقيعات	xvii

83 Text established by the author, based on MS Berlin, Staatsbibliothek, or. quart. 211, f. 1r.

Translation:[84]

	1	2
i	Signs [of the week days] of Ṣafar	6
ii	Numbers [of days] of Ṣafar	1
iii	[Numbers] of [of days] of Coptic Bashons	4
iv	[Numbers] of [of days] of Greek Nīsān	29
v	[Numbers] of [of days] of foreign[a] Abrīl	29
vi	[Numbers] of [of days] of Persian Ādar	1
vii	[Numbers] of [of days] of Hebrew Sīwān	1
viii	[Numbers] of [of days] of Jalālī Urdibihisht	22
ix	Sun	20 2 17
x	Moon	14 3 39
xi	Saturn	29 6 24
xii	Jupiter	28 7 2
xiii	Mars	15 1 31
xiv	Venus	4 4 31
xv	Mercury	12 3 22
xvi	Head [of the Dragon]	9 11 2
xvii	Executions [of actions]	Hebrew [month] Sīwān; first fertilizations, together with the sowing of the winegrowers; birthday of [king] David.

a Probably a reference to the ancient pagan Romans, as the Peoples of the Book have other calendars allocated to them here.

Appendix 2: Persian Texts

2.1 [*Ps.*-] *ʿUmar Khayyām,* Nawrūznāma

Text:[85]

بفرمود تا کبیسه کند وسال بجایگاه خویش بازآرند. حکماء عصر از خراسان بیاورند
، وهر آلتي کث رصد را بکار آید بساختند از دیوار و ذات الحلق ومانند این، ونوروز
را بفرودین بردند ولیکن پادشاه را زمانه زمان نداد و کبیسه تمام ناکرده بماند

84 Translated by the author.

85 Text: [ps.-]ʿUmar al-Khayyām, *Nawrūznāma*, p. 12.

Translation:[86]
[Malikshāh] ordered the intercalation to be made and the year to be moved to its true place. The learned people of the age were brought from Khurasan, and each needed instrument, such as the wall [quadrant], armillary sphere, and other similar ones were constructed. Nawrūz was set up at Farwardīn, but the King did not live long enough and the intercalation remained incomplete.

[illegible] *Naṣīr al-Dīn al-Ṭūsī, Zīj-i Īlkhānī*

Text:[87]

فصل پنجم در تاریخ محدث که انرا تاریخ ملکي خواهند

سلطان جلال الدوله ملکشاه بن الپ ارسلان
سلجوقي تاریخی دیگر نهاده است که اوّل
سال او روزی باشد که افتاب بحمل آمده
باشد یعنی اول نهار حقیقی و بعضي اول
هر ماه هم اول آمدن آفتاب گیرند بآن
برج که آن ماه نوبت آن برج باشد تا ماهها
شمسی حقیقی باشد وفصول سال حقیقی باشد
وماههارا نام هم نام ماههای پارسیان
باشد امّا ماههای پارسیان را بقدیم
مقیّد کنند واین ماهارا بجلالي مقیّد
کنند ومنجمان ماهها سی سی روز گیرند اسامي را تا عدد ایّام در اوراق تقویم
مختلف نباشد وپنجهء مسترقه را در آخر اسفندارمذ ماه گیرند و بهر جهار سال
يك روز کبیسه باشد وسال سیصد وشست وشش روز شود و جون هفت بار یا
هشت؟
بار بجهار سال کبیسه افتذ یکبار پنج سال کبیسه افتذ و معرفت اوائل سالها
وکبائس با ستقرا معلوم شود و ما سیصذ سال را از اول تاریخ مدخل سالها در

---

86 Translation: Sayılı, *Observatory in Islam*, p. 163.
87 Text established by the author, based on MS Oxford, Bodleian Library, Hunt. 143.

جدولی نهاذیم و همحنین؟عدد کبائس در جدول دیگر و جون تاریخي دیگر معلوم باشد و تاریخ ملکی خواهند که معلوم کند آن تاریخ با روزها کند و ماپین التاریخین از ان روزها بکاهند باقی روزها باشد از اول تاریخ ملکی آنرا بر بر سیصد وشست و پنج قسمت کند وخارج قسمت در جدول عدد کبائس طلب کند آنج؟ از عدد کبائس

بازای آن عدد یا بیشتر عددی که از خارج قسمت کمتر باشد در جدول یابند بعدد آن از ایام

باقی نقصان کند خارج قسمت سالهای تامه باشد و ایّام باقی از سال ناقصه بر سی قسمت

v

کند تا ماههای تامه حاصل آید و باقی ایّام باشد از ماه حاضر و اگر خواهند که از

تاریخ ملکی تاریخی دیگر بیرون آرند سالهای تامه در سیصد وشست ونج ضرب کند

و با زاء ان سالها تا بیستر عددی که از ان کمتر بود در جدول عدد کبائس بر حاصل ضرب

افزایند تا ایام سالها تامه شود عدد بس عدد ماهاء تام در سی ضرب کند و بر ایّام

سالها تامه افزایند و روزها از ماه حاضر تا روز مطلوب بر ان افزایند جمله ایام تاریخ ملکی باشد ما بین تاریخ ملکي و تاریخ مطلوب بروافزایند تاریخ مطلوب بروافزایند تاریخ مطلوب شود بر وجه مذکور با سال و ماه و روز مطلوب کند و ما جدول بجهت معرفت مدخل تاریخ ملکي نهاده ایم تا آنج با زای مجموعه ومبسوطه

از ان جدول یابند مدخل سال مطلوب بود وجدول انیست؟ کث بر هنح؟ دیگرست

Translation:[88]
Fifth chapter: On the new calendar which they call the Malikī calendar

The Saljuq Sultan Jalāl al-Dawla Malikshāh ibn Alp Arslan established another calendar. The beginning of its year is the day on which the sun comes into Aries, meaning the beginning of the real bright day; and some people also take as the beginning of each month the entry of the sun into the zodiacal sign, such that the month is the period of that zodiacal sign, so that the months are true solar ones. The seasons of the year are true [solar ones]. The months also have the names of the months of the [ancient] Persians. The months of the [ancient] Persians are specified as 'old' (*qadīm*), but these months are specified as 'Jalālī'. The astrologers (*munajjimān*) make the months 30 days each for the people (*asāmī*), in order that the numbers of days on the leaves of an almanac should not be different. They put the five supplementary days at the end of the month Isfandārmudh. Every four years there is one leap day, and the year has 366 [*sīṣad* rather than *sih ṣad* for 300] days. When seven times, or eight times the intercalation in four years has passed, one fifth-year intercalation takes place. The beginnings of the years and the intercalation is made known by diligent search. We have put in a table the beginnings of the year for 300 years from the epoch (*tārīkh*) onwards. And in the same way [we] have also [put] the number of intercalations into another table. If a date [in] another [calendar] is known, and one wants the Malikī date to be known, one converts (*kunand*) this date into days. One has to subtract the difference between the two epochs (*al-tarīkhayn*). The reminder are the days from the beginning of the Malikī calendar. One divides these by 365. One searches the result of the division in the table of the number of intercalations. One finds in the table … this number or a greater number which is smaller than the result of the division [?]. One subtracts this number from the remaining days. The result of the division is in complete years. One divides the remaining days of the incomplete year by 30. The remaining days are from the present month. If one wishes to find out a Malikī date in another calendar, one multiplies the complete years by 365. With this result or a greater number which is smaller than that number [?], one adds [the number] in the table of the numbers of intercalations to (?) the result of the multiplication, to the end that years are complete. One multiplies the number being the number of completed months by 30. One adds it to the days of the complete years. One adds to it the days of the present month together with the wanted days [?]. The sum are the days in the Malikī calendar. We add to it the difference between the Malikī epoch and the wanted epoch, adding [up to] the wanted date [?], being the wanted date [?]. One does it with year, month, and day in the way just mentioned. We have established a table for knowing the beginning of the Malikī

88 Translation by the author.

calendar. Thus, one finds the collected and single [years] from this table, [which] are the wanted years. The table … [?] is different.

2.3 *Horoscope of Iskandar Sultan*

Text:[89]

لدن ذكر تاريخ الولآدة المبارك

در بهترين وقي و شد نفرين ساعتي از شب دوشنبه سيوم ربيع الاول سنه ست و
ثمانين وسبعمائه

الهجريه موافق با نر دهم اردبهشت ماه جلاّلي سنه ست وثلثمائه الملكشاهيه
مناسب هفدهم مردادماه

قديم سنه ثلاث وخمسين و سبعمائه اليزدجرديه مُطابق بيست و پنجم نيسان الرومي
سنه الف و

خمس و تسعين وستمائه الاسكندريه تاريخ خطآئيان صد و چهار فنك وعشري
كد شبه ازكه هشتم از

جاع دوازدهم كه انرا نخطاتي خاتي طنغوز خوآنند از روز هشتم انه دور ستيني
كه انرآ بقاتى سن و تي

خوآنند و بتركي قوي وبحسب دور چهآرم كه اعتماد اهل قتآبران است روز چهآرم
كه انرا پِنْ خوانند

موصاف به خي كدشته از مدخل حقيقى دُورد نجآي سه روز و جهآر هزار و دو
يست و بآنزده

فنك كث ارقام ان اينست جـ اي يه و كدشته از مدخل قسم كفوُو ده روز و
هشت هرآ و شيصد و هفتاد

Translation:[90]

[…] account of the date of blessed birth […] evening of Monday, 3 Rabīʿ al-awwal of the Year 786 of the Hijra, corresponding to 15 Jalālī month Urdībihishj of the year 306 of

89 Text established by the author, based on manuscript London, Wellcome Medical Library, MS Persian 474.

90 The first part of the translation is by the author, the second part on the Chinese calendar by Elwell-Sutton, 'A Royal Timūrid Nativity Book', pp. 121–23.

Malikshāh, in accordance with 17 old month Murdādmāh of the year 753 of Yazdgird, coinciding with 25 Roman month Nīsān of the year 1695 of Alexander. The date of the Chinese [people]: 104 1/10 *fink* elapsed of the 8th *kih* of the 12th *chāgh*, which is called in Khatāʾī *khāy* [and in turkish] *ṭunghūz*, of the eight day of the sexagesimal cycle, which in Tatāʾī [sic] is called *sin wī*, and in Turkish *qūy*. According to the fourth cycle, which is favored by the people of Qatā, on the fourth day called *pin*, described as *khī*, 3 days and 4,215 *fink* (in figures 3, 1, 10, 15) past the beginning (*madkhal*) of *dūrdinj āy*, and 10 days and 8677 *fink* (in figures 10, 2, 24, 37) past the beginning of the section (*qism*) of Khavin cycle, which year is called *kā zha* [chin. *jia zi*] in Khaṭāʾī and of the *kiskū* of *sījqān* in Turkish. There had elapsed since the creation of the world 8863 complete *win*, and of the incomplete *win* 9860 years, the above-mentioned year being the incomplete year.

Appendix 3: Turkish Texts

3.1 *Ephemeris of 1450 CE*

Text.[91]

ايام اعياد ايام ايام ايام ايام فوردين ماه جلالي اوّله يوم الأربعاء
جلالي جمعات عرب روم فرس گنس اي زحل مشتري مريّخ زهره عطارد

ا گنش د كو يا يح 1018A يكو هك يكو ياد ياد جب
حمل ط ل ب لج نب نب
برجنه
گجدي

91 Text established by the author, based on MS Florence, Biblioteca Medicea Laurenziana, Or. 27.

Translation:[92]

Jalālī days	Feasts	Days of the week	Days of the Arabs	Days of the Greeks	Days of the Persians	Jalālī month Farwardīn, its beginning on Wednesday						
						Sun	Moon	Saturn	Jupiter	Mars	Venus	Mercury
1	The sun enters the zodiacal sign Aries	4	26	11	18	0 0 9	10 26 30	5 20 2	10 26 33	11 4 52	11 4 52	0 3 2

3.2 *Almanac of 1824* CE

Text:[93]

|نوروز سلطاني، اول موسم بهار | يكشنبه | ٢٠ | ٩ |قوي | سعد |

Translation:[94]
| *nevrūz-i sulṭānī*, first [day] of the season of spring | Sunday | 20 [Rajab] | 9 [March (jul.)] | fortunate |

References

R. Abdullāhī, *Tārīkh-i tārīkh dar Īrān* (Tehran: Intishārāt-i Amīr Kabīr, 1987).

A.T. Adamova and M. Bayani, *Persian Painting: The Arts of the Book and Portraiture* (London: Thames & Hudson, 2015).

S.H. Barani, 'The Jalali calendar (tarikh-i-Jalali or Maliki)', *Islamic Culture* 17 (1943), pp. 166–175.

A.D. Beihammer, *Byzantium and the Emergence of Muslim-Turkish Anatolia, ca. 1040–1130* (Birmingham Byzantine and Ottoman studies 20; London: Routledge, 2017).

S. Blair, *The Monumental Inscriptions from Early Islamic Iran and Transoxiana, Studies in Islamic Art and Architecture* (Supplements to Muqarnas 5; Leiden: Brill, 1992).

92 Translation by the author.

93 Text established by the author, based on MS Berlin, Staatsbibliothek, or. quart 1697; facsimile and German translation in Kurz, *Osmanischer Almanach*, pp. 147 (facsimile) and 89 (translation).

94 Translation by the author.

F.C. de Blois, *Persian Literature: A Bio-bibliographical Survey: Volume V, Part 2: Poetry ca. A.D. 1100 to 1225* (Leiden: Brill, 1994).

F.C. de Blois, 'The Persian Calendar', *Iran* 34 (1996), pp. 39–54.

F.C. de Blois, 'Ta'rīkh: I. Dates and Eras in the Islamic World: 1. In the sense of "date, dating", etc.', in: P.J. Bearman et al. (eds), *Encyclopaedia of Islam: New Edition*, vol. X (Leiden: Brill, 2000), pp. 258–64.

F.C. de Blois, 'The Zoroastrian Calendar: New Evidence from Achaemenid Bactria', in: N. Sims-Williams and F.C. de Blois (eds), *Studies in the Chronology of the Bactrian Documents from Northern Afghanistan* (Wien: Verlag der Österreichischen Akademie der Wissenschaften, 2018), pp. 109–12.

A. Caiozzo, 'The Horoscope of Iskandar Sultān as a Cosmological Vision in the Islamic World', in G. Oestmann, H.D. Rutkin, and K. von Stuckrad (eds), *Horoscopes and Public Spheres: Essays on the History of Astrology* (Berlin: Walter de Gruyter, 2005), pp. 115–44.

P. Casanova, 'Le montre du sutan Noûr ad Dîn', *Syria* 4 (1923), pp. 282–99.

M.S. Champion, *The Fullness of Time: Temporalities of the Fifteenth-Century Low Countries* (Chicago: The University of Chicago Press, 2017).

E. Combe, J. Sauvaget, and G. Wiet (eds), *Répertoire chronologique d'épigraphie arabe (RCEA)* (Cairo: Imprimerie de l'IFAO, 1931–).

S. Cristoforetti, *Il natale della luce in Iran: Une festa delfuoco nel cuore di ogni inverno: Ricerche sul* sada*: occorrenze rituale e temi mitologici di una celebrazione cortese tra Baghdad e Bukhara, secc. IX–XII* (Milano: Associazione Culturale Mimesis, 2002).

B. van Dalen, E.S. Kennedy, and M.K. Saiyid, 'The Chinese-Uyghur Calendar in Ṭūsī's Zīj-i Īlkhānī', *Zeitschrift für Geschichte der arabisch-islamischen Wissenschaften* 11 (1997), pp. 111–51.

B. van Dalen, 'The Malikī Calendar in the *Dustūr al-munajjimīn*', in E. Orthmann and P. Schmidl (eds), *Science in the City of Fortune: The* Dustūr al-Munajjimīn *and its World* (Bonner Islamstudien 39; Berlin: EB-Verlag, 2017), pp. 117–35.

M. Dirāyatī, *Fihristgān: nuskha-hā-yi khaṭṭī-i Īrān* (*Fankhā*), Nuskhah'shināsī 1 (Tehran: Sāzmān-i Asnād wa Kitābkhāna-i Millī-i Jumhūrī-i Islāmī-i Īrān, 2011).

G. Dohrn-Van Rossum, *History of the Hour: Clocks and Modern Temporal Orders* (Chicago: University of Chicago Press, 1996).

L.P. Elwell-Sutton, 'A Royal Timūrid Nativity Book', in R.M. Savory and D.A. Agius (eds), *Logos Islamikos: Studia Islamica in honorem Georgii Michaelis Wickens* (Toronto: Pontifical Institute of Mediaeval Studies, 1984), pp. 119–36.

S. Fang, *Zhongguo shi liri he Zhong xi liri duizhao biao* [中国史歷日和中西歷日對照表] (Shanghai: Shanghai cishu chubanshe, 1987).

H.R. Giahi Yazdi, 'The *Jalālī* Calendar: The Enigma of its Radix Date', *Archive for History of Exact Sciences* 74 (2020), pp. 165–82.

F.K. Ginzel, *Handbuch der mathematischen und technischen Chronologie: Das Zeitrechnungswesen der Völker* (Leipzig: J.C. Hinrichs, 1906–1914).

B.R. Goldstein and D. Pingree, 'Astrological Almanacs from the Cairo Geniza', *Journal of Near Eastern Studies* 38 (1979), pp. 153–75 and 231–56.

B.R. Goldstein and D. Pingree, 'Additional Astrological Almanacs from the Cairo Geniza', *Journal of the American Oriental Society* 103 (1983), pp. 673–90.

M. Gronke, *Arabische und persische Privaturkunden des 12. und 13. Jahrhunderts aus Ardabil (Aserbaidschan)* (Islamkundliche Untersuchungen 72; Berlin: Klaus Schwarz, 1982).

M. Gronke, 'The Arabic Yārkand Documents', *Bulletin of the School of Oriental and African Studies* 49 (1986), pp. 454–507.

M. Gronke, *Derwische im Vorhof der Macht: Sozial- und Wirtschaftsgeschichte Nordwestirans im 13. und 14. Jahrhundert* (Freiburger Islamstudien 15; Stuttgart: Steiner, 1993).

Ḥajjī Khalīfa, *Kash al-ẓunūn ʿan asāmī l-kutub wa-l-funūn = Keşf-el-zunun* (ed. Ş. Yaltkaya and R. Bilge; İstanbul: Milli Eğitim Basımevi, 1971).

G. Herrmann, *Persische Urkunden der Mongolenzeit: Text- und Bildteil* (Documenta Iranica et Islamica 2; Wiesbaden: Harrassowitz, 2004).

G. Herrmann, 'Urkundenfunde in Āẕerbāyǧān', *Archäologische Mitteilungn Iran* N.F. 4 (1971), pp. 249–62.

Ibn al-Athīr, ʿAlī ibn Muḥammad, *al-Kāmil fī l-taʾrīkh* (ed. Abū l-Fidāʾ ʿAbdallāh al-Qāḍī; Beirut: Dār al-kutub al-ʿilmiyya, 1987–2003).

L. Kalus, *Thesaurus d'Épigraphie Islamique* (http://www.epigraphie-islamique.org, accessed 19 August 2020).

L. Kalus and C. Guillot, 'Réinterprétation des plus anciennes stèles funéraires islamiques nousantariennes: III. Sri Lanka', *Archipel* 72 (2006), pp. 15–68.

E.S. Kennedy, 'The Chinese-Uyghur Calendar as Described in the Islamic Sources', *Isis* 55 (1964), pp. 435–43.

F. Keshavarz, 'The Horoscope of Iskandar Sultan', *Journal of the Royal Asiatic Society* 116 (1984), pp. 197–208.

F. Keshavarz, *A Descriptive and Analytical Catalogue of Persian Manuscripts in the Library of the Wellcome Institute for the History of Medicine* (London: Wellcome Institute for the History of Medicine, 1986).

Ch. Kiening, 'Mediating the Passion in Time and Space', in C. Kiening and M. Stercken (eds), *Temporality and Mediality in Late Medieval and Early Modern Culture* (Turnhout: Brepols, 2018), pp. 115–46.

D.A. King, *In Synchrony with the Heavens: Studies in Astronomical Timekeeping and Instrumentation in Medieval Islamic Civilization* (Islamic Philosophy, Theology and Science 55; Leiden: Brill, 2004–2005).

M. Kurz, *Ein osmanischer Almanach für das Jahr 1239/1240 (1824/1825)* (Islamkundliche Untersuchungen 276; Berlin: Klaus Schwarz, 2007).

J. Le Goff, 'Au Moyen Age: Temps de l'Eglise et temps du marchand', *Annales. Histoire, Sciences Sociales* 15 (1960), pp. 413–33.

J. Le Goff, 'Merchant's time and church's time', in idem, *Time, Work and Culture in the Middle Ages* (Chicago: University of Chicago Press, 1989), pp. 29–42.

J.G. Leichter, *The Zīj as-Sanjarī of Gregory Chioniades: Text, Translation and Greek to Arabic Glossary* (Ph.D. thesis, Brown University, 2004).

Ch. Melville, 'The Chinese-Uyghur Animal Calendar in Persian Historiography of the Mongol Period', *Iran* 32 (1994), pp. 83–98.

J.W. Meri (ed.), *Medieval Islamic Civilization: An Encyclopedia* (London: Routledge, 2005).

V. F. Minorsky, 'Some early documents in Persian I', *Journal of the Royal Asiatic Society* (1942), pp. 181–194.

W.H. Morley, *Description of a Planispheric Astrolabe Constructed for Sháh Sultán Husain Safawí, King of Persia, and Now Preserved in the British Museum; Comprising an Account of the Astrolabe Generally, with Nores Illustrative and Explanatory: To which are Added, Concise Notices of Twelve Other Astrolabes, Eastern and European, Hitherto Undescribed* (London: Williams and Norgate, 1856).

S.M. Mozaffari, 'The Effect of Astrological Opinions on Society: A Preliminary View', *Trames* 16 (2012), pp. 359–68.

Naba'ī, Abū l-Faḍl, *Gāhshumārī dar tārīkh*, Sāzmān-i Muṭāla'a wa Tadwīn-i Kutub-i 'Ulūm-i Insānī-i Dānishgāhhā; 565.Tārīḫ 16 (Tehran: Sāzmān-i Muṭāla'a wa Tadwīn-i Kutub-i 'Ulūm-i Insānī-i Dānishgāhhā [SAMT], 2002).

Nāṣir Ḫusrau, *Book of travels = Safarnāmah: a parallel Persian-English text* (ed. and transl. W.M. Thackston; Bibliotheca Iranica. Intellectual traditions series 6; Costa Mesa, Calif.: Mazda Publishers, 2001).

D.S. Richards, 'Written documents', in W. Kubiak and G.T. Scanlon (eds), *Fusṭāt excavation final report II* (American Research Center in Egypt Reports 2; Wiona Lake: Eisenbrauns, 1989), pp. 4–80.

B.A. Rozenfeld and A.P. Juškevič, *Omar Chajjam* (Moscow: Izd. Nauka, 1965).

F. Sarre and E. Herzfeld, *Archäologische Reise im Euphrat- und Tigris-Gebiet III* (Forschungen zur islamischen Kunst 1; Berlin: Reimer, 1911).

E. Savage-Smith, *Islamicate Celestial Globes. Their History, Construction, and Use* (Smithsonian Studies in History and Technology 46; Washington, D.C.: Smithsonian Institution Press, 1985).

A. Sayılı, *The Observatory in Islam and its Place in the General History of the Observatory* (Ankara: Türk Tarih Kurumu Basımevi, 2nd ed., 1988).

G.T. Scanlon, 'Fustat', in E.M. Meyers (ed.), *The Oxford Encyclopedia of Archaeology in the Near East* (New York: Oxford University Press, 1997), pp. 365–68.

L.P.E. Sédillot, *Prolégomènes des tables astronomiques d'Oloug-Beg, publiés avec notes et variantes et précédés d'une introduction* (Peris: Firmin Didot, 1847).

A.T. Şen, *Astrology in the Service of the Empire: Knowledge, Prognostication, and Politics at the Ottoman Court 1450s–1550s* (Ph.D. thesis, University of Chicago, 2016).

B. Spuler, *Wüstenfeld-Mahler'sche Vergleichungs-Tabellen zur muslimischen und iranischen Zeitrechnung: mit Tafeln zur Umrechnung orient-christlicher Ären* (Wiesbaden: Steiner, 1961).

S. Stern, *Calendars in Antiquity: Empires, States, and Societies* (Oxford: Oxford University Press, 2012).

J. Thomann, 'The Arabic Ephemeris for the Year 1149/1150 CE (P.Cambridge UL Inv. Michael. Chartae D 58) and the Arabic Baḫnīṭas, Greek Παχνίτης and Coptic ⲡⲁϣⲟⲛⲥ', *Chronique d'Égypte* 90.179 (2015), pp. 207–24.

J. Thomann, 'Ephemeride für das persische Jahr 413 (1044/1045 n. Chr.)', in A. Zdiarsky (ed.), *Orakelsprüche, Magie und Horoskope: wie Ägypten in die Zukunft sah* (Nilus 22; Wien: Phoibos-Verlag, 2015), pp. 138–40.

J. Thomann, 'Almanach für das koptische Jahr 707 (990/991 n. Chr.)', in A. Zdiarsky (ed.), *Orakelsprüche, Magie und Horoskope: wie Ägypten in die Zukunft sah* (Nilus 22; Wien: Phoibos-Verlag, 2015), pp. 140–41.

J. Thomann, 'A Fragment of an Unusual Arabic Almanac for 297 AH/910 CE (P.Berl. inv. 12793)', in S. Souderbala, S. Denoix, and M. Malczycki (eds), *New frontiers of Arabic papyrology: Arabic and Multilingual Texts from Early Islam* (Leiden: Brill, 2017), pp. 179–96.

J. Thomann, 'From Katarchai to Ikhtiyārāt: The Emergence of a New Arabic Document Type Combining Ephemerides and Amanacs', in A. Nodar and S. Torallas Tovar (eds), *Proceedings of the 28th Congress of Papyrology Barcelona 1–6 August 2016* (Barcelona: Publicacions de l'Abadia de Montserrat, 2019), pp. 342–54.

J. Thomann, 'Tools of Time: Devices for Organizing Public and Private Life in the Premodern Islamic World', in A. Pellitteri et al. (eds), *Re-defining a Space of Encounter. Islam and Mediterranean: Identity, Alterity and Interactions: Proceedings of the 28th Congress of the Union Européenne des Arabisants et Islamisants* (Leuven: Peeters, 2019), pp. 97–105.

J. Thomann, 'Ritual Time, Civil Time, and Cosmic Time: Three Co-Existing Temporalities in Premodern Islamic Society', *Kronoscope* 20 (2020), pp. 41–63.

S. Tourkin, 'Astrological Images in Two Persian Manuscripts' in N. Allan (ed.), *Pearls of the Orient: Asian treasures of the Wellcome Library* (Chicago: Serindia, 2003), pp. 105–9.

S. Tourkin, 'The Horoscope for Shah Tahmāsp', in S.R. Canby and J. Thompson (eds), *Hunt for Paradise: Court Arts of Safavid Iran 1501–1576* (Milan: Skira, 2003), pp. 327–331.

S. Tourkin, 'L'Oroscopo di Shāh Tahmāsp', in S.R. Canby and J. Thompson (eds), *A caccia in Paradiso: Arte di corte nella Persia del Cinquecento* (Milano: Skira, 2005), pp. 327–31.

Uluğ Beg, *Zidž: Novye Guraganovy astronomičeskie tablicy* (transl. A.A. Achmedov Taškent; Fan, 1994).

Uluğ Beg, *Uluğ Beg'in astronomi cetvelleri: Zîc-i Uluğ Beg* (Bilimin ve felsefenin doğulu öncüleri dizisi 13; Istanbul: Kültür ve Turizm Bakanlığı, 2012).

[ps.-]ʿUmar Khayyām, *Nawrūznāma* (ed. by Mojtaba Minovi; Tehran: Kaveh, 1933).

S. Vryonis, *The Decline of Medieval Hellenism in Asia Minor and the Process of islamization from the Eleventh through the Fifteenth Century* (Publications of the Center for Middle Eastern Studies 4; Berkeley: California University Press, 1971).

CHAPTER 6

The 247-Year Jewish Calendar Cycle

Origins, Diffusion, and Diversity

Nadia Vidro

By the early tenth century, a system of calendar reckoning had emerged in the Rabbinic Jewish centres of the Near East that became diffused very widely among the Jewish communities and is now universally accepted as the basis of the Jewish calendar. This system is based on a complex arithmetic calculation that must be performed regularly and requires knowledge and numerical expertise. An alternative way of setting the Jewish calendar existed alongside the standard calculation until the end of the Middle Ages and into the early modern period. This system assumes that the Jewish calendar fully repeats itself every 247 years, so that a calendar can be calculated once and re-used indefinitely. This cycle is not fully compatible with the standard Rabbinic calendar but diverges in a small number of years every century.

The use of two incompatible systems of calendar reckoning could lead to calendar differences between Jewish communities, and is significant for the study of the standardisation of the Jewish calendar. Although various aspects of the 247-year cycle have received attention in the works of historians of the Jewish calendar,[1] no comprehensive study exists that would take into account the calendrical as well as the social aspects of using this calendar. The present article is an outcome of my research on 247-year Jewish calendar cycles in the framework of the ERC Advanced Grant project 'Calendars in Late Antiquity and the Middle Ages: Standardization and Fixation' that ran at UCL between 2013–2018.[2] In this project, I collected and analysed a corpus of over two hundred medieval and early-modern sources in Hebrew, Judaeo-Arabic, Latin and Samaritan Hebrew that either contain or discuss 247-year calendars (see

1 On the 247-year cycle see Tobi, *The Jews of Yemen*, pp. 211–26. Bornstein, '*Divre yeme ha-ibbur ha-aharonim, heleq aleph*', pp. 354–58. Id., *Mahloqet*, pp. 141–44. Vidro, 'Origins'. Raviv, *Mathematical Studies*, pp. 53–111.

2 On this project see https://www.ucl.ac.uk/hebrew-jewish/research/research-projects/calendars-late-antiquity-and-middle-ages-standardization-and-fixation-0, accessed 17 May 2020. I thank my colleagues in the project, Prof Sacha Stern (PI), Dr Ilana Wartenberg, Dr Israel Sandman, and Dr François de Blois for their valuable comments and suggestions.

 | DOI:10.1163/9789004459694_008

Appendix 2). This corpus includes prayer books, Bibles, halakhic (legal) works, self-standing calendar treatises and scientific compendia.

The compiled corpus sheds light on the following research questions related to the cycle and its effects on the standardization of calendars in the Middle Ages:

- What are the origins of the 247-year cycle?
- How were 247-year calendars produced: were they copied by scribes from master copies, or each time freshly calculated and designed? How different were they from one another? How different were they from calculated calendars?
- What were the attitudes of scholars and rabbinic leaders towards the use of the 247-year cycle and its potential for calendar diversity?
- Were 247-year cycles used in practice? Did the use of these cycles lead to calendar diversity between different Jews or Jewish communities?
- Did calendar disputes occur due to the use of the 247-year cycle? Were they settled in favour of the 247-year cycle or the normative calculation?

The present article is structured as follows. In the main body of the article I attempt to answer the outlined questions on the basis of the full corpus of assembled sources. In Appendix 1, I describe 247-year cycles as they are found in manuscripts and printed sources from different geo-cultural areas. Appendix 2 lists the manuscript sources used in the project.

1 The 247-Year Calendar Cycle, Its Origins and History

In the Rabbinic calendar, the years can be of fourteen different types. The type of a year depends on three variables: the day of the week of Rosh Hashanah (the New Year), the length of the variable months Marḥeshvan and Kislev, and the number of months in that year.[3] Rosh Hashanah may fall on a Monday, Tuesday, Thursday or Saturday. The pair of months Marḥeshvan and Kislev can be 'defective' (both have 29 days), 'regular' (Marḥeshvan has 29 days and Kislev 30 days) or 'full' (both have 30 days). A Jewish year can have 12 months and be 'plain' or can have 13 months and be 'intercalated'. The inclusion or not of this additional 30-day intercalary month is a function of the 19-year cycle of intercalations in which 12 years have 12 months and seven years have 13 months. The permutations of these three variables fully determine the course of a Jewish year. Only 14 such combinations are permitted, and to set

3 For a detailed explanation of the workings of the Jewish calendar see Sar-Shalom, *Gates to the Hebrew Calendar*, esp. pp. 52, 131–40.

a calendar for a particular year ultimately means to establish which of the 14 types will apply to it.

In the standard calendar, the type of a year is established by calculating the *molad* (pl. *moladot*, mean lunar conjunctions) of a number of years and applying various other rules, a cumbersome procedure that requires knowledge and significant numerical abilities. It would ease the setting of the calendar if the calculation could be replaced by a calendar cycle, established once and used forever. In the standard calendar, the different types of year occur in a sequence which does not repeat itself exactly until it has run for 689,472 years. This period is obviously too long to serve as a cycle in practice. Of more practical value is the period of 247 years, after which the Rabbinic calendar repeats itself almost exactly: if a sequence of year types for 247 years is re-used for the following 247 years, it will deviate from a calendar calculated for these next 247 years in two to seventeen years only.[4] If the sequence is re-used again, such 'reiteration errors' accumulate, widening the gap between the standard calendar and the 247-year cycle. The approximate nature of the 247-year cycle was known already in the twelfth century.[5] Nonetheless, many medieval and early-modern Jewish sources include 247-year calendars that claim to 'repeat themselves forever' (*ḥozer ḥalilah*) accurately with no mistakes.

The 247-year cycle is best known in research literature under the title "*Iggul* ('circle') of Rav Naḥshon Gaon', attributed to Naḥshon ben Zadok, head of the Academy of Sura (in Iraq) in the second half of the ninth century, ca. 865–873.[6] However, studies of extensive corpora of calendar literature demonstrate that the cycle was unknown in the ninth century.[7] The 247-year cycle is not mentioned in the in-depth discussions of calendar procedures in the correspondence and literature relating to the calendar controversy of 921/2, and is equally absent from Abū Rayḥān al-Bīrūnī's comprehensive and well-informed section on the Jewish calendar in his work *The Chronology of the Ancient Nations* completed in 1000.[8] The earliest traceable 247-year cycles appear to have been

4 On the accuracy of the 247-year cycle see Raviv, *Mathematical Studies*, pp. 57–62.

5 See, for example, Abraham Ibn Ezra's critique of the 247-year cycle in *Shalosh She'elot* (ed. Steinschneider, *Sefer Shene ha-Me'orot*, p. 1). For another twelfth-century critique, see Vidro, 'Origins', pp. 124–26.

6 Here and in the following, all dates are CE unless otherwise specified. The precise dates of Naḥshon Gaon's incumbency are controversial; the above given dates follow Brody, *Geonim of Babylonia*, p. 344.

7 Raviv, *Mathematical Studies*, pp. 63, 86; Vidro, 'Origins'.

8 On the calendar controversy of 921/2 (previously known as the Saadia – Ben Meir dispute), see Stern, *Jewish Calendar Controversy of 921/2 CE*, superseding id., *Calendar and Community*, pp. 264–75, and Bornstein, *Maḥloqet*. See also Rustow and Stern, 'The Jewish Calendar Controversy of 921–22'. For Al-Bīrūnī, see Abū Rayḥān al-Bīrūnī, *The Chronology of Ancient*

put together in the middle of the tenth century, between the mid 920s and early 980s.[9]

The cycle may have been devised in the aftermath of the calendar controversy of 921/2. During this controversy, small differences in the calendar calculation traditions of Babylonia and Palestine caused a major rift between the two communities. It became clear that following calculated calendars did not guarantee calendar unanimity. Under these circumstances, the 247-year calendar cycle may have been proposed as an alternative for the standard calendar that was capable of preventing future calendar dissidence by establishing an unchangeable sequence of year types and eliminating the need for calculation. Had it been accepted, it would have been a relatively accurate luni-solar calendar that could have served the Jewish community almost as well as the *molad* calculation: the cycle is only slightly less astronomically accurate than the normative calculation,[10] and does not violate any of the basic Rabbinic calendation principles.[11] Instead, the 247-year cycle came to be regarded not as a replacement for the standard calendar but as an easy means of setting it, to be used alongside the more demanding calculation.

The earliest 247-year calendars, in Oriental manuscripts, are attributed to Josiah b. Mevorakh (ibn) al-ʿĀqūlī, a scholar of Babylonian descent. He is best known in connection with the calendar cycle but may have also authored Bible commentaries, one of them preserved in a Genizah manuscript dated 1009, when he was still alive.[12] The attribution of the cycle to Naḥshon Gaon first appears in late twelfth-century Ashkenazi sources and is attested in only about one third of the examined corpus.[13] This attribution is predominantly Ashkenazi but also relatively frequent in Italian manuscripts; it is rare in Sefardi, Provençal and Oriental manuscripts, and does not occur in Byzantine, Samaritan and Latin sources known to me. Clearly pseudepigraphical, the attribution of the 247-year cycle to Naḥshon Gaon, a high-ranking Rabbinic leader, must have been introduced as a means of adding authority to a deviant but

---

Nations, pp. 141–75. I thank Sacha Stern and François de Blois for drawing my attention to these sources' silence on the 247-year cycle.

9 See Appendix 1, *Oriental* and Vidro, 'Origins', pp. 120–22.

10 The normative Rabbinic calendar assumes a slightly excessive lunation, i.e. the *molad* occurs slightly later than the mean conjunction. As a result, the *molad* calculation falls behind the moon by about 25 minutes in 247 years. The 247-year cycle falls behind the moon by about 1 hour and 15 minutes (Sacha Stern, personal communication; see also Bornstein, '*Divre yeme ha-ibbur*', p. 356).

11 Such as the rule *lo ADU Rosh* that specifies that Rosh Hashanah may not fall on Sunday, Wednesday and Friday and similar rules.

12 Vidro, 'Origins', pp. 122–24.

13 See Vidro, 'Nahshon Gaon', pp. 27–28.

convenient way of setting the calendar. It is for the same reason that in some Ashkenazi manuscripts Hayye Gaon and Sherira Gaon are said to have verified the 247-year calendar's cyclicity.[14] In contrast, Josiah b. Mevorakh's authorship of the reiterative calendar appears to be historical because his name would not have added authority to it.

The reiterative calendar is most commonly found in primary sources under the titles *Thirteen Cycles* (*maḥzorim*), *Thirteen Rows* (*shurot*) or *The 247(-year) Cycle* (*maḥzor remez*).[15] In Oriental manuscripts, titles like *The Great Cycle* (*maḥzor gadol* in Hebrew or *maḥzor kabīr* in Judaeo-Arabic), *The Revolving 247* (*al-remez al-dā'ir*) and *The Great Waterwheel-like Cycle* (*maḥzor gadol dūlābī*) are also found. The title *'Iggul* (lit. 'circle'), which is regularly used in secondary literature, is found in some sources that ascribe the cycle to Naḥshon, but is never given to reiterative calendars that are not attributed to the gaon.[16] For these reasons, the 247-year cycle is not referred to in this article as the *'Iggul* of R. Naḥshon.

Although calendrical and historical considerations indicate that the 247-year cycle was first put together in the middle of the tenth century, no tenth- and eleventh-centuries copies of reiterative calendars survive.[17] In contrast, a number of early and mid twelfth-century cycles and discussions of the reiterative calendar exist, some in contemporary manuscripts, others in later copies. These sources are from many different geographical areas: Egypt and Maghreb, Byzantium, Provence, and Ashkenaz.[18] This demonstrates that by the middle of the twelfth century the cycle had spread and become widely known. The cycle's later dissemination is reflected in the large number of preserved sources that either contain or discuss 247-year calendars. In my corpus of medieval and early-modern Jewish manuscripts on the 247-year cycle, there

14 Oxford, Bodl. Opp. 614, fol. 50v (fourteenth century, Ashkenazi), Cincinnati, HUC 436, fol. 191r (1435, Ashkenazi). See Vidro, 'Origins', pp. 132–33.

15 The former two designations reflect the fact that 247 years can be represented as thirteen 19-year cycles of intercalations. See below, section 3.1: Calendars for a Particular 247-Year Period.

16 A cognate term *'iggulot remez* is used by Abraham Ibn Ezra, who does not ascribe the cycle to Naḥshon Gaon (ed. Steinschneider, *Sefer Shene ha-Me'orot*, p. 1).

17 A possible exception is a discussion of the reiterative calendar preserved in T-S Ar.29.135 and T-S Ar.29.3r, which are paleographically datable to the eleventh–twelfth century.

18 For the Orient, an early twelfth-century critique of the 247-year cycle discussed in Vidro, 'Origins', pp. 124–26, and Oxford, Bodl., Heb. e.45/45–56, with a calendar starting in 1142. For Byzantium, Vatican, Biblioteca Apostolica ebr. 303, fol. 190v, with a calendar starting in 1142. For Provence, Abraham Ibn Ezra's critique of the 247-year cycle, composed in Narbonne in ca. 1148 (ed. Steinschneider, *Sefer Shene ha-Me'orot*, p. 1). For Ashkenaz, Moscow, RSL, Guenzburg 481, fol. 102r, with a calendar starting in 1123, and Parma, Biblioteca Palatina Cod. Parm. 2295, fols 101v–103r, with a calendar starting in 1161.

are 81 texts in Ashkenazi hands, 61 in Oriental hands, 32 in Italian hands, 29 in Sefardi and Provençal hands and 4 in Byzantine hands.[19] 247-year cycles were also included in many printed works on the Jewish calendar and law between the fifteenth – nineteenth centuries. Apart from the Jews, the 247-year cycle was also known to Samaritans and to Christian scholars; the latter viewed it as a *vade mecum* of the calendar of the Old Testament and of Scriptural chronology.[20]

2 Attitudes towards the Cycle

A range of opinions about the 247-year cycle are attested in the sources. About a fifth of the sources contain a calendar for 247 years but say nothing about its cyclicity or suitability for setting the Jewish calendar. Another fifth oppose the cycle. The remaining sources support the cycle and advocate its use.

The earliest known refutations of the 247-year cycle were written already in the first half of the twelfth century, the same time as when the cycle first became widespread.[21] Among well-known medieval scholars who criticised the cycle are Abraham ibn Ezra,[22] Isaac Israeli,[23] Profiat Duran[24] and Obadiah b. David, the author of the standard commentary on Maimonides' *Mishneh Torah, Sanctification of the Moon*.[25] In addition, anonymous refutations are found in medieval and early-modern manuscripts.

All refutations are based on the following scientific argument. For a calendar cycle to be accurate, the *molad* of Tishri (first month of the year) must recur exactly at the same time and on the same day of the week at the end of its cyclicity period. However, in the Rabbinic calculation the *molad* of Tishri of year n + 247 does not equal that of year n but is 905 parts smaller. This small difference has implications for *moladot* approaching calendrical limits. Take,

19 Any of these manuscripts can contain more than one text on the reiterative calendar. The geo-cultural identification of the handwritings is based on my examination of the manuscripts, together with data provided in catalogue descriptions. This geo-cultural classication follows Beit-Arié, *Hebrew Codicology*, pp. 60–66.

20 See Appendix 1, *Oriental* and *Latin* respectively.

21 See Vidro, 'Origins', pp. 124–26.

22 See above, n. 5.

23 Isaac Israeli, *Yesod 'Olam* (1310, Toledo), 4:10; ed. Goldberg and Rosenkranz, *Liber Jesod Olam*, vol. 2, p. 23.

24 Profiat Duran, *Ḥeshev ha-Efod* (1395), chapter 22. On this unpublished work see Kozodoy, *The Secret Faith of Maestre Honoratus*, pp. 104–14.

25 Obadiah b. David, Commentary on *Sanctification of the Moon* (ca. 1340), 8:10 (first printed in the Amsterdam 1702 edition of *Mishneh Torah*).

for example, the Rabbinic rule known as *molad zaqen*. According to this rule, if the *molad* of Tishri in any year n falls after noon, year n will not start on that day but will be postponed. If the *molad* of year n exceeds the limit of noon by less than 905 parts, then 247 years later, when the molad is 905 parts less, it will fall before noon, and the year will not be postponed. As a result, the type of year n + 247 will not be the same as of year n (in some years, the limit of the *molad* is earlier than noon; an excess over this limit leads either to postponement or to other effects on the calendar type of the forthcoming year; this kind of limit can similarly lead to differences between years n and n + 247).

The cycle's critics argued that if outdated calendar information were used in such years, it would differ from the dates determined by the *molad* calculation and would result in a violation of the commandments, by eating leavened bread on Passover and not fasting on the day of Atonement. This line of argument very clearly shows that the 247-year cycle was opposed by scholars not because it was an astronomically bad calendar, but because it deviated from the standard calculated calendar and was liable to disturb calendar unanimity, an important principle of Rabbinic calendar reckoning.[26]

The scientific critique of the cycle did not sway the public's opinion: 60% of the sources in my corpus consider the cycle a perfectly acceptable way of setting the calendar, and advocate its use. Support for the 247-year cycle can be indicated in the sources through the simple phrase *ḥozer ḥalilah* ('it repeats itself forever'), or can be put more persuasively: 'all Israel in all generations should follow this order (of year types), until a priest stands to Urim and Tummim (Neh 7:65)' (i.e. until messianic times when the empirical calendar will be restored),[27] or 'one will find this correct without a shadow of a doubt,'[28] or 'so long as the earth endures, it will not cease (cf. Gen 8:22) and will not depart from the mouth of the holy seed, because it repeats itself forever.'[29] Some appear to have been aware that the 247-year cycle was not completely accurate, but still considered it a good calendar since the few mistakes that it produces could be easily corrected.[30] Others erroneously assumed that the

26 See Stern, *Calendar and Community*, pp. 241–44.

27 Moscow, RSL, Guenzburg 746, fol. 193r (Byzantine, fifteenth century): ועל הסדר הזה יתנהגו כל ישראל לדורי דורים עד עמוד כהן באורים ותומים

28 T-S Ar.2.12r (Oriental, ca. 1181): יגדה צחיח בלא שך ולא ריב

29 Oxford, Bodl. Canon Or. 1, fol. 81r (Ashkenazi, fourteenth century): עוד כל ימי הארץ לא ישבות ולא ישוב מפי זרע הקד[ש] בגלל כי הוא חוזר חלילה

30 Paris, BNF, heb. 642, fols 183v–184r (Sefardi or Provençal, fifteenth century): והוא מנה וקבע י״ג מחזורי׳ מי״ט שנים וחוזרים חלילה ואין שינוי ביניהם כי אם מעט והשנוי ההוא יש לתקנו בנקלה (He (R. Naḥshon) counted and fixed thirteen cycles of nineteen years that repeat themselves forever without many changes and those changes are easy to correct.)

247-year cycle deviated so rarely from the standard calculation that this would not happen before messianic times, when any kind of fixed calendar would be abolished and the empirical calendar restored.[31] One author went so far as to suggest that the 247-year cycle should be used in place of the empirical calendar even in messianic times 'because the Almighty agrees with it'.[32]

3 The Format of 247-Year Cycles

In the sources, the cycle exists in two different formats: a calendar for a particular 247-year period, and a cycle of remainders for any consecutive 247 years. Both formats embody the same idea of a Jewish calendar that is reckoned once and reused forever with no mistakes. A classification of calendars according to their format helps to identify separate stages in the cycle's history and to uncover regional calendar writing trends.

3.1 *Calendars for a Particular 247-Year Period (Dated Tables)*

The most common format of the 247-year cycle is a table of nineteen columns over thirteen rows (or vice versa; 13 × 19 = 247). In such tables, each column represents one 19-year cycle of intercalations, and each cell represents one year of that cycle; the cell is filled with the calendrical type of the year it stands for. The tables are dated by numbering the 19-year cycles, where cycle 1 stands for years 1–19 Anno Mundi (henceforth AM), cycle 2 for years 20–38 AM, and so on.[33] Individual years' dates are also occasionally given, by original scribes or by later users. Numbered 19-year cycles and individual years' dates clearly indicate which 247 years are covered by the table. Such dated calendars are

---

31 Moscow, RSL, Guenzburg 481, fol. 102r (Ashkenazi, fifteenth century): והם חוזרין כן חלילה לעולם עד עמוד כהן לאורים ותומים אלא שיש הפסק כמעט וטרם יפסוק נאכל ימות המשיח ברחמי יוצרינו (They repeat themselves forever until a priest stands to Urim and Tummim but with a small discrepancy. However, before this discrepancy occurs we will eat in Messianic times by the mercy of our Creator.)

32 Oxford, Bodl., Mich. 330, fol. 11r (Ashkenazi, 1559): ואפילו עמוד הכהן הגדול לאורים ותומים אין צריך לעדים על חידוש הלבנה כי על עיגול דרב נחשון עליו השלום יש לסמוך כל זמן שהשמש וירח קיימם עד עולם כי בזה מסכים הקבה (Even when a priest stands to Urim and Tummim, there is no need for witnesses of the new moon because one should rely on the *ʿIggul* of R. Naḥshon, peace be upon him, all the time that the sun and the moon are in existence, forever, because the Almighty agrees with it.)

33 Despite this dating system, tables never begin at 1 AM and only rarely at 1 AM *modulo* 247 (see Appendix 1, *Latin*).

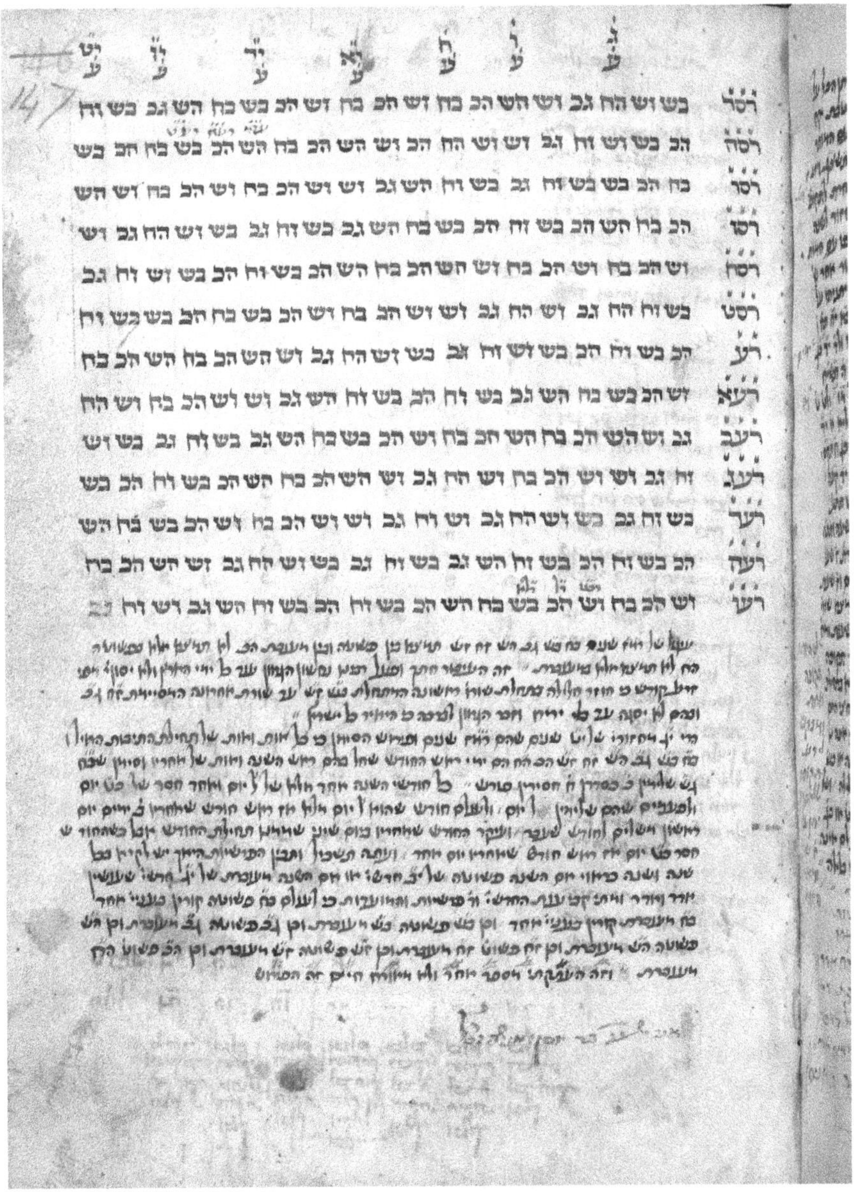

FIGURE 6.1 Oxford, Bodleian Library, Ms. Laud. Or. 166, fol. 147r. A dated 247-year calendar table for 19-year cycles nos. 264–276 (4998–5244 AM, 1237/8–1483/4 CE). The table was copied in Fritzlar, Germany in 1470. In the last row of the table, years 5229–5231 AM (1468/9–1470/71 CE) were marked by the scribe as they corresponded to the time of copying. In the second row of the table, years 5277–5279 AM (1516/7–1518/9 AM) were marked by a later user, as they corresponded to the second iteration of the cycle.

attested in all Jewish geo-cultural areas and in Latin manuscripts.[34] They are sometimes anachronistically associated with R. Naḥshon Gaon.

Tables for a particular stretch of 247 years are not intrinsically cyclical unless explicitly so stated. Their intended reiterativeness can be indicated by that statement that the table repeats itself forever (*ḥozer ḥalilah*) or through instructions to go back to the first line and start using the table all over again.[35] In some tables, reiterativeness is signaled by noting 19-year cycle numbers for more than one iteration of the 247-year cycle.[36]

To use a reiterative 247-year table in any given year, one must take that year's date and calculate its 19-year cycle and its number within that cycle. The cell that corresponds to the established 19-year cycle and year number contains the year type for the sought year. If the required 19-year cycle is not explicitly covered by the table, one must look thirteen 19-year cycles earlier, or a multiple of thirteen 19-year cycles earlier, and take the data from there.

3.2 *Cycles of Remainders (Undated Calendars)*

Another attested format of the reiterative calendar are sequences of 247 year types that are not dated to specific years. In these calendars, years are counted not from an established era (such as the era of Creation or the Seleucid era) but from an arbitrarily chosen epoch, and are referred to by their position within the cycle, from 1 to 247. To calculate the position of a year in a 247-year cycle counted from any epoch, one must take that year's date according to the era with which the cycle operates and subtract from it the years of the epoch minus one. One must then calculate the remainder after casting out 247s from the obtained number. This remainder determines the position of the year in the 247-year cycle. For example, years 1 AM, 248 AM and 495 AM all correspond to year 1 of a 247-year cycle that takes the year of Creation as its epoch. Years 1006 SE, 1253 SE and 1500 SE are year 6 of the 247-year cycle with an epoch in 1001 SE. A reiterative calendar in the shape of a cycle of remainders specifies to which year type each remainder is assigned. In order to use this cycle in any given year, one must follow the algorithm described above, find the calculated

34 Dated calendars for a particular 247-year period are not always tabular and can have a variety of other, less common formats. Since the exact details of their layout are less important for the present article than the fact that they cover a particular set of thirteen 19-year cycles, I shall refer to all such calendars as tables.

35 E.g., Oxford, Bodl., Opp. 59, fol. 164r (Ashkenazi, ca. 1308).

36 See, for example, Paris, BNF heb. 263, fol. 57v (Italian, 1480/1–1481/2), Cesena, Biblioteca Malatestiana Pluteo sinistro XXIX 4, fol. 3v (Sefardi, fourteenth century), Frankfurt, UB, Hebr. Oct. 31, fols 38v–39r (Ashkenazi, 1662/3), Nicholas Trevet, *Compotus Hebreorum* (in Nothaft, *Medieval Latin Christian Texts*, p. 363).

remainder in the cycle of remainders, and see to which of the fourteen year types this remainder is assigned.

The simplest cycle of remainders operates with dates from Creation and takes 1 AM as its epoch; the position of a year in this cycle equals its AM date *modulo* 247. Descriptions of the algorithm based on this epoch survive, but no actual calendars.[37] Cycles of remainders that start from other epochs are common in Oriental and Italian manuscripts and are described in Appendix 1, *Oriental* and *Italian* (see Figures 6.2 and 6.3).

It appears that cycles of remainders are the original format of the reiterative calendar, and tables for a particular set of thirteen 19-year cycles represent a later development. The earliest cycle of remainders was devised by Josiah b. Mevorakh (ibn) al-ʿĀqūlī in the middle of the tenth century. In contrast, the earliest surviving dated table starts in 1123/4,[38] and the earliest such table implied by the sources must have started in 1104/5.[39] This is not merely a corollary of the surviving manuscripts: calendar data in 247-year tables indicates that such tables did not exist before the eleventh century. As can be seen in Table 6.1 below, the 247-year cycle deviates from the standard *molad* calculation in a number of years, so that using a reiterative calendar in these years will create mistakes. Significantly, reiteration errors in the years between 984/5–1252/3 are never attested in the tables. Between 1252/3 and 1354/5 there are no errors either, but that is because in these years the cycle does not differ from the standard calculation.[40] But in 1354/5–1355/6 and later, reiteration errors are very common in the tables. This indicates that calendar data for years prior to and including 1005/6 (247 years earlier than 1252/3) was never incorporated in 247-year tables. Had there been 247-year tables that covered years before 1005/6, they would have left a trace in later tables in the form of reiteration errors reflecting the early year types. This leads to the conclusion that the first dated 247-year

37 I am aware of three sources, from different geo-cultural areas: T-S Ar.29.135 and T-S Ar.29.3r (Oriental, ca. eleventh–twelfth centuries); Cambridge, Trinity College F 12 22, fol. 6v (Ashkenazi, fourteenth century); and Moscow, RSL, Guenzburg 365, fol. 163v (Provençal, fifteenth century).

38 Moscow, RSL, Guenzburg 481, fol. 102r (Ashkenazi, fifteenth century). The earliest fragmentary table of a similar format (Manchester, Rylands B 4471), which may or may not have covered the entire 247 years, takes 1048/9 as its starting point.

39 This is inferred from a marginal note in Munich, BSB, Cod. hebr. 128, fol. 28r (Provençal, fifteenth century) that speaks of 19-year cycle 270 as the beginning of a next iteration of the 247-year cycle, indicating that its previous iteration started thirteen cycles earlier in the beginning of the 19-year cycle 257, i.e. 1104/5.

40 Reiteration errors would have been possible in 1276/7–1277/8 if year types for 782/3–783/4 (two 247-year periods earlier) were used in 1029/30–1030/31 (one 247-year period earlier) and then again in 1276/7–1277/8. This, however, is not attested in any surviving tables.

calendars must have been put together between 1006/7 and 1107/8 (which is 247 years earlier than 1354/5) or possibly 1104/5 (which is the estimated starting date of the earliest table implied by the sources). Cycles of remainders may have been re-formatted into dated tables for the sake of convenience, since an undated calendar is clearly less intuitive than a dated one. Parallels for this conjectured development may be found in some copies of the Oriental cycle of remainders, where pre-calculated remainders for a number of years are noted down in order to help users find their position within the cycle,[41] and in an Italian manuscript where a part of a cycle of remainders has been converted into a dated calendar for the years close to the time of copying.[42]

4 The Production of 247-Year Cycles

A calendar for 247 years can be copied from an earlier *Vorlage* or can be freshly calculated. To determine how a calendar was produced, one must look at particular years that in the standard *molad* calculation are fixed differently from corresponding years 247 years earlier. A calendar copied from an earlier *Vorlage* will contain outdated year types, correct for 247 (or a multiple of 247) years earlier, in years susceptible to reiteration errors; a calendar produced by a fresh calculation will have no such mistakes. Needless to say, all calendars, either copied or calculated, can have mistakes unrelated to the use of the 247-year cycle. These mistakes can be easily distinguished from reiteration errors since they usually apply to single years and generate random results that are incorrect in any iteration of a 247-year cycle. On the contrary, reiteration errors occur in groups of two to three consecutive years and produce year types that are incorrect in one iteration of the cycle but correct in another.

The earliest identifiable 247-year cycle, put together by Josiah b. Mevorakh, correctly represents the standard calendar for years 689/90–935/6.[43] In its further iterations the cycle deviates from the standard *molad* calculation in the years that are specified here in Table 6.1 (the years in which the 247-year cycle remains identical to the standard calculation are not included in the Table). In this table, the column '1st iteration' shows the types of year that are fixed in Josiah b. Mevorakh's 247-year cycle of remainders; they conform identically to

41 T-S AS 144.111 and the folio formed by T-S AS 203.216, T-S AS 144.286, T-S AS 144.228.

42 Frankfurt, UB, Hebr. Oct. 142, fols 80r–81v (Italian, fifteenth century).

43 The starting date of 689/90 is not indicative of the time when this cycle was conceived. The year 689/90 CE corresponds, in fact, to 1001 SE, the beginning of the second millennium of the Seleucid era. See Appendix 1, *Oriental*.

the standard calculation in the cycle's first iteration. Columns '2nd–5th iteration' show the types of years occurring 247 (or a multiple of 247) years later according to the standard calculation. Where these year types, in the 2nd–5th iterations, differ from those in the 1st iteration, reiteration errors can occur if data is copied from an earlier *Vorlage*.

The type of a year is commonly presented in the form of a shorthand notation consisting of three letters. The first letter of the year type stands for the day of the week of Rosh Hashanah, which can be ב (2) for Monday, ג (3) for Tuesday, ה (5) for Thursday and ז (7) for Saturday. The second letter of the year type stands for the length of the variable months Marḥeshvan and Kislev, and can be ח for חסרים 'defective' (both have 29 days), כ for כסדרן 'regular' (Marḥeshvan has 29 days and Kislev 30 days); and ש for שלמים 'full' (both have 30 days). The third letter represents the day of the week of Passover, which can be א (1) for Sunday, ג (3) for Tuesday, ה (5) for Thursday and ז (7) for Saturday. There are also other ways of representing year types in medieval manuscripts. In Table 6.1, the days of the week of Rosh Hashanah and Passover are represented by numbers, from 1 (for Sunday) to 7 (for Saturday). The length of the variable months is represented by the letters D, R, F for 'defective', 'regular' and 'full' respectively.

Reiteration errors in a dated calendar for a particular stretch of 247 years can be detected by collating it with a standard calculated calendar for the same years. The procedure is more complicated with cycles of remainders, which are undated. In such cycles one must determine whether a given sequence of year types corresponds to the standard calendar for any single 247-year period counted from the cycle's epoch. If all year types assigned to remainders fit the same 247-year period and that period ended before the time of copying, or if year types do not all fit the same 247-year period, the cycle of remainders can be assumed to have been copied from an earlier *Vorlage*.[44]

Of the fourteen surviving copies of Josiah b. Mevorakh's cycle, only four contain a version of the calendar in which all year types pertain to one and the same iteration of the 247-year cycle. Of these four, two are copies of a scientific critique of the 247-year cycle, in which remainders were fully updated for the second iteration of the cycle (see Appendix 1, *Oriental*). The other two were copied 100–200 years after the end of the iteration which they represent, and would have created reiteration errors if used. In most copies, year types for lower remainders fit a later iteration of the cycle than those for higher remainders.[45] This indicates that Josiah b. Mevorakh's calendar was commonly

44 See Vidro, 'Origins', pp. 110, 124–28.
45 Ibid., pp. 112–19, 126–28.

TABLE 6.1 Deviations of Josiah b. Mevorakh's 247-year cycle from the standard calculation, late tenth–early twentieth centuries

Josiah b. Mevorakh's 247-year cycle	Standard calculation									
Remainder	1st iteration		2nd iteration		3rd iteration		4th iteration		5th iteration	
	CE date	year type	CE date	year type	CE date	year type	CE date	year type	CE date	year type
6	694/5	7F3	941/2	7F3	1188/9	7F3	1435/6	7D1	1682/3	7D1
7	695/6	5R7	942/3	5R7	1189/90	5R7	1436/7	3R5	1683/4	3R5
8	696/7	2D5	943/4	2D5	1190/1	2D5	1437/8	7F5	1684/5	7F5
49	737/8	7F5	984/5	7D3	1231/2	7D3	1478/9	7D3	1725/6	7D3
50	738/9	7D1	985/6	5F1	1232/3	5F1	1479/80	5F1	1726/7	5F1
53	741/2	7F3	988/9	7D1	1235/6	7D1	1482/3	7D1	1729/30	7D1
54	742/3	5D1	989/90	3R7	1236/7	3R7	1483/4	3R7	1730/1	3R7
55	743/4	3R5	990/1	2F5	1237/8	2F5	1484/5	2F5	1731/2	2F5
65	753/4	2F7	1000/1	2F7	1247/8	2D5	1494/5	2D5	1741/2	2D5
66	754/5	2D3	1001/2	2D3	1248/9	7F3	1495/6	7F3	1742/3	7F3
69	757/8	2F5	1004/5	2F5	1251/2	2D3	1498/9	2D3	1745/6	2D3
70	758/9	7D1	1005/6	7D1	1252/3	5F1	1499/1500	5F1	1746/7	5F1
84	772/3	5F3	1019/20	5F3	1266/7	5F3	1513/4	5F3	1760/1	5D1
85	773/4	5R7	1020/1	5R7	1267/8	5R7	1514/5	5R7	1761/2	3R5
86	774/5	2D3	1021/2	2D3	1268/9	2D3	1515/6	2D3	1762/3	7F3
89	777/8	2F5	1024/5	2F5	1271/2	2F5	1518/9	2F5	1765/6	2D3
90	778/9	7D3	1025/6	7D3	1272/3	7D3	1519/20	7D3	1766/7	5F3
94	782/3	5F1	1029/30	5R7	1276/7	5R7	1523/4	5R7	1770/1	5R7
95	783/4	3R7	1030/1	2F7	1277/8	2F7	1524/5	2F7	1771/2	2F7
147	835/6	7F5	1082/3	7F5	1329/30	7F5	1576/7	7D3	1823/4	7D3
148	836/7	7D1	1083/4	7D1	1330/1	7D1	1577/8	5F1	1824/5	5F1
151	839/40	7F3	1086/7	7F3	1333/4	7F3	1580/1	7D1	1827/8	7D1
152	840/1	5D1	1087/8	5D1	1334/5	5D1	1581/2	3R7	1828/9	3R7
153	841/2	3R5	1088/9	3R5	1335/6	3R5	1582/3	2F5	1829/30	2F5
172	860/1	5F1	1107/8	5F1	1354/5	5R7	1601/2	5R7	1848/9	5R7
173	861/2	3R5	1108/9	3R5	1355/6	2F5	1602/3	2F5	1849/50	2F5
192	880/1	5F1	1127/8	5F1	1374/5	5F1	1621/2	5R7	1868/9	5R7
193	881/2	3R7	1128/9	3R7	1375/6	3R7	1622/3	2F7	1869/70	2F7
225	913/4	7F5	1160/1	7F5	1407/8	7F5	1654/5	7F5	1901/2	7D3

TABLE 6.1 Deviations of Josiah b. Mevorakh's 247-year cycle (*cont.*)

Josiah b. Mevorakh's 247-year cycle	Standard calculation									
Remainder	1st iteration		2nd iteration		3rd iteration		4th iteration		5th iteration	
	CE date	year type	CE date	year type	CE date	year type	CE date	year type	CE date	year type
226	914/5	7D1	1161/2	7D1	1408/9	7D1	1655/6	7D1	**1902/3**	**5F1**
233	921/2	5F3	1168/9	5D1	1415/6	5D1	1662/3	5D1	1909/10	5D1
234	922/3	5R7	1169/70	3R5	1416/7	3R5	1663/4	3R5	1910/1	3R5
235	923/4	2D3	1170/1	7F3	1417/8	7F3	1664/5	7F3	1911/2	7F3
238	926/7	2F5	1173/4	2D3	1420/1	2D3	1667/8	2D3	1914/5	2D3
239	927/8	7D3	1174/5	5F3	1421/2	5F3	1668/9	5F3	1915/6	5F3

Note: Grey background represents years when Josiah b. Mevorakh's cycle deviates from the standard calculation, and hence reiteration could lead to errors. In bold: years for which reiteration errors are attested in dated 247-year tables. In non-bold: years for which reiteration errors are not attested in dated 247-year tables, but only in cycles of remainders.

copied from earlier *Vorlagen* and was gradually updated in order to bring it in line with the standard calendar.

Two different cycles of remainders are attested in Italian manuscripts. The first cycle correctly represents the standard calendar for years 1237/8–1483/4 (19-year cycles 264–276); the second one correctly covers years 1256/7–1502/3 (19-year cycles 265–277). All copies of the cycles known to me are found in manuscripts produced in the fourteenth–fifteenth centuries, within the first iteration of the sequences and too early for reiteration errors to occur.

The 247-year calendars most susceptible to reiteration errors are dated calendar tables. Scribes often strove to start such tables from the 19-year cycle in which they worked. To do this they could either produce or copy a table that started in the desired cycle, or they could take a reiterative table that started in an earlier 19-year cycle and move all 19-year cycles that have already passed to the end of the table, renumbering them but not updating the year types. Although reusing a calendar table in the described way inevitably creates reiteration errors, this method of creating calendar tables appears to have been popular in the Middle Ages and early-modern times.

In the overall corpus of manuscript tables for 247-years ca. 70% have reiteration errors. The percentages are highest in Ashkenazi and Italian tables (80% and 100% respectively) and lowest in tables from the Orient (24%, with just 13% of Yemenite tables having reiteration errors). In tables starting before the end of the fifteenth century, 66% have reiteration errors. In tables starting in the beginning of the sixteenth century and later, the percentages rise to 85%. It is evident that with the exception of the Orient, dated tables for thirteen 19-year cycles were most commonly produced by reusing old tables. Some *Vorlage* tables were iterated more than once. Thus, year types given in many tables for 1601/2–1602/3 were last correct in 1107/8–1108/9, two iterations of the 247-year cycle earlier.

The state of reiterative tables did not improve in the age of printing: a 247-year calendar printed in Guadalajara (ca. 1482) as well as those included in *She'erit Yosef* by Joseph ben Shem Ṭov (Salonika, 1521), in the printed *Ṭur Oraḥ Ḥayyim* by Jacob ben Asher (first table printed in Constantinople, 1540 edition), and in many *Sifre Evronot* (first printed in Riva di Trento, 1561), have reiteration errors and must have been based on earlier calendars. Despite the errors, these printed tables did not remain tucked away in books but were displayed, a fact that highlights their authoritative status. Thus, a visiting scholar in Salonika in 1538/9 described seeing a faulty reiterative calendar on a wall of a synagogue and published a list of its mistakes.[46]

Copying calendars from earlier *Vorlagen* was so widespread that outdated year types are often found in tables that are not otherwise recognisable as produced by reiteration – they cover less than thirteen 19-year cycles, do not have cyclicity statements and are not associated with R. Naḥshon. For example, Toronto, MS Friedberg 3–016, fol. 91r (Ashkenazi, fourteenth century) has a calendar for the three 19-year cycles numbered 268–270 (1313/4–1369/70) with a reiteration error in 1354/5–1355/6. Munich, BSB Cod. hebr. 109, fols 36r–36v (Sefardi, fifteenth century) has a calendar for the six 19-year cycles 274–279 (1427/8–1540/1) with a reiteration error in 1435/6–1437/8. A striking example is Zürich, Jeselsohn 16, fol. 146r (Ashkenazi, sixteenth century) where a calendar for the five 19-year cycles 281–285 (1560/1–1654/5) has reiteration errors in all but one cycle, and includes year types that were last correct in the early twelfth century. It is clear that these and similar tables effectively represent truncated

46 Issachar Ibn Susan, *Tiqqun Issachar* (Venice, 1578/9, fols 10v–11v). The described calendar was most probably a copy of the table in *She'erit Yosef* by Joseph ben Shem Ṭov, as is demonstrated by the identicalness of the reiteration and printing mistakes in this table with those discussed in *Tiqqun Issachar*, and is further corroborated by Ibn Susan's statement that the described calendar was printed in Salonika.

247-year cycles. Of tables of up to ca. 100 years that cover years susceptible to reiteration errors, every third contains outdated year types; of shorter tables of up to ca. 60 years, it is every fourth. If calendars of any length are considered, almost two thirds of those that can have reiteration errors do so. This means that Jews relying on ready-made calendars were more likely to use one with reiteration errors than without them.

Outdated year types can be found in tables that provide *moladot* – a hallmark of the calculated calendar.[47] In such cases, it is evident that the year types were not calculated on the basis of the *moladot*, but that these two kinds of calendrical data were uncritically put together for the sake of completeness. Even more surprising is the presence of outdated calendar information in scientific works on the calendar that explicitly refute the 247-year cycle. Examples include *Yesod ʿOlam*, a comprehensive monograph on mathematics, astronomy and calendar by Isaac Israeli composed in 1310, and *Ḥeshev ha-Efod*, a calendar monograph by Profiat Duran composed in 1395. Tables in all surviving copies of *Yesod ʿOlam* and *Ḥeshev ha-Efod* have mistakes linked to reusing old calendrical information from thirteen or a multiple of thirteen 19-year cycles earlier.[48] At the same time, both authors explicitly state that the Jewish calendar does not recur after 247 years and warn that relying on calendrical information for thirteen 19-year cycles earlier leads to mistakes.[49] As I have suggested elsewhere, these tables may have been put together by the scholars' less skilled amanuenses.[50]

5 Corrections and Use

Evidence from different parts of the Jewish world indicates that 247-year calendars were widely used in practice. Some sources explicitly state the cycle's

---

47 See e.g. Oxford, Bodl., Poc. 262, fols 249r–251v (Oriental, 1202), 12 cycles; London, BL Or 2674, fols 131v–135r (Italian, fifteenth century), 8 cycles; Frankfurt, UB, Hebr. Oct. 142, fols 80r–81v (Italian, fifteenth century), ca. 5 cycles, New York, JTS 2435, fols 50v–54r (Byzantine, fifteenth century), ca. 8 cycles.

48 The table in *Yesod ʿOlam* covers 19-year cycles 268–300 and has generic mistakes linked to using outdated year types in cycles 274, 281, 282, and 287 but not in cycles 284, 291, 294, 295, 298, 299, and 300, where such mistakes are also possible (see, for example, London, BL Add 15977, fol. 178r). The table in *Ḥeshev ha-Efod* covers cycles 272–281 and has generic mistakes linked to using outdated year types in cycle 281 (see, for example, Parma, Biblioteca Palatina Cod. Parm. 2776, fol. 130v).

49 See above, nn. 23–4.

50 Vidro, 'Calendar tables', p. 74.

popularity 'because it is simple and easy to grasp'.[51] Elsewhere there is evidence of people warning others about imminent mistakes, which implies that 247-year tables were expected to be used and cause trouble.[52] We also find practical advice on how to keep track of the position of the year in a 247-year table by adding the AM date to the relevant cell at the start of each year.[53]

AM dates in secondary hands are found in many tables, suggesting that the tables were consulted in those years. Especially interesting are cases where secondary hand dates and 19-year cycle numbers are marked not for the initial iteration of the cycle as it is found in the manuscript, but for a later period. For example, Vatican, BAV, ebr. 318, fol. 259v (Ashkenazi, end thirteenth century?) contains a 247-year calendar table for 19-year cycles 261–273 (1180/1–1426/7). In this table some original cycle numbers are crossed out and new cycle numbers, thirteen 19-year cycles later, are written instead, e.g. 277 for 264, 278 for 265, etc. In addition, AM dates for years 1458/9–1619/20, in the next iteration of the calendar, are added in at least two secondary hands. Such marks clearly indicate that users of the manuscript believed in the reiterative nature of the 247-year cycle and implemented it in practice.[54]

Cycles of remainders, too, have user marks that indicate their being used in years beyond the 247-year period that they correctly cover. Sixteenth-century dates are marked in Italian cycles of remainders that correctly cover 1237/8–1483/4.[55] Similarly, a user of the Judaeo-Persian manuscript London, BL,

---

51 T-S NS 98.2 (Oriental, thirteenth – fourteenth century): ונצ̇רנא כתרה אעתמאד אלנאס עלי עבור לרבינו יאשיהו בן מבורך ז״ל לאנה סהל קריב אלמאכד (We saw that people often rely on the calendar of R. Josiah b. Mevorakh of blessed memory because it is simple and easy to grasp.)

52 Moscow, RSL, Guenzburg 365, fol. 147r (Provençal, fifteenth century): זאת התעייה ... לא שנת קצו וקצז לבד כמו שהעירוך טארשקון אבל גם שנת קצח (This error ... is not just in the years 1435/6 and 1436/7 as (the people of) Tarascon warned you, but in the year 1437/8, too.)

53 *Sefer Minhage Maharil*, cited here according to Frankfurt, UB, Hebr. Oct. 94, fol. 219v (Ashkenazi, 1460): בכל תחילת השנה תכתוב הפרט על הלוח ולא תשגה (at the beginning of each year write down its date in the table and you will not make mistakes.)

54 Additional examples can be found in Zürich, Jesselssohn 17, fols 263v–264r (Ashkenazi, end twelfth–end thirteenth century), Cincinnati, HUC 436, fol. 191v (Ashkenazi, 1435), Cambridge, Trinity College F 12 21, fol. 10v (Ashkenazi, fourteenth century), Oxford, Bodl. Heb. e.60, fol. 462r (Persian, 1485), Oxford, Bodl. Laud Or. 166, fol. 147r (Ashkenazi, 1470), Paris, BNF, heb. 1032, fol. 1v (Ashkenazi, fourteenth century), London, BL Or 2389, fols 141v–142r (Yemenite, 1635), Vatican, BAV ebr. 423, fol. 30r (Italian, fifteenth century), Jerusalem, Kapah 16, fol. 12v (Yemenite), London, BL Or 4104, fol. 21v (Yemenite, fifteenth century).

55 See, e.g., Jerusalem, NLI Heb. 38°4281, fol. 298v (Italian, 1391), Paris, BNF, heb. 620, fol. 281v (Italian, fourteenth–fifteenth century), Vatican, BAV Ross 437, fol. 395v (Italian, 1448), Parma, Biblioteca Palatina, Cod. Parm. 2467, fols 46v–49v (Italian, ca. 1410).

Or 2451, fol. 369v (1482/3–1483/4) determined year types for eighteenth-century dates on the basis of a cycle that correctly covers 1183/4–1429/30. However, since cycles of remainders are not inherently dated, their users cannot be said to have consciously reiterated the calendar.

As demonstrated above, a large proportion of 247-year calendars in medieval and early modern Jewish sources were produced by copying from earlier *Vorlagen*. This affected the accuracy of the cycles in more than one way, since in addition to reiteration errors, such calendars were more likely to have scribal mistakes.[56] Manuscript readers occasionally attempted to update outdated year types and correct scribal mistakes. Such corrections rarely cover whole tables, but apply to separate years or to relatively short stretches of years, presumably close to the time when a reader consulted the cycle. While some corrections are based on a calculation, sometimes performed in a manuscript's margins,[57] others are a result of collating different calendar tables. When collating, scribes and readers often noted divergent year types as alternative versions, without making a final decision which ones to follow.[58]

Corrections did not always improve a table's readings. Whether calculating or collating different calendars, readers were as likely to correct mistakes as to replace originally correct data with year types that were either outdated or wrong for other reasons. In a secondary note in Zürich, Jesselssohn 17, fol. 264r (Ashkenazi, end twelfth–end thirteenth centuries), a user calculated *moladot* and incorrectly inferred from them the year types for years 1462/3 and 1463/4, suggesting them instead of the correct data in the table. The table in Jerusalem, NLI, Heb. 8°1997, fol. 73r (Italian, late fifteenth century) underwent two sets of interventions. One reader collated it with a much earlier calendar and glossed it with year types that were last correct two 247-year periods earlier. Another reader acted to update the table and crossed out all outdated year types, both those in the main body of the table and in the glosses.

---

56 Scribes were aware of this danger and insisted that a copied calendar should be checked and rechecked. See Paris, BNF, heb. 646, fol. 139v (Ashkenazi, fourteenth century): מאד יש לו לאדם ליזהר שלא יטעה כי יכול לבא לידי קילקול גדול לקלקל כל המועדים ולאכול חמץ בפסח ולחילול יום הכיפורים על כן כל המעתיק יש לו לקרות ב׳ פעמים או ג׳ פעמי׳ ולדקדק שלא יטעה אחרי שיעתיק (One should be very careful not to make mistakes because one can come to a big distortion and distort all the festivals, eat leavened bread on Passover and profane the day of Atonement. This is why a copyist must read (what he copied) twice or three times and examine carefully after he had copied that he made no mistakes).

57 See, e.g. Zürich, Jesselssohn 17, fols 263v–264r (Ashkenazi, end twelfth–end thirteenth century).

58 See, e.g., New York, JTS Rab. 689, fol. 122r (Ashkenazi, 1437), Jerusalem, The Israel Museum, 180/051, fol. 471r (Italian, fourteenth century).

A calendar used in practice should be reflected in dates found in colophons, legal documents, and other dated texts. The use of the 247-year cycle, however, cannot be verified from such sources. Only one date has so far been discovered which is wrong in the standard calendar but compatible with the cycle.[59] This unique example may reflect the use of a 247-year cycle, but the error could equally have been caused by other factors. A number of reasons can be suggested for the apparent lack of evidence of use of the 247-cycle in documentary dates. Firstly, the cycle produces only few characteristic errors of reiteration. With prolonged use, reiteration errors multiply and are more likely to be detected, but since 247-year calendars other than the earliest Oriental cycles of remainders do not deviate from the standard calendar before the middle of the fourteenth century, reiteration errors do not begin to accumulate before the last third of the sixteenth century, when year types of nine out of the 46 years 1576–1622 differ from those 247 or a multiple of 247 years earlier (see Table 6.1). However, in this period printed wall calendars for a single year began to be produced and gain popularity, and the use of manuscript calendar tables may have lessened.[60] Secondly, reiteration errors can only be detected in documents with a full date that includes the day of the week, the day of the month and the year. Given the rarity of fully dated documents, and the paucity of affected years, the chances of finding a date with a reiteration error are slim. Thirdly, there may have been some awareness among users of the 247-year cycle as to when it was unsafe to rely on it. Some theoretical discussions of the 247-year cycle give isolated examples of years or 19-year cycles when using an outdated calendar produces erroneous results.[61] This information may

59 New York, JTS Rab. 689 (Ashkenazi), dated in the colophon on fol. 194r Thursday, 23 Marḥeshvan 5198 AM (1437 CE). In 5198, 23 Marḥeshvan fell on a Tuesday. In contrast, in 4951 AM (1190 CE), 247-years earlier, 23 Marḥeshvan fell on a Thursday.

60 Carlebach, *Palaces of Time*, pp. 59–66. I have not analysed printed wall calendars for this article.

61 See Isaac Israeli, *Yesod 'Olam*, 4:10, where the author explicitly writes that years 9–11 of 19-year cycle no. 274 (1435/6–1437/8) are not identical to corresponding years in 19-year cycle no. 261 (ed. Goldberg and Rosenkranz, *Liber Jesod Olam*, vol. 2, p. 23). One reiteration error in year 10 of 19-year cycle no. 274 (1436/7) is pointed out in Obadiah b. David's commentary on Maimonides' *Sanctification of the Moon*, vol. 8, p. 1. Surprisingly, neither work mentions possible reiteration errors in 1354/5–1355/6 (years 4–5 of 19-year cycle 270), closer to their composition times in 1310 and 1340 respectively. Paris, BNF, heb. 642, fol. 184r (fifteenth century, Sefardi or Provençal), includes a calculation of *moladot* that demonstrates that some types of years in 19-year cycle no. 282 differ from corresponding years in cycle no. 269.

have been available to some of the cycle's users. A small number of reiterative tables are provided themselves with a full or partial list of errors that a table produces, either in its initial or its second iteration.[62] Another small group of tables alerts users to potential problems by not supplying the AM dates in some years with reiteration errors, whereas all other years are provided with a date.[63] Significantly, there is never a comprehensive list of all potential reiteration errors, so that users could only receive a patchy idea of when it was unsafe to rely on a 247-year cycle.[64] Finally, corrections found in the calendars, and public enquiries into the validity of the cycle which were occasionally initiated in the years leading up to possible years of calendar divergence, may have prevented mistakes in particular years.

6 Public Enquiries

The widespread use of the 247-year cycle led to a number of public enquiries into its adequacy as a means of fixing the Jewish calendar. These enquiries

62 Munich, BSB Cod. hebr. 343, fol. 167v (Sefardi, fifteenth century) includes an accurate 247-year table and comments on all the reiteration errors that the table will produce in its second iteration. The same can be observed in a fourteenth-century Yemenite work *Ner Israel* by Joseph ben Yefet ha-Levi (*Ner Israel*, chapter 13, see, e.g., London, BL, Or 4104, fol. 11r). Zürich, ZB Heid. 145, fol. 45v (Ashkenazi, 1340/1) has a calendar with reiteration errors in 1354/5–1355/6 and in 1435/6–1437/8 accompanied by a reader's gloss that points out the former set of mistakes, but not the latter one. In Parma, Biblioteca Palatina, Cod. Parm. 3266, fols 19r–19v (Ashkenazi, fourteenth century) different notations are used for year types in 19-year cycles that can and cannot have reiteration errors.

63 In Cincinnati, HUC 436, fol. 191v (Ashkenazi, 1435), AM dates are supplied by the scribe in all years starting from 1239/40, except in years 1435/6–1437/8 which are fixed incorrectly in the included table; years 1354/5–1355/6, which also have reiteration errors but had already passed at the time of copying in 1435, are not marked in this way. In Moscow, RSL, Guenzburg 1068, fol. 47r (Italian, sixteenth century) and Oxford, Bodl., Mich. 484, fol. 169r (Italian, sixteenth century) an identical table is included, in which AM dates are missing in years 1601/2–1602/3 and 1681/2–1683/4 where reiteration mistakes are possible. Interestingly, the table has correct data in these years, but makes reiteration errors in other places without indicating them. It seems clear that the table was collated with a different table with reiteration errors in 1601/2–1602/3 and 1681/2–1683/4 and dates were not added in those years in which the tables differed.

64 Thus, the reiteration error in year 10 of 19-year cycle no. 274 (1436/7) pointed out by Obadiah b. David (see above, n. 61) appears to have been the only mistake known to a nineteenth-century maskil Reuven Joseph Wunderbar (see Wunderbar, *Immerwährender Kalender*, p. 5).

never settled the matter in a final way, but may have prevented mistakes in the years involved. Three such cases are known to me: in fourteenth-century Yemen, in fifteenth-century Avignon, and in the early twentieth-century Russian Empire.[65]

The earliest enquiry and dispute took place in Yemen and was reported in a calendar treatise by Maʿūda al-Lidānī, otherwise unknown, composed in the first half of the fourteenth century.[66] Maʿūda al-Lidānī reported that for a long time the Jews of Yemen relied on ready-made calendars including the reiterative calendar by Josiah ben Mevorakh. In 1336 the reiterative calendar was checked against the standard calculation as it was laid out in Maimonides' *The Sanctification of the Moon*, and was found to produce mistakes in years 1354/5–1355/6 and in various other years. The mistakes in years 1354/5–1355/6, described by Maʿūda al-Lidānī, were reiteration errors due to their year types differing from those 247 years earlier; these year types must have been not updated in the calendar in circulation, a state of affairs observed in most surviving copies of Josiah ben Mevorakh's calendar.[67] A confrontation ensued between those who checked the cycle by calculation, established its faultiness and pushed for relying on the standard calculation, and those who supported the 247-year cycle even though, in the words of Maʿūda al-Lidānī, they had no arguments for it[68] and could only put forth an old claim that it was not appropriate for anyone to fix the calendar outside Palestine.[69] It is unknown how the years 1354/5–1355/6 were fixed by the communities involved, but later sources indicate that a preoccupation with the cycle continued in Yemen for centuries to come. Whereas most surviving Yemenite sources refute the cycle and emphasize that once a calendar for 247 years is finished, a new one must be prepared by calculation,[70]

65 Abraham Ibn Ezra's critique of the 247-year cycle can be seen as an outcome of another such enquiry, initiated by David b. Joseph Narboni who found mistakes (דברים שאינם הגונים) in reiterative tables. See ed. Steinschneider, *Sefer Shene ha-Meʿorot*, p. 1.

66 On this dispute see Tobi, *The Jews of Yemen*, pp. 211–26; id., 'The dispute over the 247-year cycle in Yemen', in Morag, Ben-Ami and Stillman (eds), *Studies in Judaism and Islam*, pp. 193–228 (an edition and Hebrew translation of Maʿūda al-Lidānī's account are on pp. 210–21).

67 See Vidro, 'Origins', pp. 118–19 (remainders 172 and 173). No Yemenite copies of Josiah b. Mevorakh's calendar are known to me.

68 מן גיר ביאן (Tobi, 'The Dispute', p. 211).

69 Tobi, 'The Dispute', pp. 199–201, 210–11.

70 For example, Paris, BNF, heb. 1331, fol. 123r (Yemenite, a seventeenth-century copy of a fifteenth-century text): ומתי מא פרג' אלמחזור אלכביר עמל דור מתלה עלי אלחסאב (When the big cycle is used up, make a similar cycle by calculation). Cincinnati, HUC 765, fol. 139v (Yemenite, a seventeenth-century copy of a fifteenth-century text): ומתי פרג'ת

at least one treatise, composed in 1397 and still copied in the eighteenth and nineteenth centuries, advocates using it reiteratively.[71]

The second enquiry was linked to the next stretch of years when the 247-year cycle deviated from the normative calculation, in 1435/6–1437/8. We learn about it from a letter by a fifteenth-century Avignon physician Mordecai Nathan[72] to his uncle Don Boniac Astruc Nasi, composed between 1427/8–1435/6 and preserved in a later copy.[73] Mordecai Nathan's letter is a reply to an earlier letter from Don Boniac Astruc, which is not preserved but some of its contents can be reconstructed. Don Boniac Astruc was warned by Tarasconians that the years 1435/6–1436/7 were fixed incorrectly in the 247-year table.[74] Alarmed and presumably unable to check and correct the calendar by himself, Don Boniac Astruc asked his nephew for a correct 247-year calendar, considering it 'impossible to be without it'.[75] Instead, Mordecai Nathan wrote an epistle on the reiterative calendar, pointing out a third imminent mistake in 1437/8 and explaining why the 247-year calendar is inaccurate. He reported that the faultiness of the reiterative calendar had been sensed in their community already some eighty years earlier, but had since been forgotten because the table produced no more mistakes until the time of the letter. Anticipating the disturbance in 1435/6–1437/8, Mordecai Nathan did all he could to alert people around him and prevent them from being 'lead astray by the table'.[76] Realising that one man's efforts were not enough, Mordecai Nathan proposed an

---

פאנצם לך מחאזיר גירהא ואעתמד עלי אלקואנין אלתי ביינתהא לך פי מא תקדם ולא יכֹטר בבאלך קול מן קאל אן תֹם דור לאלסנין אצלא לא רֹמֹז ולא גֹירה (When it is used up, arrange for yourself further 19-years cycles. Rely on the rules that I explained to you above and do not at all let it come to your mind what they said about there being a cycle of years, not of 247 (years) and not any other!).

71 New York, JTS 5543, fol. 37v (Yemenite, fifteenth century): פאדֹי בלֹגת אכֹרה עדת אלי אולה בלא תעב ולא ענא דאיר ידור אבדא לאלדהר כלה (If you reach its end, go back to the beginning without difficulty and without anguish, it repeats itself forever for all eternity.). On this work see Tobi, *The Jews of Yemen*, pp. 213, 221–22; id., 'The Dispute', pp. 202–3, 222–25; Langermann, 'What does *Sefer Yesira* have to do with the Jewish calendar?'

72 On Mordecai Nathan see Simon and David, 'Nathan, Mordecai'. Accessed on Gale Virtual Library on 23 March 2018.

73 Moscow, RSL, Guenzburg 365, fols 147r–148r (Provençal, ca. 1487).

74 See, e.g., Munich, BSB, Cod. hebr. 128, fol. 28r (late fifteenth century) for a surviving Provençal table with these mistakes.

75 Moscow, RSL, Guenzburg 365, fol. 148r: כי אי אפשר בלעדיו.

76 Moscow, RSL, Guenzburg 365, fol. 147v: ואני כבר נתעוררתי בתעייה המשלשת הזאת זה כמה והעירותי הנ׳ הנבונים אחי וגבירי קרשקש נתן ואחיו בונגואש יא׳ בדרוש ושחרתי להעיר כה וכה לבל יתעו בלוח (I have become alert to this triple error some time ago and alerted the wise ones, my brother and lord Crescas Nathan and his brother Bonjudas of(?) Béziers. And I sought to comment on it this way and that way so that people are not lead astray by the table.)

ingenious solution for preventing reiteration errors while not giving up 247-year tables completely. He formulated criteria to know when it was safe to carry over year types from an earlier table, and suggested that every scholar should write these criteria under every 247-year table that he encounters. He urged Don Boniac Astruc to use his authority to encourage this course of action. Mordecai Nathan did not fulfill his uncle's request for a correct 247-year calendar. Instead, he recommended using a different, non-reiterative, table that provided *moladot* alongside year types, which at the time had been in circulation for about 70 years but contained calendar data for another century and a half.

The last of the three public enquiries known to me happened in 1902, when Jews in the Russian Empire suspected their almanacs of containing incorrect year types for years 1901/2 and 1902/3.[77] Their main concern was that the almanacs disagreed with the table in *Ṭur Oraḥ Ḥayyim*, the first part of the highly influential legal code *Arbaʿah Ṭurim* ('The Four Pillars') by Jacob ben Asher (Cologne, ca. 1270–Toledo, after 1340). In all printed editions of *Ṭur Oraḥ Ḥayyim* the calendar is presented as a cycle of 247 years that repeats itself four times.[78] This table provides correct year types for 1541/2–1787/8[79] but produces errors in other iterations of the cycle, including in years 1901/2–1902/3. Its deficiency had already been exposed in the seventeenth century,[80] but many Rabbis in the Russian Empire still considered the *Ṭur* table authoritative two hundred years later.[81] In a bid to clarify the matter, calendar experts Zvi Hirsch Yaffe and Rafael Gordon published articles in the pages of the periodical *Ha-Melits*,[82] confirming the year types given in the almanacs, refuting the 247-year cycle and dissuading people from relying on the table in *Ṭur Oraḥ Ḥayyim*.

77 On this dispute see also Raviv, *Mathematical Studies*, pp. 105–7. Yaffe, *Qorot Ḥeshbon ha-ʿIbbur*, pp. 163–64.

78 On this calendar and its history see Vidro, 'Calendar tables'.

79 In some editions, mistakes are found in this period, too (Vidro, 'Calendar tables', p. 81).

80 Ḥezekiah da Silva (ca. 1656–1695), *Peri Ḥadash*, first published in Amsterdam 1706, fols 5r–6v.

81 Gordon, 'Further on the matter of year types', p. 3. Available online at http://www.jpress.nli.org.il/Olive/APA/NLI/?action=tab&tab=browse&pub=HMZ#panel=browse, accessed 23 March 2018. In *Ha-Melits*, Julian dates were used, and Gregorian dates were given in brackets; I have followed the same system in the bibliography. The online archive is organised by Gregorian date.

82 See Yaffe, 'Intercalation in its correct time', and Gordon, 'Further on the matter of year types'.

7 The Status of the 247-Year Cycle in Relation to the Calculated Calendar

Ample evidence confirms that the 247-year cycle, although incompatible with the standard *molad* calculation, was not regarded as an alternative Jewish calendar but as a means of setting the standard calendar. 247-year cycles are often found in treatises on the standard calendar, in close proximity with an explanation of the *molad* calculation. All identified refutations of the cycle oppose it for inaccurately representing the standard calendar, and 247-year calendars themselves are relatively frequently corrected in the manuscripts to bring them in line with the standard calculation.

Why did Jews use a calendar scheme that was known to yield erroneous calendrical information and, in some years lead to celebrating festivals on the wrong days, eating leavened bread on Passover and profaning the day of Atonement? The answer appears to be linked to the general level of knowledge in the field of calendar reckoning. An introduction to a work on the calendar preserved in T-S Ar.54.7b distinguishes between people who have mastered the science of calendar reckoning and can calculate the calendar from first principles; those who studied less but can use pre-calculated data to set the calendar; and those whose knowledge is so insufficient that they must use a ready-made calendar.[83] It appears that most medieval and early-modern Jews belonged to the latter group of people, and did not know how to construct or correct a calendar.[84] This included the educated elites: in the Avignon correspondence described above, Don Boniac Astruc, who bore the title 'the prince' (heb. *nasi*) and must have been a high-standing member of the community, found himself unable to verify calendar data. In another episode, the elders of Toledo struggled to fend off an accusation that they had celebrated Passover on a wrong date and turned to a calendar expert, Isaac Israeli, for justifying their date.[85]

A long-term, ready-made calendar ensured that a community without calendar experts would not find itself without a means of setting the calendar. A 247-year cycle, which could be reused and only rarely produced mistakes, must have been especially attractive for those lacking calendar expertise. The sources frequently connect the use of the reiterative calendar with lack of knowledge:

---

83 T-S Ar.54.7b recto: ויכון בתא הדא עלי ג̇ טבקא[ת] פמן אכתר יתעלם אלעבור פיקף עלי אלפצול ומן לם יכתאר יתעב פי דלך פיחסב מא אשתהא מן אלגדאול אלמנצובה ומן צעב עליה דלך איצא פיעלם אלסנה או אלשהר מן אלסנין אלמנצובה עלי אלנסק

84 See also Carlebach, *Palaces of Time*, p. 6.

85 Isaac Israeli, *Sefer Yesod 'Olam*, 4:18 (ed. Goldberg and Rosenkranz, *Liber Jesod Olam*, vol. 2, p. 36).

> … and the degradation and the lack of knowledge and deficiency of learning. And we saw that people often rely on the calendar of R. Josiah b. Mevorakh of blessed memory because it is simple and easy to grasp.[86]
>
> Blessed be our Rock who poured out his spirit from on high (cf. Isa 23:15) on the enlightener of the Diaspora, R. Naḥshon. Were it not for him, our feet would collapse in our long exile and through our lack of knowledge, but he counted and fixed thirteen cycles of nineteen years that repeat themselves forever without many changes and the few changes are easy to correct.[87]

The popular nature of the 247-year cycle is also further highlighted by the fact that hardly any attempts were made to prove the cycle's validity, and those that are attested are scientifically weak. In Paris, BNF, heb. 646 it is erroneously stated that the 247-year calendar recurs because *moladot* 'go in sevens' after 247 years, which appears to mean, because the same *molad* recurs after 247 years, and not earlier.[88] In Oxford, Bodl. Heb. e.45/45–56 (Oriental, 1151/2), the year types for fifteen 19-year cycles, nos. 259–273 (1142/3–1426/7), are listed in order to demonstrate repetition after the first thirteen, a demonstration that only worked due to a lucky choice of the years covered by the cycle: had the table covered just one more 19-year cycle, it would have produced mistakes in 1435/6–1437/8. In addition, some manuscripts state that whereas *moladot* do not repeat themselves after 247 years, year types do,[89] a claim that cannot be sustained in the framework of the standard calendar.

---

86 T-S NS 98.2 recto (Oriental, thirteenth–fourteenth century): ואלדל וקלה אלמערפה ונקץ אלעלם ונצ̇רנא כתרה אעתמאד אלנאס עלי עבור לרבינו יאשיהו בן מבורך ז̇ל̇ לאנה סהל קריב אלמאכד

87 Paris, BNF, heb. 642 (fifteenth century, Sefardi or Provençal), fols 183v–184r: ברוך צורינו אשר הערה רוחו ממרום על מאור גלותינו רב נחשון שאלמלא הוא נתמוטטו רגלינו באורך גלותינו וחסרון ידיעתינו והוא מנה וקבע י״ג מחזורי׳ מי״ט שנים וחוזרים חלילה ואין שינוי ביניהם כי אם מעט והשנוי ההוא יש לתקנו בנקלה (the manuscript erroneously reads מאוד גלותינו).

88 Paris, BNF, heb. 646, fol. 139v (Ashkenazi, fourteenth century): והנה לך הטעם למה חוזרים לרמז שנים כשתמנה המולדות מרמז שנים ילכו כולם שביעיות לא יבואו לשביעיות עד רמז שנים. Interestingly, this statement is followed by what appears to be a later gloss incorporated into the main text that explains that the cycle is inaccurate but is not harmful up to a certain point in the calendar.

89 New York, JTS Rab 689, fol. 121v (Ashkenazi, 1437): זה הלוח שלפנינו תקן רב נחשון גאון והק־בועים חוזרי׳ חלילה אבל המולדות אינ׳ חוזרי׳ שיש חילוק מן תתק״ה חלקי׳ (This table here was fixed by R. Naḥshon Gaon. The year types repeat themselves forever, but the *moladot* do not repeat themselves because there is a difference of 905 parts.)

8 Conclusions

This article has analysed a large corpus of medieval and early modern sources on the 247-year calendar cycle, a system of calendar reckoning that gained considerable popularity in all corners of the Jewish world despite being incompatible with the standard Rabbinic calendar. The corpus demonstrates that the 247-year cycle was put together in the middle of the tenth century by a Babylonian scholar Josiah b. Mevorakh (ibn) al-ʿĀqūlī. The cycle may have been proposed in the aftermath of the calendar controversy of 921/2, as an alternative to the standard calendar that would be capable of preventing future calendar dissidence caused by differences in the calendar calculation procedures practiced in Babylonia and Palestine. Later, the 247-year cycle came to be regarded not as a replacement for the standard calendar but as an easy means of setting it, used alongside the more demanding calculation. By the middle of the twelfth century, the cycle spread to different geo-cultural areas and became widely known, often under the title *Thirteen Cycles*. In contrast, the title *ʿIggul* and the attribution of the 247-year calendar cycle to R. Naḥshon Gaon were predominantly Ashkenazi and did not appear before the late twelfth century. This attribution was probably introduced in order to give authority to a way of setting the Jewish calendar that had already become popular.

Although scholars regularly refuted the 247-year cycle on scientific grounds, demonstrating its inaccuracy as a means of setting the standard calendar, the cycle enjoyed considerable support throughout the studied period. A large proportion of calendar tables in medieval and early modern Jewish sources were produced by copying from master copies for 247 years earlier, with many shorter calendars effectively being truncated 247-year cycles. This caused reiteration errors and scribal mistakes and undermined the accuracy of the tables. As a result, many conflicting and partially erroneous calendars were in circulation. Evidence from users' glosses and public disputes indicates that at least some Jewish communities relied on 247-year cycles. Reiterative tables continued to be used even after mistakes in them were discovered and their imprecise nature exposed, due to the relative rarity of the errors. The generally low level of calendar knowledge even among educated elites meant that it was difficult for non-specialists to fix or check the calendar by calculation, and this created a demand for long-term ready-made calendars. In this situation, the 247-year cycle filled the niche of a non-technical way of setting the Jewish calendar that was not limited to a set range of years. Some of the cycle's users may not have been aware of its deficiency, while others may have known that it is not entirely accurate but found themselves unable to set the calendar in a more scientific way. Yet despite the cycle's popularity, the present research did not reveal any cases of communities celebrating festivals on different dates due to the use of the cycle. This suggests that there may have been some awareness, among users of the 247-year cycle, of when it was unsafe to rely on it.

Appendix 1: 247-Year Cycles in Different Geo-Cultural Areas

1 *Manuscripts in Oriental Hands*[90]

My corpus includes 61 Oriental manuscripts, penned between the eleventh and nineteenth centuries, that contain or discuss 247-year calendars. These include Genizah fragments and complete manuscripts, the latter mostly from Persia and Yemen. In Persian manuscripts, 247-year calendars are usually found in Bible codices; in Yemenite manuscripts they are often discussed in calendar treatises within prayer books. The popularity of the 247-year cycle in the Orient is highlighted by a reference to it in a poem on the Samaritan calendar.[91] Since the 247-year cycle plays no role in the Samaritan calendar calculation, its mention must reflect a tradition picked up from local Jews.

The earliest identifiable 247-year cycle was put together in the Orient in the middle of the tenth century and has the shape of a cycle of remainders. Attributed to Josiah b. Mevorakh al-ʿĀqūlī, it is known from Genizah fragments and Judaeo-Persian manuscripts.[92] This cycle operates with Seleucid dates (SE), has the epoch of 1001 SE (689/90), and is not synchronised with the 19-year cycle of intercalations (it starts in year 4 of the cycle). The epoch of 689/90 CE is not indicative of the time when the cycle of remainders was conceived. Indeed, data for Josiah b. Mevorakh's cycle must have been originally put together using the standard *molad* calculation,[93] which in all likelihood has been known since the ninth century but not before.[94] This makes a construction date in the seventh century very unlikely. More probable is that the cycle was constructed around the end of the 247-year period from 689/90–935/6 or the beginning of the next one, using calendrical records for the near past in combination with data calculated retrojectively for more remote years, or by retrojective calculation alone. The most likely *terminus post quem* for the creation of Josiah b. Mevorakh's calendar is the calendar controversy of 921/2, when methods of calendation were discussed in fine detail but the 247-year cycle was not mentioned. The decade of 980s may be suggested as the cycle's *terminus ante quem*. Josiah b. Mevorakh's cycle was based

---

90 The geo-cultural area called 'Oriental' includes territories in the Middle East such as Egypt, Palestine, Iraq, Iran, and Yemen.

91 Powels, *Der Kalender der Samaritaner*, pp. 55, 197, 254. I thank Sacha Stern for drawing my attention to this source.

92 For a detailed description of this cycle, see Vidro, 'Origins'.

93 Vidro, 'Origins', p. 108.

94 Stern, *Calendar and Community*, pp. 200–10.

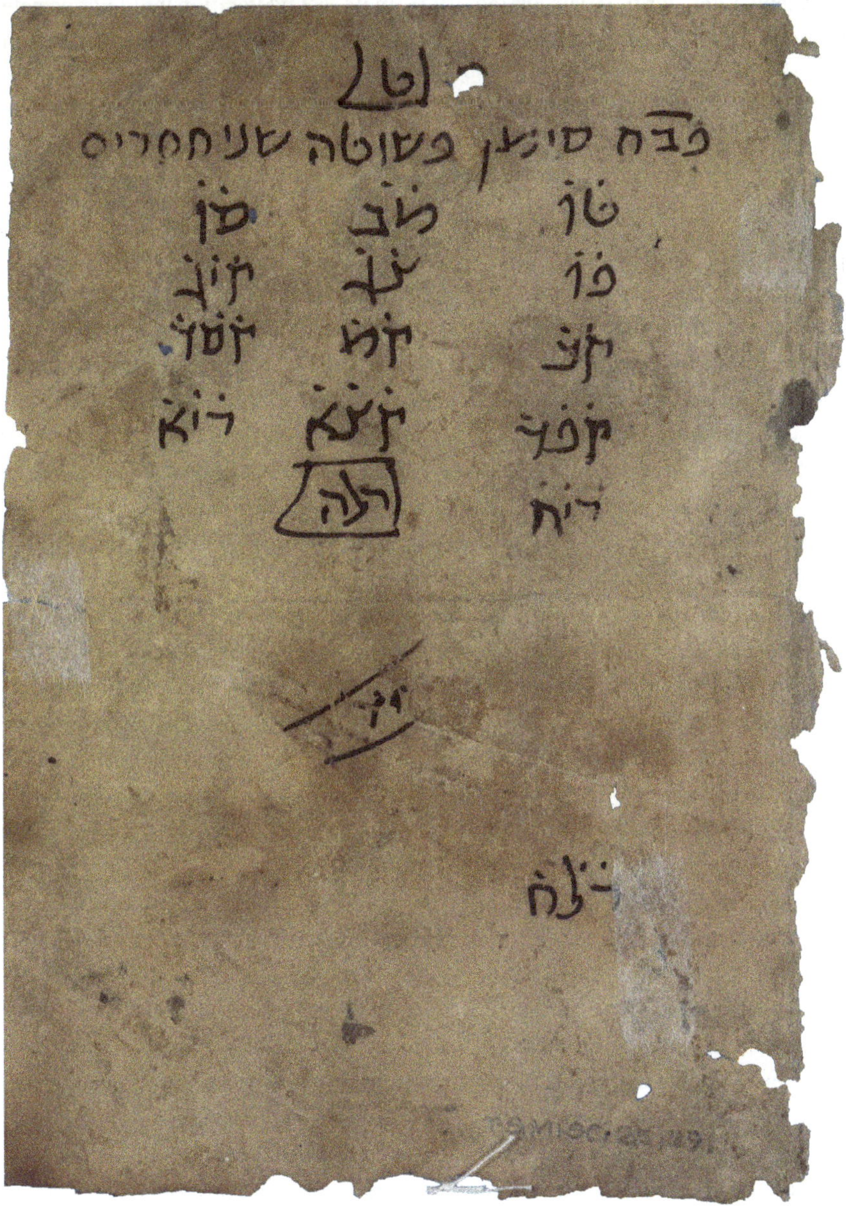

FIGURE 6.2 Cambridge University Library, T-S Misc.25.29v. A page from an Oriental 247-year cycle of remainders, listing remainders assigned to the year type 'בח (2D) plain' (second half of the twelfth century – first half of the thirteenth century). The manuscript includes a graphic indication of the errors created by the cycle if it is reused for a second iteration: remainders that deviate from the standard calculation are framed, and those correct at the time of copying are added below the main text. WITH PERMISSION OF THE SYNDICS OF THE CAMBRIDGE UNIVERSITY LIBRARY

on calendar data for the years 689/90–935/6. In its second iteration it did not deviate from the calculated calendar until the 980s, when the two calendars disagreed first in 984/5–985/6 and then 988/9–990/1 (see Table 6.1). If the cycle had been put together in the 980s, it would have been immediately apparent that it differed from the standard calendar, and therefore data for 689/90–935/6 would not have been used to construct the cycle.

Josiah b. Mevorakh al-ʿĀqūlī's treatise consists of fourteen chapters, each dedicated to one of the fourteen possible types of the Jewish year. Each chapter is made up of three elements: 1) the year type described in that chapter, 2) a set of remainders assigned to this year type, and 3) a fuller roster of the calendar, including the beginning of months, the festivals and the fasts. In order to use the cycle in any given year, one must establish its remainder by taking that year's SE date, subtracting 1000 and casting out 247s (in modern notation, (SE date – 1000) *modulo* 247). One must then look for the remainder in one of the fourteen chapters. The chapter that lists the sought remainder will describe the correct course of the year for that date.

The earliest version of Josiah b. Mevorakh's cycle correctly covered years 689/90–935/6. In later versions, the remainder lists were gradually updated in order to bring them in line with the standard calendar. In a number of manuscripts, the cycle of remainders is embedded in a twelfth-century critique of the 247-year cycle authored by Joseph bar Āraḥ, in which remainders were fully updated for the second iteration of the cycle, viz., 936/7–1182/3. Due to this update, year types for the first iteration of the cycle survive in only a small number of copies.[95] Through a series of further, partial updates, most surviving copies of Josiah b. Mevorakh's cycle were made to agree with the standard calendar up to and including 1353/4 but would have caused mistakes in 1354/5–1355/6 and later years susceptible to reiteration errors.[96]

Dated tables for a particular 247-year period also circulated in the Oriental geo-cultural area from the twelfth century onwards. Reiteration errors are infrequent in surviving tables from the Orient (5 out of 23 tables) but some tables have been glossed with secondary notes re-dating them for future iterations of the cycle, when they would have produced mistakes.[97] Marks and notes in secondary hands often found in Oriental manuscripts demonstrate that these calendars were consulted (20 out of 49 calendars, both dated 247-year tables and cycles of remainders).

Different attitudes to the cycle are expressed in Oriental manuscripts. Here a distinction must be made between manuscripts from Yemen and from other regions

95 Vidro, 'Origins', pp. 120–21, 126.

96 Compare this to the earliest reiteration errors attested in dated 247-year tables: see above, p. 255.

97 Jerusalem, Kapah 16, fol. 12v (Yemenite), London, BL Or 4104, fol. 21v (Yemenite, fifteenth century), Oxford, Bodl. Heb. e.60, fol. 462r (Persian, 1485).

within the Oriental geo-cultural area. In non-Yemenite manuscripts, the cycle is explicitly refuted in only six out of thirty-nine identified manuscripts, half of which are copies of Joseph bar Āraḥ's critique of the cycle of remainders. In Yemen, the overwhelming tendency is to refute the cycle and to stress that each new 247-year calendar must be constructed by calculation.[98] It is likely that the fourteenth-century dispute caused by the use of the 247-year calendar in Yemen sensitized calendar-makers and scribes to the problems involved in using the cycle.

2 *Manuscripts in Ashkenazi Hands*[99]

The 247-year cycle was more popular in Ashkenaz than in all other geo-cultural areas. In my corpus there are 81 Ashkenazi-hand manuscripts copied between the twelfth and the eighteenth centuries, and later discussions in printed sources can also be identified. In medieval Ashkenaz, 247-year calendars were most often included in prayer books and halakhic works. In later centuries, they became a staple component of early modern books on the calendar known as *Sifre Evronot*.[100]

All but one calendar cycle in Ashkenaz are tables dated for a particular stretch of 247 years. One description of a cycle of remainders starting from Creation is also attested.[101] The majority of Ashkenazi 247-year cycles contain data carried over from earlier calendars (61 calendars with reiteration errors, and 13 calendars without). These statistics go well with the fact that only few Ashkenazi manuscripts refute the 247-year cycle (12 manuscripts). The refutations first appear in fourteenth century manuscripts, much later than in other areas, but are most common in *Sifre Evronot* composed in or after the sixteenth century. Nearly half of all Ashkenazi 247-year calendars have user marks, showing that the cycles were widely consulted.

3 *Manuscripts in Italian Hands*

My corpus contains 32 Italian manuscripts on the 247-year cycle, which include both cycles of remainders and dated calendar tables for a particular 247-year period. Cycles of remainders most commonly occur in prayer books of the Italian rite, whereas dated tables are mainly found in calendar sections of various compendia together with other scientific, medical or halakhic texts.

Unlike in the Orient, Italian cycles of remainders explicitly divide the 247 years into thirteen 19-year cycles and start at the beginning of a 19-year cycle. The 19-year cycles are numbered 1–13 and years are numbered 1–247. The calendars are formatted as a sequence of 247 year types, followed by a description of the fourteen types of the

---

98 See above n. 70.

99 'Ashkenaz' refers mainly to Germany and Northern France.

100 On *Sifre Evronot*, see Carlebach, *Palaces of Time*.

101 Cambridge, Trinity College, F 12 22, fol. 6v (Ashkenazi, fourteenth century).

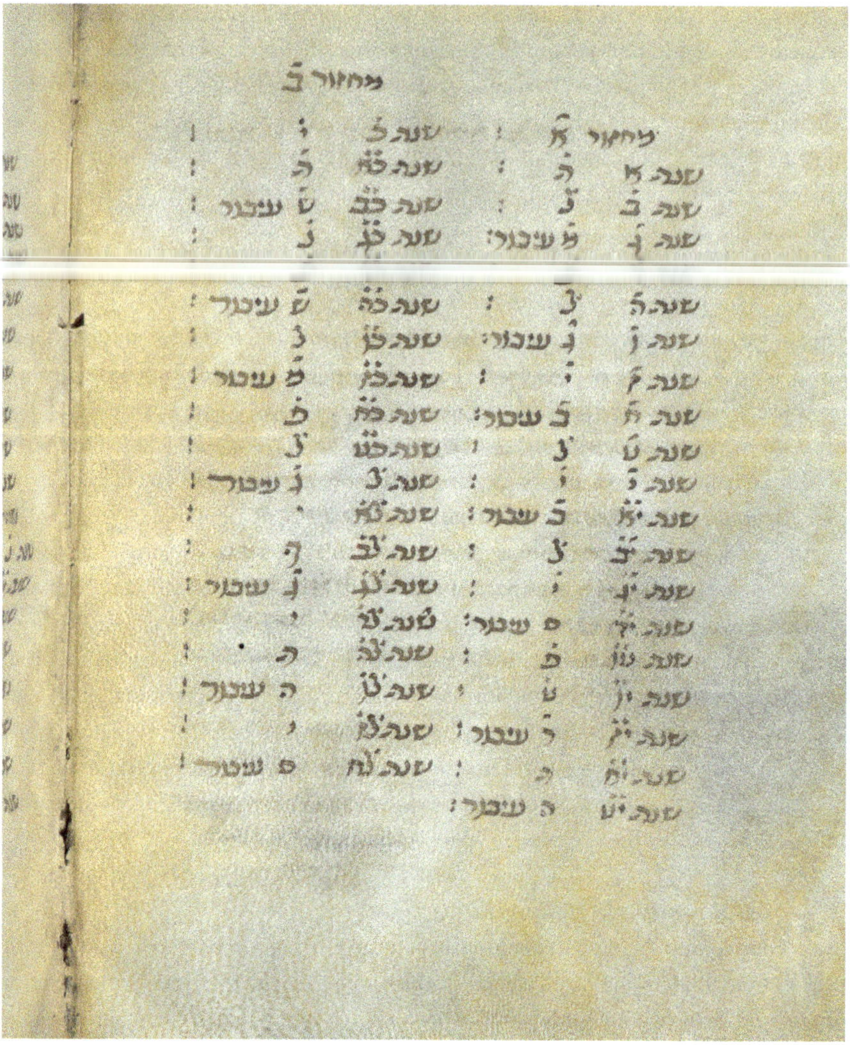

FIGURE 6.3 Cambridge University Library, Ms Add.642, fol. 186v. A page from an Italian 247-year cycle of remainders (fourteenth–fifteenth centuries). The 19-year cycles are numbered 1–13 and years are numbered 1–247. Year types corresponding to remainders are given in code.

Jewish year. Two different sequences of year types are attested, with epochs in 4998 AM (1237/8) and 5017 AM (1256/7) respectively. The algorithm for determining the place of a year within the first sequence can be written in modern notation as (AM date-5000 + 3) *modulo* 247.[102] The algorithm for the second sequence is (AM date-5016) *modulo* 247.[103] The first sequence gives correct year types for 1237/8–1483/4 (19-year cycles nos. 264–276). The second sequence correctly covers years 1256/7–1502/3 (19-year cycles nos. 265–277). Judging by their epochs and their fully correct calendar data, both sequences were probably put together in the thirteenth century on the basis of a calculation rather than by copying an existing calendar.

The first Italian sequence of remainders was used in practice, as is demonstrated by dates from the Creation that are added in secondary hands. Although some of the added dates fall in the second iteration of the cycle, the sequence has not been updated in any copies known to me. User marks have not been found in manuscripts of the second sequence. In most copies of this sequence, the years are written out of order, making the calendar difficult to use.

Dated Italian tables for a particular 247-year period are generally later than the sequences of remainders, and most begin in the fifteenth–sixteenth centuries. All Italian calendar tables have reiteration errors and many start before the manuscript's time of copying. It is clear that these calendars are based on earlier *Vorlagen*. Many Italian tables have user marks in secondary hands such as added dates and corrected mistakes. Critique of the 247-year cycle is rare in Italian sources: I am aware of only one manuscript that refutes the cycle, a seventeenth–eighteenth century calendar treatise in Moscow, RSL, Guenzburg 365 (see fol. 134r).

4 *Manuscripts in Sefardi*[104] *and Provençal Hands*

Copies, mentions and discussions of the 247-year cycle are relatively common in Sefardi and Provençal astronomical-calendrical miscellanies and Bibles. In my corpus there are 29 manuscripts from this area. The majority of the cycles are dated tables, although there are two Sefardi-hand copies of the first Italian sequence of remainders. Some of the Sefardi 247-year cycles are derivative from a non-reiterative calendar in the astronomical treatise *Luḥot ha-Nasi* by the twelfth-century mathematician

102 Jerusalem, NLI, Heb. 38°4281, fol. 298r (Italian, 1391): הרוצה לידע קביעות ראשי חדשים ומועדים וצומות וקריאת פרשיות והפסקות שלכל שנה ושנה יחשוב שנות הפרט שעל חמשת אלפים שנה ויוסיף עליהן ג̇ שנים ויכללם יחד ויוציאם ר̇מ̇ז̇ ר̇מ̇ז̇ והנמצא בידו הוא החשבון שעומד בו וימצאהו באילו י̇ג̇ מחזורים שלפנינו

103 Cambridge, CUL Add. 491, 2, fol. 157v (Italian, fifteenth century): הרוצה לידע קביע׳ ראשי חדשי׳ והמועדים והצומ׳ וקריאת פרשיות של כל שנה ושנה יחשוב השנים שעל חמשת אלפי׳ וששה עשר לפרט ויוציאם ר̇מ̇ז̇ ר̇מ̇ז̇ והנמצא בידו הוא החשבון שעומד בו וימצאהו באלו י̇ג̇ מחזורים

104 'Sefarad' refers mainly to the Iberian Peninsula.

and astronomer Abraham bar Ḥayya.[105] Bar Ḥayya's original table covers 65 19-year cycles and is laid out in 24 columns, each representing between one and five 19-year cycles.[106] In this table, 19-year cycles that share the same sequence of year types are presented in one column, whereas cycles that do not share a sequence of year types with any other cycle covered by the table are given a separate column. In a number of fourteenth-century Sefardi manuscripts of *Luḥot ha-Nasi*, this unusually structured table was turned into a reiterative calendar for five iterations of the 247-year cycle.[107] This reformatting of a non-reiterative calendar into a reiterative one demonstrates Sefardi Jews' close familiarity with the 247-year cycle.

The 247-year cycle is refuted in a number of works from Sefarad and Provence, including *Shalosh She'elot* by Abraham Ibn Ezra (Narbonne, 1148), *Yesod 'Olam* by Isaac Israeli (Toledo, 1310), *Ḥeshev ha-Efod* by Profiat Duran (Castille, 1395), and a number of anonymous refutations found in manuscripts. More than a half of all Sefardi and Provençal copies of the 247-year cycle exhibit reiteration errors but only few have users' glosses. The paucity of the glosses may indicate that the cycle was relatively rarely used. At the same time, the Avignon correspondence between Mordecai Nathan and his uncle Don Boniac Astruc Nasi provides direct evidence of the cycle's being used in the fifteenth century by some Jewish communities in the area.[108]

5 *Manuscripts in Byzantine Hands*

Only four reiterative calendars in Byzantine hands could be identified.[109] Three are dated calendars and one is a unique cycle of remainders. Oxford, Bodl. Poc 368, fol. 219r (fifteenth century) contains Josiah b. Mevorakh's cycle of remainders translated into Hebrew and dated according to the era of Creation, which was in use in the Byzantine Empire, rather than the Seleucid era common in the Near East.[110] Three out of four Byzantine 247-year cycles in the corpus would produce reiteration errors; none has users marks.

---

105 On this work see Millás Vallicrosa, *La Obra Séfer Hešbón Mahlekot ha-Hokabim*, pp. 109–16; Langermann 'Science in the Jewish Communities', pp. 15–16; Garstein, *Relationship*.

106 See, for example, Paris, BNF, heb. 1046, fol. 2r. For a description of the table in *Luḥot ha-Nasi* see Vidro, 'Calendar tables', pp. 71–73, 76 n. 38.

107 Cesena, Biblioteca Malatestiana, Pluteo sinistro XXIX 4, fol. 3v, Parma, Biblioteca Palatina, Cod. Parm. 3821, fol. 10v and Oxford, Bodl., Marsh. 114, fol. 22r.

108 See above, pp. 267–68.

109 This may be in proportion with the low number of Byzantine Hebrew manuscripts that are generally extant.

110 See Vidro, 'Origins', pp. 134–35. On the use of the era of Creation and the Seleucid era see Krakowski and Stern, 'The "oldest dated document of the Cairo Genizah" (Halper 331)' (forthcoming).

6 *Latin Manuscripts*

247-year cycles accompanied by descriptions of the fourteen types of the Jewish year have been identified in four Latin expositions on the Jewish calendar: *Compotus philosophicus* by an unidentified Friar John (ca. 1273, Germany),[111] Robert of Leicester's *Tractatus de compoto Hebreorum aptato ad kalendarium* (1294, England),[112] John of Pulchro Rivo's *Compotus novus* (1297, Germany)[113] and Nicholas Trevet's *Compotus Hebreorum* (1310, England).[114] In addition, Roger Bacon (ca. 1214/20–ca. 1292) is known to have possessed a Hebrew manuscript containing a 247-year table, which he sent to Pope Clement IV commenting that it was 'a wonderful work of astronomical art and highly useful for the understanding of the [Mosaic] Law and the feasts prescribed by it'.[115]

Understanding the calendrical background of the feasts and other events mentioned in the Old Testament and reconstructing the chronology of the Scriptures appears to have been among the main interests that Christian scholars had in the Jewish calendar.[116] Another, more practical, concern was establishing the date of Easter in accordance with the precepts of the Mosaic Law. The turn to the contemporary Jewish calendar was based on a notion accepted by many Hebraists that the calendar used by medieval Jews was the same as that presupposed by the Hebrew Scriptures. A composition consisting of a 247-year cycle and a description of the fourteen possible types of the Jewish year was particularly well suited for elucidating the feasts prescribed in the Hebrew Bible and reconstructing Biblical chronology. A description of the fourteen year types contained all necessary information on the Jewish months, feasts and fasts. The 247-year cycle allowed working out the calendar for years in the distant past without complicated calculations. The application of the 247-year cycle for reconstructing Biblical chronology is explicit in *Tractatus de compoto Hebreorum aptato ad kalendarium* by Robert of Leicester, who used calendrical data obtained from the 247-year cycle to work out the internal chronology of the year of the Flood and the time of the Exodus from Egypt.[117] To simplify the application of data in a 247-year table to Biblical events, calendars in Robert of Leicester's and Nicholas Trevet's works start at the beginning of a 247-year cycle if counted from Creation, i.e. year 1 of their tables corresponds *modulo* 247 to 1 AM. Robert of Leicester's table starts in 19-year cycle no. 261, precisely twenty 247-year cycles from Creation. Nicholas Trevet's table is dated for four

111 Nothaft, *Medieval Latin Christian Texts*, pp. 573–78, 586.
112 Ibid., pp. 139–335, esp. 161–62, 299.
113 Ibid., pp. 570–611, esp. 586–87.
114 Ibid., pp. 336–77, esp. 349–50, 363.
115 Ibid., p. 31. Nothaft, *Dating the Passion*, pp. 183–84.
116 Nothaft, *Medieval Latin Christian Texts*, pp. 11–12, 14.
117 Ibid., pp. 174–76, 180–81.

iterations of the cycle, starting in 19-year cycles nos. 222, 235, 248 and 261 respectively, 17–20 full 247-year cycles from Creation.[118]

In Jewish manuscripts there is very little evidence of 247-year cycles' being used for chronology. Tables usually start near the time of copying and are not purposefully synchronised with the era of Creation. I am aware of only one manuscript, London, BL Or 11594, fol. 2v (Sefardi, fifteenth century), where the position in the 247-year cycle of the Exodus (פקודה) and of the destruction of the Temple (חרבן) are marked in a table [illegible] represent the start of Jewish chronological eras, it is most likely that they were noted in the table to help people date events according to these eras.

7 *Printed Works*

247-year cycles continued to be a staple of calendar literature in the age of printing. A reiterative table was first printed in Guadalajara around 1482, possibly by Solomon Alkabiz.[119] A number of sixteenth-century printed books include 247-year cycles, among them the calendar treatise *She'erit Yosef* by Joseph ben Shem Ṭov ben Jeshua Hai printed in Salonika in 1521, and the distinct work of the same title by Daniel b. Peraḥya ha-Kohen (Salonika 1568). The first *Sefer Evronot* printed for a Jewish readership was issued by Jacob Marcaria in Riva di Trento in 1561, and has a reiterative calendar entitled *'Iggul de-Rav Naḥshon*. Such tables continued to be included in printed *Sifre Evronot* and other printed works on the Jewish calendar until the nineteenth century.[120] In all these works, the calendar is explicitly said to repeat itself forever (*ḥozer ḥalila*) and is frequently attributed to Naḥshon Gaon. In many of the printed works, the included tables contain outdated calendar data carried over from earlier iterations of the cycle.

Apart from calendar literature, 247-year cycles can be found in printed halakhic works. A calendar for four iterations of the 247-year cycle was included in printed editions of the legal codes *Arba'ah Ṭurim* (*Ṭur Oraḥ Ḥayyim*) by Jacob ben Asher and *Tsedah la-Derekh* by Menaḥem ben Aharon ibn Zeraḥ, although the original tables

118 In contrast, the table in the Latin works composed in Germany, *Compotus philosophicus* and *Compotus novus*, start near the time of the works' composition and are not synchronised with the start of a 247-year period counted from Creation.

119 *Gesamtkatalog der Wiegendrücke* (online), number M1603320, accessed on 26 March 2018.

120 For the nineteenth century, see Bendavid, *Zur Berechnung und Geschichte des juedischen Kalendar* and Wunderbar, *Immerwährender Kalender*. A review of Bendavid's treatise in the *Leipziger Literatur-Zeitung* (no. 280, 4 November 1817, cols 2237–2239) referred to the included reiterative calendar as 'eine dem Verfertiger eines Judenkalenders nützliche and nothwendige Zugabe'. For a discussion of Bendavid's views on the reiterative calendar and his nineteenth-century critics, see Carlebach, 'When does the modern period of the Jewish calendar begin?', pp. 46–47, 50–51.

found in manuscript copies of these works are non-cyclical.[121] A 247-year calendar is also included in *Levush Malkhut* by Mordecai Yoffe (1530–1612), a commentary on *Arba'ah Ṭurim, Ṭur Oraḥ Ḥayyim* published in 1590. Mordecai Yoffe stressed that this cycle did not recur, but included a correct table for 247 years in his commentary in order to provide readers with an error-free table of the familiar format.

Refutations of the 247-year cycle can be found in a number of early-modern printed works. To give just two examples, the mathematician and calendar scholar Issachar Ibn Susan published a critique and a list of mistakes in a reiterative calendar that he witnessed in Salonika, most probably a copy of the table in *She'erit Yosef* by Joseph ben Shem Ṭov.[122] A refutation of the reiterative calendar in *Arba'ah Ṭurim* was written by Hezekiah da Silva, who in his commentary on the compendium entitled *Peri Ḥadash* criticised the use of the 247-year cycle, pointed out mistakes in the reiterative calendars in *Arba'ah Ṭurim* and *She'erit Yosef*, and provided a non-cyclical calendar 'up to the end of the world'.[123]

Appendix 2: List of Manuscript Sources

Appendix 2a: Manuscripts that either contain or discuss the 247-year cycle.

Where the cycle is part of a known work, the title of the work is given in brackets. Genizah fragments that belong to the same manuscript are recorded as one source. Some manuscripts contain more than one 247-year cycle. Some cycles are incomplete.

Berlin, SBB, Or. Fol. 1198, 14th–15th c., Ashkenazi, fol. 42r
Berlin, SBB, Or. Oct. 352, ca. 1300, Ashkenazi, fols 13r, 17r–17v, 20r
Berlin, SBB, Or. Oct. 3150, 1649, Ashkenazi, fols 44v–45v
Berlin, SBB, Or. Qu 682 (Joseph b. Yefet ha-Levi, *Ner Israel*), 1455, Yemenite, fols 88v–89v, 93v
Berlin, SBB, Or. Qu. 826, 16th c., Italian, fol. 36r
Bet Nekofa, Kaplan 1, 18th c., Persian, fols 62r–62v
Budapest, MTAK, Kaufmann A 370, end 14th–15th c., Ashkenazi, p. 442
Budapest, MTAK, Kaufmann A 399, 14th c., Ashkenazi, pp. 460–461, 500
Budapest, MTAK, Kaufmann A 418, 16th–17th c., Italian, p. 127
Budapest, MTAK, Kaufmann A 513, 16th c., Italian, p. 209

---

121 See Vidro, 'Calendar tables', pp. 79, 81. Compare this with the table in some copies of *Luḥot ha-Nasi* (see above, Appendix 1, *Sefarad and Provence*).

122 Issachar Ibn Susan, *Tiqqun Issachar* (Venice, 1578/9, fols 10v–11v). See above, n. 46. See also Carlebach, *Palaces of Time*, pp. 51–53.

123 First published in Amsterdam 1706, fols 5r–6v (commentary on *Arba'ah Ṭurim, Ṭur Oraḥ Ḥayyim* chapter 428).

Budapest, MTAK, Kaufmann A 515, 1598, Ashkenazi, p. 90
Budapest, MTAK, Kaufmann A 520, 16th c., Ashkenazi, pp. 115–116
Cambridge, CUL, T-S 6K2.1, late 12th–early 13th c., Oriental
Cambridge, CUL, T-S 10K20.2 and T-S K19.12, 13th–14th c., Oriental
Cambridge, CUL, T-S K2.8, ca. 1297, Oriental
Cambridge, CUL, T-S K2.41, 13th–14th c., Oriental
Cambridge, CUL, T-S K2.82 and T-S AS 144.72, 13th–14th c., Oriental
Cambridge, CUL, T-S K[illegible], Oriental
Cambridge, CUL, T-S Ar.2.7, 13th c., Oriental
Cambridge, CUL, T-S Ar.2.12, late 12th c., Oriental
Cambridge, CUL, T-S Ar.29.31 and T-S Ar.29.3v, 13th–14th c., Oriental
Cambridge, CUL, T-S Ar.29.135 and T-S Ar.29.3r, 11th–12th c., Oriental
Cambridge, CUL, T-S NS 98.2 and T-S AS 144.118, 13th–14th c., Oriental
Cambridge, CUL, T-S NS 98.40, T-S Misc.25.29 and T-S AS 144.164, Oxford, Bodl. Heb e. 100.46, second half of the 12th–first half of the 13th c., Oriental
Cambridge, CUL, T-S NS 98.95, Oriental
Cambridge, CUL, T-S NS 312.94, 13th–14th c., Oriental
Cambridge, CUL, T-S AS 144.32, 13th c.?, Oriental
Cambridge, CUL, T-S AS 144.46 and T-S AS 144.166, 12th–14th c., Oriental
Cambridge, CUL, T-S AS 144.111, 12th c., Oriental
Cambridge, CUL, T-S AS 144.228, T-S AS 144.286 and T-S AS 203.216, 12th c., Oriental
Cambridge, CUL, Add.491.2, 15th c., Italian, fols 157r–160v
Cambridge, CUL, Add.635 (Zalman of St. Goar, *Sefer Maharil*), 1465, fol. 210r
Cambridge, CUL, Add.642, 14th–15th c., Italian, fols 186r–194v
Cambridge, CUL, Add.1200, 16th–17th c., Yemenite, fol. 166v
Cambridge, CUL, Add.1727, 16th–17th c., Yemenite, fols 252v–253r
Cambridge, CUL, Add.3127, 1399, Ashkenazi, fol. 330v
Cambridge, CUL, Add.3203, 15th c., Sefardi, 6v–7r, 9v
Cambridge, Trinity College, F 12 21, 14th c., Ashkenazi, fol. 10v
Cambridge, Trinity College, F 12 22, 14th–15th c., Ashkenazi, fol. 6v
Cesena, Biblioteca Malatestiana, Pluteo sinistro XXIX.4 (Abraham Bar Ḥayya, *Luḥot ha-Nasi*), 14th c., fol. 3v
Cincinnati, HUC 436, 1435, Ashkenazi, fols 191r–191v
Cincinnati, HUC 717, 17th c., Ashkenazi, fol. 9r
Copenhagen, Det Kgl. Bibliotek, Cod. Sim. Hebr. 37, 1749, Ashkenazi, fols 61r–61v
Florence, BML, Laur. Plut.I.08, 1553, Sefardi, fol. 43
Frankfurt, UB, Hebr. Oct. 31, 17th c., Ashkenazi, fols 38v–39r
Frankfurt, UB, Hebr. Oct. 35, end 13th c., Ashkenazi, fols 29v–41r
Frankfurt, UB, Hebr. Oct. 94 (Zalman of St. Goar, *Sefer Maharil*), 1460, Ashkenazi, fol. 219v
Frankfurt, UB, Hebr. Oct. 120, 17th–18th c., Ashkenazi, fols 19v–20r

Frankfurt, UB, Hebr. Oct. 142, 15th c., Italian, fols 58r–72r
Hamburg, SUB, Cod. Hebr. 246 (Jacob ben Asher, *Arbaʿah Ṭurim*), 1464, Ashkenazi, fol. 56r
Hamburg, SUB, Cod. Hebr. 249, 15th c., Ashkenazi, fol. 30r
Hamburg, SUB, Cod. Hebr. 37, 1434, Ashkenazi, fols 122r, 125v
Holon, Nahum 116, 1580, Yemenite, fols 58v–60r
Holon, Nahum 177, 17th c., Yemenite, fols 151v–155r
Imola, Biblioteca Comunale, A A 3 23 Ms. N 77, 15th c., Sefardi, fol. 335r
Jerusalem, Ben Zvi Institute 1236, 17th c., Yemenite, fols 2r, 10r, 11v–12r, 14v–15r
Jerusalem, Mossad ha-Rav Kook, OL 246, 1850, Oriental, fols 1v–2v
Jerusalem, Kapah 16 (Joseph b. Yefet ha-Levi, *Ner Israel*), Yemenite, 9r–10r, 12v, 13v–14r
Jerusalem, NLI, Heb. 8°1997, late 15th c., Italian, fol. 73r
Jerusalem, NLI, Heb. 8°2380, 1716, Ashkenazi, fols 51r–52r
Jerusalem, NLI, Heb. 8°3857, 15th c., Ashkenazi, fols 60r–63r
Jerusalem, NLI, Heb. 38°4281, 1391, Italian, fols 298r–314v
Kiev, OPI 753, 17th c., Yemenite, fols 149v–150r
Leiden, UB, Cod. Or. 4730, ca. 1465, Italian, fols 14v–15v
London, BL, Add 18684, 1392, Ashkenazi, fol. 48r (marginal note in a secondary 15th-c. hand)
London, BL, Add 26970, 1308, Ashkenazi, fol. 182r
London, BL, Add 27150 (Jacob ben Asher, *Arbaʿah Ṭurim*), 1492, Italian, fol. 99r
London, BL, Add 27205, late 12th–13th c., Ashkenazi, fol. 175r
London, BL, Harley 5716 (Jacob ben Asher, *Arbaʿah Ṭurim*), 1475, Sefardi, fol. 100r
London, BL, Harley 5584, 14th–15th c., Ashkenazi, fols 173r–174r
London, BL, Or 2227, 1540, Yemenite, fols 199r–201v
London, BL, Or 2389, 1635, Yemenite, fols 140v, 141v–145r
London, BL, Or 2451, 1483–1484, Persian, fols 362v–375v, 378r
London, BL, Or 2674, 15th c., Italian, fols 138r–140r
London, BL, Or 4104 (Joseph b. Yefet ha-Levi, *Ner Israel*), 15th c., Yemenite, fols 10r–11r, 13r–15v, 20r, 21v
London, BL, Or 5866, 15th c., Sefardi, fols 3v–4r
London, BL, Or 9884, 15th c., Persian, fols 308r–317r, 321r–321v
London, BL, Or 10576, 16th–17th c., Persian, fols 153r–159r
London, BL, Or 10702, 15th c., Persian, fol. 30r
London, BL, Or 10765, 16th c., Yemenite, fol. 126r
London, BL, Or 11594, 15th c., Sefardi, fol. 2v
Manchester, Rylands B 3390 and Rylands B 5508, 11th–13th c., Oriental
Milan, Biblioteca Ambrosiana, X 123 sup (Jacob ben Asher, *Arbaʿah Ṭurim*), ca. 1479, Ashkenazi, fol. 167v
Moscow, RSL, Guenzburg 107, 1329, Sefardi/ Provençal, fol. 264r

Moscow, RSL, Guenzburg 365–1, 17th–18th c., Italian, fol. 134r

Moscow, RSL, Guenzburg 365–2, 15th–16th c., different Provençal hands, fols 147r–148r, 163v, 164v, 231r–231v, 271v

Moscow, RSL, Guenzburg 481, late 12th–early 13th c., Ashkenazi, fols 101v–102r

Moscow, RSL, Guenzburg 705, 15th–16th c., Italian, fol. 48r

Moscow, RSL, Guenzburg 746, 15th c., Byzantine, fols 186v–193r

Moscow, RSL, Guenzburg 821, 14th–15th c., Sefardi, fol. 230v

Moscow, RSL, Guenzburg [illegible], [illegible] c.?, Ashkenazi, fol. [illegible]

Moscow, RSL, Guenzburg 1068, 16th c., Italian, fol. 47r

Moscow, RSL, Guenzburg 1379, 17th c., Ashkenazi, fols 12r–12v

Moscow, RSL, Guenzburg 1380, 1717, Ashkenazi, fol. 9r

Moscow, RSL, Guenzburg 1451, 1536, Italian, fol. 94r

Munich, BSB, Cod. hebr. 109, 15th c., Sefardi, fols 38r–38v

Munich, BSB, Cod. hebr. 128, 15th c., Provençal, fol. 28r

Munich, BSB, Cod. hebr. 343, 15th c., Sefardi/Provençal, fol. 167v

Munich, BSB, Cod. hebr. 394–1, 1677, Ashkenazi, fols 11r–12r

Munich, BSB, Cod. hebr. 394–2, 1566, Ashkenazi, fol. 72v

New York, JTS, 2435, 15th c., Byzantine, fols 50r–50v

New York, JTS, 2540, 1631, Ashkenazi, fols 32r, 41v–42r

New York, JTS, 2548, 16th c., Ashkenazi, fols 4r, 41r–v

New York, JTS, 2590, 15th c., Italian, fols 6r–16r

New York, JTS, 2641, 16th c., Italian, fols 16v–17r

New York, JTS, 4246, end 13th–beginning 14th c., Ashkenazi, fols 1r–11r

New York, JTS, 4463, 18th c., Yemenite, fols 98r–98v

New York, JTS, 4660, 1531, Ashkenazi, fol. 70v

New York, JTS, 5543, 15th c., Yemenite, fols 37r–40v

New York, JTS, 8892, 1490, Italian, fols 441r–442r, 443v–450r

New York, JTS, 9227, 15th–16th c., Yemenite, fols 1r–3r

New York, JTS, 9487, 16th c., Ashkenazi, fol. 32v

New York, JTS, ENA 1640.5 and ENA 3329, 13th–14th c., Oriental

New York, JTS, ENA 2752.21, 14th c.(?), Oriental

New York, JTS, ENA 3082.4, 1266 or later, Oriental

New York, JTS, ENA 3616.32, 1266 or later, Oriental

New York, JTS, Rab. 532 (Zalman of St. Goar, *Sefer Maharil*), 1473, Ashkenazi, fols 459r, 462v

New York, JTS, Rab. 689 (Jacob ben Asher, *Arba'ah Ṭurim*), 1437, Ashkenazi, fol. 122r

Oxford, Bodl., Canon Or. 1, 14th c., Ashkenazi, fols 81r, 88r

Oxford, Bodl., Heb. d. 11, 14th c., Ashkenazi, fol. 4r

Oxford, Bodl., Heb. e. 45.45–56, 1152, Oriental

Oxford, Bodl., Heb. e. 60, 1485, Persian, fols 450r–462r

Oxford, Bodl., Laud. Or. 166 (Jacob ben Asher, *Arba'ah Ṭurim*), 1470, Ashkenazi, fol. 147r
Oxford, Bodl., Marsh 114 (Abraham Bar Ḥayya, *Luḥot ha-Nasi*), 14th c., Sefardi/ Provençal, fol. 22r
Oxford, Bodl., Mich. 74, 14th c., Ashkenazi, fols 3r, 12v
Oxford, Bodl., Mich. 260 (16th c., Ashkenazi): 247-year cycle (fol. 83r)
Oxford, Bodl., Mich. 292 (Zalman of St. Goar, *Sefer Maharil*), 1450, Ashkenazi, fol. 152r
Oxford, Bodl., Mich. 328, 14th c., Ashkenazi, fol. 294r
Oxford, Bodl., Mich. 330, 1559, Ashkenazi, fols 10r–11r, 21r
Oxford, Bodl., Mich. 369 (Jacob ben Asher, *Arba'ah Ṭurim*), 1444, Ashlenazi, fol. 71r
Oxford, Bodl., Mich. 484, 16th c., Italian, fol. 169r
Oxford, Bodl., Mich. 569, 13th c., Ashkenazi, fols 89v–94v
Oxford, Bodl., Opp. 59, 14th c., Ashkenazi, fol. 164r
Oxford, Bodl., Opp. 332, 16th c., Ashkenazi, fol. 162v
Oxford, Bodl., Opp. 614, 14th c., Ashkenazi, fols 50v–55r
Oxford, Bodl., Opp. 642, 14th c., Ashkenazi, fols 308v–309v
Oxford, Bodl., Opp. 701, 1586, Ashkenazi, fol. 33r
Oxford, Bodl., Opp. 712, 16th c., Ashkenazi, fol. 295r
Oxford, Bodl., Opp. 758, 14th c., Ashkenazi, Liturgy, 338r–345
Oxford, Bodl., Opp. Add. Qu. 157 (Joseph b. Yefet ha-Levi, *Ner Israel*), 15th c., Yemenite, fols 12r–12v, 17v–18v
Oxford, Bodl., Poc. 182–1 (Obadiah b. David, Commentary on Maimonides' *Sanctification of the Moon*), 14th c., Oriental, fols 68v–71v
Oxford, Bodl., Poc. 182–2, 14th c., Oriental, fol. 169v
Oxford, Bodl., Poc. 368, 15th c., Byzantine, fol. 219r
Oxford, Bodl., Regg. 43 (Profiat Duran, *Ḥeshev ha-Efod*), 14th–15th c., Sefardi, fol. 35v
Paris, Cluny Museum of the Medieval World, 12290, 14th c., Ashkenazi, fol. 294v
Paris, AIU, H 21 A (Zalman of St. Goar, *Sefer Maharil*), 1475, Ashkenazi, fol. 142r
Paris, BNF, heb. 20, 1300, Sefardi, fols 3r–6v
Paris, BNF, heb. 21, 14th c., Sefardi, 2r–3v
Paris, BNF, heb. 263, 1481, Italian, fols 57v–60r
Paris, BNF heb. 426 (Jacob ben Asher, *Arba'ah Ṭurim*), 1455, Ashkenazi, fol. 102v
Paris, BNF, heb. 429 (Jacob ben Asher, *Arba'ah Ṭurim*), 15th c., Ashkenazi, fol. 48r
Paris, BNF, heb. 605, 1401, Italian, fols 323v–331r
Paris, BNF, heb. 609, 1348, Italian, fols 298r–308r
Paris, BNF, heb. 620, 14th–15th c., Italian, fols 280r–285r
Paris, BNF, heb. 642, 15th c., Sefardi/ Provençal, fols 183v–184v (marginalia)
Paris, BNF, heb. 646, 14th c., Ashkenazi, fols 138v–139v
Paris, BNF, heb. 1032, 14th c., Ashkenazi, fols 1r–11r
Paris, BNF, heb. 1064 (Abraham Bar Ḥayya, *Luḥot ha-Nasi*), 15th c., Sefardi, fol. 81r
Paris, BNF, heb. 1077, 15th–16th c., Ashkenazi, fol. 51r

Paris, BNF, heb. 1089, 15th–16th c., Italian, fols 34v–36v
Paris, BNF, heb. 1120, 15th c., Ashkenazi, fol. 133v
Paris, BNF, heb. 1311, 15th c., Italian, fols 91v–105v
Paris, BNF, heb. 1331, 17th c., Yemenite, fols 123r–126v
Paris, BNF, heb. 1480, 15th c. secondary hand, Ashkenazi, fol. 203r
Parma, Biblioteca Palatina, Cod. Parm. 350, 14th–15th c., Italian, fols 385v–392v
Parma, Biblioteca Palatina, Cod. Parm. 1902, 1470, Ashkenazi, fol. 276v
Parma, Biblioteca Palatina, Cod. Parm. 2305, 13th–14th c., Ashkenazi, fols 101v–106r
Parma, Biblioteca Palatina, Cod. Parm. 2467, 15th c., Sefardi, fols 46v–54v
Parma, Biblioteca Palatina, Cod. Parm. 3004, 15th c. secondary hand, Ashkenazi, fol. 317v
Parma, Biblioteca Palatina, Cod. Parm. 3134, 14th–15th c., Italian, fols 348r–355v
Parma, Biblioteca Palatina, Cod. Parm. 3228, 16th c. secondary hand, Italian, fol. 243v
Parma, Biblioteca Palatina, Cod. Parm. 3518–1, 14th c., Ashkenazi, fol. 16v
Parma, Biblioteca Palatina, Cod. Parm. 3518–2, 15th c. secondary hand, Sefardi, fols 38v–39r
Parma, Biblioteca Palatina, Cod. Parm. 3821 (Abraham Bar Ḥayya, *Luḥot ha-Nasi*), 14th–15th c., Sefardi, fol. 10v
Sassoon Collection, 368 (The Farhi Codex), 1366–1383, Sefardi, fol. 12
St. Petersburg, RNL, Evr. I 237, 14th c., Oriental, fols 11v–12r
Vatican, BAV, ebr. 152 (Jacob ben Asher, *Arbaʿah Ṭurim*), 15th c., Ashkenazi, fol. 158r
Vatican, BAV, ebr. 303, 14th c., Byzantine, fol. 190v
Vatican, BAV, ebr. 318, 14th c.?, Ashkenazi, fol. 259v
Vatican, BAV, ebr. 323, 14th–15th c., Ashkenazi, fol. 198r
Vatican, BAV, ebr. 387, 15th c., Italian, fol. 162v
Vatican, BAV, ebr. 423, 15th c., Italian, fols 30r–36v
Vatican, BAV, Ross. 437, 1448, Italian, fols 395r–403v
Vienna, ÖNB, Cod. hebr. 75 (Zalman of St. Goar, *Sefer Maharil*), 1468?, fol. 154r
Vienna, ÖNB, Cod. hebr. 175 (Zalman of St. Goar, *Sefer Maharil*), 1470, fols 6r–6v
Zürich, Braginsky, 247, 17th c., Ashkenazi, fols 45v–46r
Zürich, Jeselsohn, 16, 14th c., Ashkenazi, fols 147r–151v
Zürich Jeselsohn, 17, 1280?, Ashkenazi, liturgy: siddur, 263v–264r
Zürich, ZB, Heid. 51, 14th–15th c., Ashkenazi, fols 3r–5v
Zürich, ZB, Heid. 145, 1348, Ashkenazi, fol 45v (marginal note in a secondary 15th-c. hand)

Appendix 2b: Manuscripts that contain calendars with mistakes due to the use of the 247-year cycle despite covering less than thirteen 19-year cycles, having no cyclicity statements, and not being associated with R. Naḥshon or Josiah b. Mevorakh al-ʿĀqūlī.

Bern, Bürgerbibliothek, Cod. 253 (Jacob ben Asher, *Arbaʿah Ṭurim*), 15th c., Ashkenazi, fol. 84r

Budapest, MTAK, Kaufmann A 520, 16th c., Ashkenazi, pp. 117, 139
Cambridge, CUL, T-S K2.42, 14th–15th c., Sefardi
Cambridge, CUL, Add.391, 4 (Profiat Duran, *Ḥeshev ha-Efod*), 16th c., Italian, fols 246r, 261r
Cambridge, CUL, Add.667.1, 13th–14th c., Ashkenazi, fol. 185v
Cincinnati, HUC 439, 15th c., Ashkenazi, fols 197r–198v
Frankfurt, UB, Hebr. Oct. 31, 17th c., Ashkenazi, p. 36r
Hamburg, SUB, Cod. Hebr. 91, 1412, Ashkenazi, fol. 195v
Jerusalem, NLI, Heb. 8°1282, 14th c., Ashkenazi, fols 17v–18r
Jerusalem, NLI, Heb. 8°3857, 15th c., Ashkenazi, fols 21r–22r
Jerusalem, NLI, Heb. 34°1114, 1419, Ashkenazi, fols 242r–242v
London, BL, Add 11639, 13th c., Ashkenazi, fol. 536v
London, BL, Or 2674, 15th c., Italian, fols 131v–137v
Mantua, Comunita Israelitica, ebr. 10 (Profiat Duran, *Ḥeshev ha-Efod*), 15th c., Sefardi, fol. 97v
Moscow, RSL, Guenzburg 421, 18th c., Oriental, fol. 78r
Moscow, RSL, Guenzburg 453 (Profiat Duran, *Ḥeshev ha-Efod*). 15th–16th c., Italian, fols 20v, 30v
Munich, BSB, Cod. hebr. 299 (Profiat Duran, *Ḥeshev ha-Efod*), 15th–16th c., Sefardi, fols 29r, 35r
New York, JTS, 2435, 15th c., Byzantine, fols 50v–55v
New York, JTS, 2540, 1631, Ashkenazi, fols 40r, 42v
New York, JTS, 2569, 19th c., Sefardi, fol. 124v
New York, JTS, 2590, 15th c., Italian, fols 22r–22v
New York, JTS, 2641, 16th c., Italian, fols 27r–28v
Oxford, Bodl., Canon Or. 1, 14th c. secondary hand, Ashkenazi, fol. 92v
Oxford, Bodl., Opp. 156, 15th c., fol. 82r
Oxford, Bodl., Poc. 262, 1202, Oriental, fols 249r–251v
Oxford, Bodl., Regg. 49, 1491, Byzantine, fols 71r–74v
Paris, BNF, heb. 1047, 15th c., Byzantine, fols 156r–157r
Paris, BNF, heb. 1311, 15th c., Italian, fol. 114r
Parma, Biblioteca Palatina, Cod. Parm. 2018, 1484, Sefardi, fol. 5v
Parma, Biblioteca Palatina, Cod. Parm. 2198, 15th–16th c., Italian, fols 45r–51v
Parma, Biblioteca Palatina, Cod. Parm. 2412, 14th c., Ashkenazi, fols 24r–28r
Parma, Biblioteca Palatina, Cod. Parm. 2776 (Profiat Duran, *Ḥeshev ha-Efod*), 15th c., Sefardi, fols 113r, 130v
Parma, Biblioteca Palatina, Cod. Parm. 3266, 14th c., Ashkenazi, fols 19r–22r
Toronto, University of Toronto Libraries, Fr 3–016, 14th c., Ashkenazi, fols 90v–91r
Vatican, BAV, ebr. 318, 15th c. secondary hand, Ashkenazi, fols 275r–277r
Vatican, BAV, Ross. 555 (Jacob ben Asher, *Arbaʿah Ṭurim*), 1436, Italian, fol. 78r

Vienna, ÖNB, Cod. hebr. 127 (Jacob ben Asher, *Arbaʿah Ṭurim*), 1436, Ashkenazi, fol. 61v
Zürich, Jeselsohn, 16, 16th c. secondary hand, Ashkenazi, fol. 146v
Zürich, ZB, Heid. 145, 14th c., Ashkenazi, fols 45v–46r, 167r

Appendix 2c: Manuscripts that contain calendars for more than but not a multiple of thirteen 19-year cycles with mistakes due to the use of the 247-year cycle

Berlin, SBB, Or. Qu. 649 (Abraham Bar Ḥayya, *Luḥot ha-Nasi*), 15th c., Sefardi, fol. 1v
Cambridge, CUL, Add.548 (Jacob ben Asher, *Arbaʿah Ṭurim*), 15th c., Sefardi, fol. 148r
Cambridge, CUL, Add.1199.1 (Jacob ben Asher, *Arbaʿah Ṭurim*), 1432, Byzantine, fol. 150v
Cambridge, CUL, Oo.6.65 (Isaac Israeli, *Yesod ʿOlam*), 16th–17th c., Sefardi, fol. 21r
Chicago, Newberry Library, 2 (Abraham Bar Ḥayya, *Luḥot ha-Nasi*), 15th c., Sefardi
Florence, Laurentian Library, Or. 491 (Abraham Bar Ḥayya, *Sefer ha-ʿIbbur*), 14th–15th c., Italian
Frankfurt, UB, Hebr. Oct. 31, 17th c., Ashkenazi, fol. 39v
Hamburg, SUB, Cod. Hebr. 34 (Jacob ben Asher, *Arbaʿah Ṭurim*), 14th–15th c., Ashkenazi, fol. 260r–260v
Leipzig, UBL, B.H. fol. 8 (Jacob ben Asher, *Arbaʿah Ṭurim*), probably before 1412, Ashkenazi, fol. 133v
London, BL, Add 15977 (Isaac Israeli, *Yesod ʿOlam*), 15th c., Sefardi, fol. 178r
London, BL, Add 26899 (Abraham Bar Ḥayya, *Sefer ha-ʿIbbur*), 1316/1317, Italian, fol. 63r
London, BL, Or 10583 (Abraham Bar Ḥayya, *Sefer ha-ʿIbbur*), 14th–15th c., Italian, fol. 87r
London, BL, Or 11796 (Abraham Bar Ḥayya, *Luḥot ha-Nasi*), 15th c., Sefardi, fol. 3r–4r
Modena, Estense University Library, a.W.8.10 (Isaac Israeli, *Yesod ʿOlam*), 14th–15th c., Sefardi, fol. 184r
Moscow, RSL, Guenzburg 571 (Isaac Israeli, *Yesod ʿOlam*), 1770, Italian, fol. 116r
Munich, BSB, Cod. hebr. 421 (Jacob ben Asher, *Arbaʿah Ṭurim*), 14th–15th c., Ashkenazi, fols 112v, 113v
New York, JTS, 5512 (Abraham Bar Ḥayya, *Sefer ha-ʿIbbur*), 14th c., Italian, fol. 45r
New York, JTS, 8188 (Jacob ben Asher, *Arbaʿah Ṭurim*), 15th c., Ashkenazi, fol. 32r
New York, JTS, Rab. 527 (Jacob ben Asher, *Arbaʿah Ṭurim*), 1380, Sefardi, fol. 184v
New York, JTS, Rab. 1147 (Jacob ben Asher, *Arbaʿah Ṭurim*), 1450, Ashkenazi, fol. 106v
Oxford, Bodl., Hunt. 299 (Isaac Israeli, *Yesod ʿOlam*), 1484, Sefardi, fol. 239r
Oxford, Bodl., Hunt. 327 (Abraham Bar Ḥayya, *Luḥot ha-Nasi*), 13th–15th c., Provençal, fols 3r–4r
Oxford, Bodl., Mich. 127 (Jacob ben Asher, *Arbaʿah Ṭurim*), 15th c., Ashkenazi, fol. 170r–170v
Oxford, Bodl., Opp. 51 (Jacob ben Asher, *Arbaʿah Ṭurim*), 1456, Ashkenazi, fol. 169v
Oxford, Bodl., Opp. 53 (Jacob ben Asher, *Arbaʿah Ṭurim*), 14th–15th c., Ashkenazi, fol. 93r

Oxford, Bodl., Poc. 368 (Isaac Israeli, *Yesod ʿOlam*), 14th–15th c., Byzantine, fol. 206r
Paris, BNF, heb. 422 (Jacob ben Asher, *Arbaʿah Ṭurim*), 1487, Ashkenazi, fol. 70v
Paris, BNF, heb. 430 (Jacob ben Asher, *Arbaʿah Ṭurim*), 15th c., Ashkenazi, fol. 1r
Paris, BNF, heb. 1038 (Abraham Bar Ḥayya, *Luḥot ha-Nasi*), 14th–15th c., Sefardi, fols 5v–6r
Paris, BNF, heb. 1045 (Abraham Bar Ḥayya, *Luḥot ha-Nasi*), 15th c., Sefardi, fols 1v, 13v
Paris, BNF, heb. 1046 (Abraham Bar Ḥayya, *Luḥot ha-Nasi*), 15th c., Sefardi, fol. 2r
Paris, BNF, heb. 1068 (Isaac Israeli, *Yesod ʿOlam*), 15–16th c., Sefardi, fol. 123v
Paris, BNF, heb. 1069 (Isaac Israeli, *Yesod ʿOlam*), 1491, Sefardi, fol. 137v
Paris, BNF, heb. 1070 (Isaac Israeli, *Yesod ʿOlam*), 16th c., Italian, fol. 172r
Paris, BNF, heb. 1072 (Isaac Israeli, *Yesod ʿOlam*), 15th–16th c., Sefardi/Byzantine, fol. 88r
Parma, Biblioteca Palatina, Cod. Parm. 1741 (extract from Isaac Israeli, *Yesod ʿOlam*), 15th c., Sefardi, fols 4v–5v
Parma, Biblioteca Palatina, Cod. Parm. 3262, 1459, Sefardi, fol. 85v
St. Petersburg, RNL, Evr. I 209 (Jacob ben Asher, *Arbaʿah Ṭurim*), 1419, Ashkenazi, fol. 118v
St. Petersburg, RNL, Evr. I 210 (Jacob ben Asher, *Arbaʿah Ṭurim*), 14th–15th c., Ashkenazi, fol. 66r
St. Petersburg, RNL, Evr. I 211 (Jacob ben Asher, *Arbaʿah Ṭurim*), 15th c. (ca. 1456?), Ashkenazi, fol. 54v
Toronto, University of Toronto Libraries, Fr 5–014 (Jacob ben Asher, *Arbaʿah Ṭurim*), 14th c.?, Sefardi, fol. 46v
Vatican, BAV, Neofiti 30 (Abraham Bar Ḥayya, *Sefer ha-ʿIbbur*), 15th c., Italian, fol. 99v
Vatican, BAV, Neofiti 31 (Isaac Israeli, *Yesod ʿOlam*), 14th–15th c., Sefardi, fol. 146r
Vatican, BAV, Ross. 600 (Jacob ben Asher, *Arbaʿah Ṭurim*), 16th c.? secondary hand, Italian, fol. 76v
Vatican, BAV, Urb. 48 (Abraham Bar Ḥayya, *Sefer ha-ʿIbbur*), 13–14th c., Italian, fol. 56v

References

Primary Sources (Printed)

Abraham Ibn Ezra, *Shalosh Sheʾelot*; ed. M. Steinschneider, *Sefer Shene ha-Meʾorot* (Berlin: Zarentsanski, 1846).
Ḥezekiah da Silva, *Peri Ḥadash* (Amsterdam, 1706).
Issachar Ibn Susan, *Tiqqun Issachar* (Venice, 1578/9).
Isaac Israeli, *Yesod ʿOlam*; ed. B. Goldberg and L. Rosenkranz, *Liber Jesod Olam*, 2 vols (Berlin: Sumtibus editorum, 1848).
Obadiah b. David, commentary on Maimonides' *Sanctification of the Moon* (first printed in the Amsterdam, 1702 edition of Maimonides, *Mishneh Torah*).

Secondary Literature

L. Bendavid, *Zur Berechnung und Geschichte des juedischen Kalendar: aus den Quellen geschoepft* (Berlin: Nicholaische Buchhandlung, 1817).

H.Y. Bornstein, *Mahloqet Rav Saadia Gaon and Ben Meir* (in Hebrew; Warsaw, 1904).

H.Y. Bornstein, *'Divre yeme ha-ibbur ha-aharonim'*, part 1, *Ha-Tequfah* 14–15 (1922), pp. 321–72 (in Hebrew).

R. Brody, *The Geonim of Babylonia and the Shaping of Medieval Jewish Culture* (New Haven, London: Yale University Press, 1998).

E. Carlebach, 'When does the modern period of the Jewish calendar begin?', in L.B. Strauss and M. Brenner (eds), *Mediating Modernity: Challenges and Trends in the Jewish Encounter with the Modern World: Essays in Honor of Michael A. Meyer* (Detroit: Wayne State University Press, 2008), pp. 43–54.

E. Carlebach, *Palaces of Time: Jewish Calendar and Culture in Early Modern Europe* (Cambridge, Mass., London: Belknap, 2011).

N. Garstein, *The Relationship between Abraham Bar Hiyya's Astronomical Tables and his Treatise "Calculation of the Stellar Motions"* (MA thesis, Bar Ilan University, 2016).

R. Gordon, 'Further on the matter of year types', *Ha-Melits* (27 March [9 April], 1902), pp. 3–4 (in Hebrew).

M. Kozodoy, *The Secret Faith of Maestre Honoratus: Profayt Duran and Jewish Identity in Late Medieval Iberia* (Philadelphia, Pa.: University of Pennsylvania Press, 2015).

E. Krakowski and S. Stern, 'The "oldest dated document of the Cairo Genizah" (Halper 331): The Seleucid era and sectarian Jewish calendars' (forthcoming).

Y.T. Langermann 'Science in the Jewish Communities of the Iberian Peninsula: An Interim Report', in idem, *The Jews and the Sciences in the Middle Ages* (Aldershot: Ashgate, 1999), 1–54.

Y.T. Langermann, 'What does *Sefer Yesira* have to do with the Jewish calendar? The place of *Sefer Yesira* in an introduction of a work on the 247-year cycle', *Ale Sefer* 26–27 (2017), pp. 9–16 (in Hebrew).

J.M. Millás Vallicrosa, *La Obra Séfer Hešbón Mahlekot ha-Hokabim* (*Libro del Cálculo de los Movimientos de los Astros*) *de R. Abraham bar Ḥiyya ha-Bargeloní* (Barcelona: Consejo Superior de Investigaciones Cientificas, 1959).

C.P.E. Nothaft, *Dating the Passion: The Life of Jesus and the Emergence of Scientific Chronology* (*200–1600*) (Leiden: Brill, 2011).

C.P.E. Nothaft, *Medieval Latin Christian Texts on the Jewish Calendar: A Study with Five Editions and Translations* (Brill: Leiden, 2014).

S. Powels, *Der Kalender der Samaritaner anhand des Kitāb hisāb as-sinīn und anderer Handschriften* (Berlin, New York: De Gruyter, 1977).

E. Raviv, *Mathematical Studies in the Hebrew Calendar* (unpublished Ph.D. thesis, Bar Ilan University, 2015), pp. 53–111.

M. Rustow and S. Stern, 'The Jewish Calendar Controversy of 921–22: Reconstructing the Manuscripts and their Transmission History', in S. Stern and Ch. Burnett (eds), *Time, Astronomy, and Calendars in the Jewish Tradition* (Leiden: Brill, 2014), pp. 79–95.

C.E. Sachau, *Abū Rayḥān al-Bīrūnī, The Chronology of Ancient Nations* (London: W.H. Allen & Co., 1879).

R. Sar-Shalom, *Gates to the Hebrew Calendar* (in Hebrew; Netanyah: R. Sar-Shalom, 1984).

I. Simon and A. David, 'Nathan, Mordecai,' in *Encyclopaedia Judaica*, vol. 15, pp. 10–11.

S. Stern, *Calendar and Community. A History of the Jewish Calendar, 2nd cent. BCE–10th cent. CE* (Oxford: Oxford University Press, 2001).

S. Stern, *The Jewish Calendar Controversy of 921/2 CE* (Leiden: Brill, 2019).

Y. Tobi, 'The dispute over the 247-year cycle in Yemen', in Sh. Morag, I. Ben-Ami and N.A. Stillman (eds), *Studies in Judaism and Islam Presented to Shelomo Dov Goitein on the Occasion of His Eightieth Birthday by His Students, Colleagues and Friends* (in Hebrew; Jerusalem: Magnes Press, 1981), pp. 193–228.

Y. Tobi, *The Jews of Yemen. Studies in Their History and Culture* (Leiden: Brill, 1999), pp. 211–26.

N. Vidro, 'The origins of the 247-year calendar cycle', *Aleph* 17 (2017), pp. 95–137.

N. Vidro, 'Calendar tables in manuscript and printed *Arbaʿah Ṭurim: Ṭur Oraḥ Ḥayyim*, chapter 428', *Journal of Jewish Studies* 69 (2018), pp. 58–85.

N. Vidro, 'Nahshon Gaon: calendar scholar or pseudo-author', *Jewish Studies Quarterly* 26 (2019), pp. 17–34.

R.J. Wunderbar, *Immerwährender Kalender der Juden. Deutsch und Hebräisch. Nach den Quellen bearbeitet und herausgegeben* (Dessau: Baumgarten und Co, 1854).

Z.H. Yaffe, 'Intercalation in its correct time', *Ha-Melits* (4 [17] March 1902), pp. 3–4 (in Hebrew).

Z.H. Yaffe, *Qorot Ḥeshbon ha-ʿIbbur* (in Hebrew; Jerusalem: Darom, 1930).

Index of Calendars and Related Terms

Calendars

Days of the Week

Month Names

General Index

www.ingramcontent.com/pod-product-compliance
Lightning Source LLC
LaVergne TN
LVHW020634300926
849161LV00004B/692